IoT System Testing

Jon Duncan Hagar

IoT System Testing

An IoT Journey from Devices to Analytics
and the Edge

Apress®

Jon Duncan Hagar
Hot Sulphur Springs, CO, USA

ISBN-13 (pbk): 978-1-4842-8275-5   ISBN-13 (electronic): 978-1-4842-8276-2
https://doi.org/10.1007/978-1-4842-8276-2

Managing Director, Apress Media LLC: Welmoed Spahr
Acquisitions Editor: Spandana Chatterjee
Development Editor: James Markham
Coordinating Editor: Mark Powers

Cover designed by eStudioCalamar
Cover image by Denny Muller on Unsplash (www.unsplash.com)

Distributed to the book trade worldwide by Apress Media, LLC, 1 New York Plaza, New York, NY 10004, U.S.A. Phone 1-800-SPRINGER, fax (201) 348-4505, e-mail orders-ny@springer-sbm.com, or visit www.springeronline.com. Apress Media, LLC is a California LLC and the sole member (owner) is Springer Science + Business Media Finance Inc (SSBM Finance Inc). SSBM Finance Inc is a **Delaware** corporation.
For information on translations, please e-mail booktranslations@springernature.com; for reprint, paperback, or audio rights, please e-mail bookpermissions@springernature.com.
Apress titles may be purchased in bulk for academic, corporate, or promotional use. eBook versions and licenses are also available for most titles. For more information, reference our Print and eBook Bulk Sales web page at http://www.apress.com/bulk-sales.
Any source code or other supplementary material referenced by the author in this book is available to readers on GitHub (https://github.com/apress). For more detailed information, please visit http://www.apress.com/source-code.

Printed on acid-free paper

The dedication of this book is to my mother, who never expected her son to write books when he was in high school crying about English classes...

Contents

About the Author

 **Jon Duncan Hagar** is a senior tester with 40 years' experience in software development and testing. He has supported software product design, integrity, integration, reliability, measurement, verification, validation, and testing on various projects and software domains (environments). He has an M.S. degree in Computer Science with specialization in Software Engineering and Testing from Colorado State University and a B.S. degree in Math with specialization in Civil Engineering and Software from Metropolitan State College of Denver, Colorado. Jon has worked in business analysis, systems, and software engineering, specializing in testing, verification, and validation. The projects he has supported include the domains of embedded and mobile devices, IoT, PC/IT systems, and test lab and tool development. Currently, Jon works as a consultant for Grand Software Testing, LLC.

Jon has taught hundreds of classes and tutorials in software engineering, systems engineering, and testing throughout the industry and universities. He has published numerous articles on software reliability, testing, test tools, formal methods, and mobile and embedded systems. He is the author of the book *Software Test Attacks to Break Mobile and Embedded Devices* and contributor to books on agile testing and test automation. Jon makes presentations regularly at industry working groups and conferences. He most recently has been working on combinatorial testing, test automation, handheld mobile devices, IoT security testing, and error taxonomies for IoT/embedded systems. Currently, Jon is serving on various IEEE and ISO international standard working groups and executive committees addressing software engineering, testing, measurement, estimation, and AI.

About the Technical Reviewer

Atonu Ghosh is a Ph.D. scholar at the Indian Institute of Technology Kharagpur, West Bengal, India. He holds a B.Tech. and an M.Tech. in Computer Science and Engineering from the Maulana Abul Kalam Azad University of Technology (MAKAUT), West Bengal, India. IoT, IIoT, and multiagent systems are his domains of research. Atonu has been building IoT solutions for over seven years now. He is an active reviewer for several journals. He welcomes you to connect with him through LinkedIn (www.linkedin.com/in/atonughoshcse/), Facebook (www.facebook.com/atonu.ghosh/), and email (atonughosh@outlook.com).

Acknowledgments

I would like to thank and recognize my many mentors, colleagues, and teachers who have taught me so much through the years. There are many ideas in this book from nearly all of them, which I cite throughout the book. Many of us are standing on each other's shoulders to advance software tech. My most significant acknowledgment goes to my reviewers and to my wife, Laura M. Hagar – software-systems geek and chief reviewer. Laura's dedication to my writing made this book possible. With all of these people, the references in this book, and the book itself, I hope that the IoT device world becomes safer, more secure, and fun to use.

I particularly would like to thank the following people who, over the years, have helped me improve my critical thinking skills: Cem Kaner, James Whittaker, James Bach, Lisa Crispin, Becky Fielder, and again, my wife, Laura M. Hagar.

Part 1
Getting Started

In this book, I provide information on testing IoT software systems. I begin by introducing IoT, testing, and then their place within a software system framework. IoT should be thought of as different from many pure software systems because testing must include the unique aspects of IoT hardware and systems. IoT technology is part of the gradual evolution of software from just information processing on mainframes and PCs to actually controlling and interacting with technology, people, and the world.

IoT is projected to be a growth area for software, so teams need to be engaged with the changes IoT brings. Even experienced software information technology testers will find differences in IoT. This makes a new challenge for testers, and so I share lessons learned for the embedded and IoT software space to help in planning, designing, securing, and understanding IoT. My goal is to make you a better IoT tester and have the projects you support to be successful.

Part 1

Getting Started

Chapter 1
The Internet of Things, V&V, and Testing

Before reviewing how this book is organized and who will best benefit from it, let's first understand what exactly the Internet of Things (IoT) is and what it is composed of. With that foundation in place, we can begin the testing questions in subsequent sections of this book.

IoT at a Glance

IoT is a collection of various "smart" devices that communicate with the Internet or other networks as shown in Figure 1-1. For our reference purposes, smart means there are software, firmware, and data processing abilities. The proliferation of these devices in a sensing-communicating-actuating network creates the IoT, wherein computers, sensors, and actuators blend seamlessly (hopefully) with the stakeholder environments. The information is then shared across platforms. The "smartness" of these devices comes from various computers running software at different levels, for example, the edge, fog, and cloud. The communication of computers, software, and hardware makes dependability a quality which includes security, functionality, and other qualities [1]. The testing of a small IoT device may initially seem simple, but put into a larger context, as most IoT devices must be, *their very use becomes a global risk*. Assessing these quality elements requires verification and validation (V&V) and testing to one degree or another. This book deals with the testing, verification, and validation of IoT devices and their associated network systems.

The quality we first think of is functionality, and certainly it is a top priority from a company profit perspective. However, the data, security, dependability, and performance of any IoT system cannot be assigned a lower priority. Knowing these important points leads us into testing and V&V to understand more about IoT product qualities and how they can affect the viability and security of IoT products, systems, and services.

- Verification checks whether the development products of a cycle conform to the previous cycle's activity (e.g., the code is verified to design and requirements).
- Validation assesses if a product and system satisfy user's needs, going beyond any written requirements.

J. D. Hagar, *IoT System Testing*, https://doi.org/10.1007/978-1-4842-8276-2_1

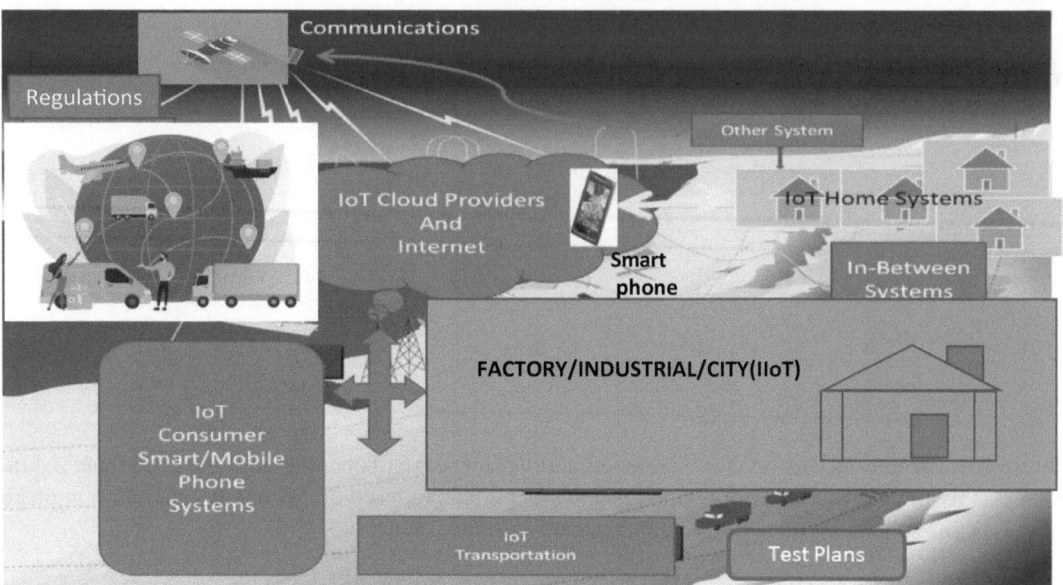

Figure 1-1. The IoT system of systems world

Composite: Jon Hagar

V&V approaches used include assessment, analysis, evaluation, review, inspection, demonstration, and testing of products and processes. V&V is performed during and with development, often by the development staff, and not "at the end" as an afterthought (e.g., *you cannot test quality into a product once the product is done*).

Test is the activity in which a system or component is executed under specified conditions, the results are observed and recorded, and an evaluation is made of some aspect of the system or component [2].

IoT is part of the world of technology that continually moves from one hot area to another. We had computers, then personal computers, then the web/.com systems, and more recently we have mobile/smart devices. Each jump in technology represents an expansion of software and system functionality along with the growing pains of errors as well as many company failures. However, for every Microsoft or Google, there are tens or hundreds of companies that did not succeed in each of these market "explosions."

To date, many of these explosions have involved companies that were already technology based or startup organizations where the people involved had high-tech backgrounds. Additionally, there were influxes of people that had little or no knowledge of technology but showed some ability to learn quickly. These people were often "new hires" who were tasked to sink or swim within these companies. Like the companies, some people succeeded, and some did not.

This book offers a reference point for people and companies working on IoT testing/V&V. These entities will be working on the development, operations of IoT systems, and testing of the systems or components, and be worried about security as well as other qualities that make a system viable. This book is organized into parts addressing significant areas of interest. The book also includes references to historical works and industry standards. These references are provided for those readers wanting more details and assistance in understanding.

The book presents the related ideas of test automation, artificial intelligence (AI), and data analytics. While not being a detailed AI or data analytics test book, due to the nature of IoT systems, these are important topics for testers to begin to consider to improve their work and to add skills to their career.

Understanding Chaos, Security, and Other Product Qualities

After some introduction to IoT, this book deals with team leadership for IoT planning, test, strategy, and architecture. However, the more critical later parts of the book deal with test design, security attacks, architectures, and support environments. My experience is these parts can make or break test/V&V efforts, particularly as teams move from the IoT device itself on to a larger system testing view.

IoT grew out of the embedded software environments (that I worked in for decades) and incorporated the larger network and data analytics–centric world. This large world leads to the ideas of chaos engineering and trustworthy dependability. Also, this larger world complicates testing, hence why many smaller IoT teams may wish to limit their test scope. The larger a system is, the more traditional testing (like manual testing and human inspections of products) becomes unachievable. This situation leads to teams cutting corners or relegating testing/V&V to other stakeholders.

A better solution, advocated in the later sections of this book, is the use of data analytics, automation in test environments, and AI.

Trustworthy means that an IoT software product or component is safe, dependable, reliable, and secure. There is a direct correlation between trustworthiness and reliability. As reliability of a design improves, the trustworthiness of the design also improves, and vice versa. Reliability, safety, testing, and security can be measured [1]. Trustworthy software is a composite of these measurable items. Composite trustworthiness infers that the IoT system consistently performs what it is supposed to do, repeating required functions, dependability qualities, and actions time after time, always producing the same results or outputs from the same stimuli or inputs. Trustworthy IoT software systems demonstrate design robustness in the face of unexpected uses and lifecycle changes. If the software experiences a break in functionality, causing a transition from an operable state to a failed state, it is optimal if the system is able to recover from the failed state in an expeditious and tolerable manner [3].

The National Institute of Standards and Technology (NIST) defines trustworthiness as "software that can and must be trusted to work dependably in some critical function, and failure to do so may have catastrophic results, such as serious injury, loss of life or property, business failure or breach of security" [4, 5].

For many IoT systems, the quality assessment and testing activities must address the preceding concepts. Just testing the written requirements will not be enough, and so V&V will become a common practice, if not for the IoT development team, then for the IoT user-stakeholders.

Why Test IoT Beyond the Device Itself

Many IoT projects will wish to test only the IoT device itself. For some teams, this will be acceptable, but in these cases, I advise a team to disclose the level of testing (e.g., only the IoT device itself in certain environments). This disclosure allows stakeholders incorporating a "small" IoT device into their large system to understand what additional V&V may be needed to keep the product viable. This approach is in line with much of the IoT world, where software, hardware, and system are subject to integrated system testing by the stakeholders or a representative of the stakeholders.

The reason to test beyond the IoT device interface is discussed in later chapters. Industrial, government, and commercial stakeholders will typically understand such limitations. However, the less experienced home or private users may not understand the limitations. Such latter users may be disappointed in an IoT system and stop using it – so much for product viability. While this does not sound like a large problem, in the days of social media and unhappy users, reviews can kill many systems with bad posts. IoT development companies wishing to stay in business, make sales, and expand should be aware of where and how to stop the assessment of IoT device qualities by less experienced assessors, as well as having an adequate operation and help desk center to assist customers, before they become totally unhappy and post that bad review.

Agile vs. Traditional vs. "Who Cares About It" Development

For years now, there has been a push to be agile. Traditional software projects were looked down on. Teams moved into development-test-security-operation modes (a.k.a. DevOps). Each methodology worked for some but failed for other projects. This leads to the question of "Who is doing the work: the method or the people?" I believe it is the latter because the teams do not understand or care enough to do a good job.

> **DIFFERENT PRACTICES FOR A TEAM**
> A customer representative who was of a "traditional" mindset went to a software requirements inspection meeting, where Agile-DevOps was being practiced. The involvement of continuous DevOps, testing, integration, and passionate engineering with the customer rep having an open-door policy to all engineering meetings is an agile concept. At first, the customer was "alarmed" at the strong engineering debates between the development staff and tester. The customer rep was about to complain to upper management. As the test manager, I explained to the customer rep that their being there for agile was new but good and that the heated debates were also good. The whole team involvement reduced the chance of failures in early product configurations and testing. This was a sign of sound engineering. They listened and later became a big supporter of the agile development team "debate" practice.

Is there a perfect methodology or approach to building IoT software systems? *No.*

Software and IoT systems are difficult to create, test, secure, operate, and maintain. I have been on successful teams, and I have been called in to help teams needing improvement. There is no one single cause or solution like "just be agile." The team must care, understand the many different options in testing, and plan, then replan to get the testing right.

What I do believe is this: if dedicated people who are willing to learn, find what works, and fix what doesn't work are given a chance, IoT product successes tend to happen. Good enough IoT has to be accepted over trying for perfection. However, I never stop trying to reach perfection – whatever that is. When teams or management cuts corners on a methodology, to the point of impacting the project, bad things usually happen. Likewise, just adding layers of heavy actions into a methodology can also result in problems.

IoT devices will be part of a system or system of systems. Different communities and stakeholders will be involved in many cases. Test leaders will need to watch for dogma from any methodology camp and be willing to listen to stakeholders. Applying some of the ideals of this book combined with other references can help teams be successful. Testers should not get hung up on labels or names of methodologies, but look for hybrid solutions of mixing and matching. Can you use some test standards? Sure. Can you use some ideals from the context test community? Absolutely, I do. Does agile have good practices? I have used many of them. Most projects will be a hybrid of methodologies.

It has to be said: *ethics counts* in developing software, hardware, systems, networks, and many other fields. For example, I subscribe to a code of ethical conduct in IEEE and ISO. I stand up for the concepts in this book, but only as a starting point for test planning and execution. I have been willing to go to stakeholders or management and give them "the bad news." I learned from them that they did not like the news, but most of them respected me, because with the bad news I gave them ways to move the project forward. This often meant hard work for the developers and testers to recover from the bad news. A few times, I needed to walk away. These are all part of ethical and legal engineering.

In Chapter 13, I present information on security testing, and this information could be misused. What Chapter 13 covers could be considered illegal, if used improperly. **I advise you to stay legal!** To be ethical, you must try to break software and security to know where others (black hats, who are *not legal*) will try to break through. *I advise you to use the information I have given you to remain within the bounds of good ethics.* For example, being ethical or having good ethics (morals) means that you are acting for the greater good of mankind vs. acting to the benefit of a singular person or group.

SolarWinds happened because malware got into a software build. Did someone violate ethics or did software security testing fail? We may never know. What we do know is that it cost a *lot of time and money to many companies* [6].

This Book's Audience

Development-operations-testing-security has many facets. Most readers of this book will spend periods of time building tester knowledge and skill. I am still learning new testing skills even with over 40 years' experience. Further, readers should keep in mind that there is no "best" path to the development and testing of software, systems, or networks. There are many options for testing; each has positives, negatives, as well as cost and schedule impacts. Software development and testing are technical skills based on knowledge that is practiced over a lifetime. Once knowledge and experience are gained, a tester can make informed decisions about what testing or V&V should be used to achieve the desired end, in accordance with a company, project test architecture, strategy, and test plans. Keep in mind that *context matters.*

This book is written for organizations and people that are new to testing and IoT. The industry value of the IoT world is expected to be trillions of US dollars in the following years on tens of billions of devices. Almost every traditional industrial company will enter into IoT, and there will be hundreds of new startups each year. Driven by the market for such devices, some estimates have IoT projected to reach $1.5 trillion by 2027 [7].

The following kinds of organizations and people may find helpful information in this book:

- Companies that have never developed software that want to expand into IoT within their existing product lines. These might include industrial/medical companies (e.g., medical devices, health monitoring systems, heating control systems, transportation systems, etc.) and consumer companies (e.g., wearables, clothing, home entertainment, etc.). While it is hard to imagine any company that offers technology products could be a novice at software development and testing, the software in IoT devices requires special considerations, and companies would be wise to do some homework on exactly what is entailed in IoT development. IoT development (Dev) and test are hard, *really hard.*
- Companies with experience in electronics and limited aspects of software but looking to expand their software footprint, while lacking networked software testing backgrounds (e.g., television, audiovisual devices, automotive, etc.).
- Startup companies looking for a good beginning reference into IoT development and testing.
- Testers looking to learn more about IoT software testing to enhance their careers.

- Groups and government officials looking to use IoT during procurement using contracts, standards, and regulations.
- White hat security testers, who wish to make products more secure.
- Anyone, managers, developers, and support staff, interested in IoT testing.

How to Use This Book

The book can be quickly skimmed or read end to end. It is probably best to read this book by jumping around to topics of interest in IoT vs. reading it in detail from cover to cover.

The table of contents can be used to index topics that are of interest to readers. I suggest the reader assess if they understand basic concepts of testing found in Chapters 1 and 2 first. Then, use the index or table of contents to find specific topics of interest (e.g., IoT planning, design, security, attacks, patterns, or test environments). I suggest that testers refer back to this book and other references to test an IoT device, as the system progresses. Remember, learning is a constant, ongoing activity.

Part 1 focuses on introducing testing concepts as related to IoT systems. New and inexperienced staff will benefit from this introduction of IoT concepts, while experienced testers may want to skip ahead to other parts and chapters of interest.

Reference Standards, Books, and the Internet: Context Matters

A friend of mine said years ago that many great testers had an epiphany at some point in their test career where they realized testing is hard, context of the test problem matters, and one needs many sources of test information. I had my epiphany more than 30 years ago. I started learning all I could about software testing, and I am still learning today. I quickly found that there were many voices on the subject of testing, and often they did not agree. This lack of agreement is where context comes into play. Context varies wildly in IoT devices and systems. You may have IoT devices that need regulations and standards to be considered, for example, when they are life critical. Another team may have an IoT "game" device that just needs to get to market as soon as possible, and so regulations are of less or no interest. The proper books, references, and Internet information on each project are driven by the context.

Throughout the book, I provide reference information. IoT testers must consider which ones to use given the context. Many testers and software people dislike and dismiss industry standards like ISO and IEEE. This can be a mistake. Testers may come to believe that one book or idea on testing is "best." However, there is no "right and perfect" answer to the test problem except by the test team considering context. If you have not had your awakening epiphany, maybe you should read more in this book and look for ideas on the Internet. In any case, you should question and test what each author, including me, says. If software and testing were easy, anyone could create them. Context matters.

Valuable References to Start Your IoT Test Library

Test and development teams should use this book with other technical books, optionally software testing standards and web references. No single book or reference can answer all your questions to fit all of the multitude of unique scenarios. You must have the references, standards, and Google searches to continue your thinking and never stop learning about software and testing. I have 44 years of learning, and I'm still not done.

Here are some of my current favorite references (as a starting point) that I use often:

- *The Art of Software Testing* by Myers (the grandfather of them all)
- *Software Test Attacks to Break Mobile and Embedded Devices* by Jon D. Hagar
- *A Practitioner's Guide to Software Test Design* by Lee Copeland
- *How to Break Software Security* and "How to Break" test books by James Whittaker
- *The Domain Testing Workbook* by Cem Kaner
- *Lessons Learned in Software Testing* by Kaner, Bach, Pettichord
- *How to Break Software: A Practical Guide to Testing* by James Whittaker
- *Software Test Automation* by Fewster and Graham
- *Agile Testing* by Crispin and Gregory
- *More Agile Testing* by Crispin and Gregory
- *Test Driven Development* by Beck
- *Testing Computer Software* by Kaner, Falk, and Nguyen
- *Software Verification and Validation* (V&V) by Deutsch
  Safeware by Leveson
- *Systematic Software Testing* by Craig and Jaskiel
- *Software Reliability Engineering* by Musa
- ISO/IEC/IEEE 29119 Software Testing Series of Standards
- IEEE 1012 Standard for Software Verification and Validation (includes system and hardware)
- ISO Standard 15288 Systems and software engineering – System lifecycle processes
- ISO Standard 12207 Systems and software engineering – Software lifecycle processes
- IEEE 982 Standard Dictionary of Measures of the Software Aspects of Dependability
- ISO 9000 series – Quality Management
- ISO/IEC 20246 – Systematic Review (of software, hardware, and system information)

> **Note**   For large IoT systems, only critical element human review may be possible; otherwise, automated tools may be needed.
>
> *Many of these books and standards do not exactly agree with each other all the time on testing/V&V.* However, the context of any project matters. Therefore, *testers need to be aware of a wide range of views and opinions.* I have worked on many of these reference standards and contributed to several of the listed books. The references for testing are large and have an ever-growing list on context. Context can be based on the following:

1. The nature of the project including management and development staff
2. The IoT device
3. The users
4. The risks
5. The cost and schedule
6. Any standards and/or regulations that impact the project
7. The project and tester history

We will address and identify other references later in the book.

Summary

This chapter is a quick overview and introduction to the world of IoT and this IoT testing book and covered the following topics:

Standards and references
Verify – Did we build the product right (compared to a reference document)?
Validate – Did we build the right product (compared to any user expectations)?
Audience and how to use this book
Important references to have in your professional library to be an informed tester

The next chapter explores IoT technology, testing, and projects.

References

1. IEEE 982.1-2005 IEEE Standard Dictionary of Measures of the Software Aspects of Dependability, in revision as of 2022
2. "IEEE/ISO/IEC 29119, ISO/IEC/IEEE International Standard – Software and systems engineering – Software testing" – Parts 1 to 5 series
3. Design for Reliability, by Raheja and Gullo, Wiley, 2012, chapter on Design for Trustworthiness
4. Committee on Information Systems Trustworthiness, NIST, Trust in Cyberspace, National Research Council, Washington, DC, 1999
5. Trustworthy Systems Through Quantitative Software Engineering, Bernstein, L., and Yuhas C. M., Wiley
6. www.channele2e.com/technology/security/solarwinds-orion-breach-hacking-incident-timeline-and-updated-details/ – accessed spring 2022
7. www.fortunebusinessinsights.com/press-release/internet-of-things-iot-market-9155 – accessed spring 2022

Figure Reference

1. https://image.freepik.com/free-vector/illustration-satellite_53876-8504.jpg; https://image.freepik.com/free-vector/internet-store-goods-international-shipment_335657-2454.jpg

Chapter 2
IoT Technology in Time and Space

In the first chapter, I introduced testing, verification, and validation since I use those terms constantly. This chapter starts by considering where IoT fits within the world of software, systems, and computers. Hence, we gain an understanding of why testing is essential even when an IoT project might wish otherwise.

IoT at a Glance

Today, we see the increased merging of physical and cyber systems to create the Internet of Things (IoT). IoT became possible because of maturing cyber and physical systems, including

- Embedded computer software devices
- Advanced physical systems
- Cyber IT, data analytics, and artificial intelligence (AI) computer systems
- Integrated communication technologies (Wi-Fi) and networks (e.g., the Web/Internet), which allow for components to easily communicate
- Smart-mobile cyber-physical devices

The IoT is, in part, an extension of an existing product area called embedded software devices that have been around almost since the beginning of computers and software. Technology companies have been using small computers, often called microprocessors or integrated circuit (IC) chips, for over 40 years. These devices typically had small amounts of storage, computing power, and software. They controlled systems but often did not have a standard computer user interface – a keyboard and screen. The embedded software was often used to develop "advanced physical systems" where the combination of hardware and software provided limited features and functions. Often, users of the embedded advanced devices were not aware they were working with software systems, for example, IT security researchers who found a major embedded software virus/worm; when they found the worm, they did not even know the kind of computer software the worm was targeting.

Over the last decades, mobile-smart devices using cellular technology now dominate the cyber market. Companies are putting "smarts" and IoT everywhere (Figure 2-1).

J. D. Hagar, *IoT System Testing*, https://doi.org/10.1007/978-1-4842-8276-2_2

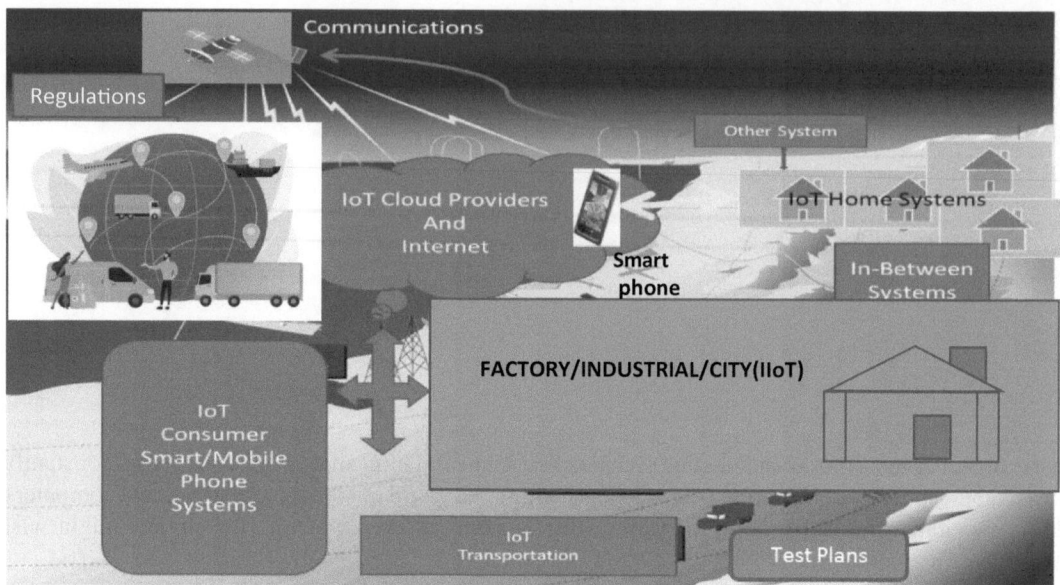

Figure 2-1. Example IoT systems and advanced physical systems moving to IoT

Composite by Jon Hagar from references and classes

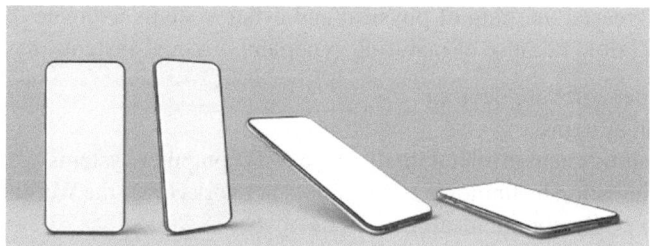

Figure 2-2. 2000s and 2010s mobile-smart cell system now many times part of the IoT Edge

The number of mobile-smart devices now outnumbers classic personal computers (Figure 2-2). The small devices are small, hand-carried, battery-powered, have more complex user interfaces, and have connectivity to computer networks (e.g., cellular systems and the Internet). The devices are close "kin" to physical systems since they have sensors, and many are now "connected" to other physical systems (e.g., cars, medical devices, etc.). This blending continues into the world of IoT.

With IoT, we see a merger of the traditional physical and cyber systems. IoT devices have become more viable now because IT, communication, data, hardware, and mobile technology have matured and blended. Some chips have processors, memory, sensor suites, and communication nodes. The cost has become very low compared to historical larger computers. These newer IoT devices are small and interfaced to what many would consider classic embedded hardware and fully connected to networks, including satellite networks. This blending offers new product features and capabilities. The number of IoT devices has driven the need for IPv6, a newer numbering scheme for IP addresses (`https://en.wikipedia.org/wiki/IPv6`), since the older Internet addresses have "run out" (IPv4) of indicators within the scheme. Besides, the number of IoT devices connected to the Internet is expected to be in the billions or more very soon.

Table 2-1. IoT Market Segment Examples

Industrial IoT 4.0	Mixed IoT	Consumer IoT
Government	Vehicles/robots	Home
- Health	- Driverless	- Security
- Safety	- Monitors	- Control and monitor
- Data analytics	- Infotainment	- Infotainment
Transport/robots	Office	Human
- Navigation	- Security	- Health
- Optimization	- HVAC	- Fitness
- Logistics	- Worker monitor	- Data analytics
Workplace/factories/robots	Retail	Travel
- Operations and control	- Ordering	- Fun
- Data analytics	- Point of sale	- Location
- Automation	- Advertising	Robots - Who knows where this will go?

Computer inputs and outputs have evolved in many ways, almost all of which affect software testing. Early inputs were notoriously clumsy and prone to error. No one wants a return to paper tape and punched Hollerith cards. A giant leap came with terminals and their familiar typewriter layout. However, more importantly, terminals are windows interacting with shell scripts and what became the start of automated testing. For outputs, computer users get immediate feedback on the screen. With IoT, these trends continue – as long as you do black box testing. You tickle the IoT device just like a consumer or another IoT device and observe the behavior. However, for more sophisticated testing, the absence of a keyboard and mouse (for inputs or outputs) and a large monitor requires new hardware/software configuration skills. We now need to consider where IoT is taking us.

IoT Market Segments – Where Is IoT Now and in the Future?

I see three IoT segments as a general abstraction: industrial (Table 2-1), consumer, and middle (mixed). Industrial IoT devices control and communicate about our cities, offices, factories, transportation, utilities, and everything that makes modern life possible. Consumer devices will be things at home and used (usually) by a person (in the future – possibly robots). The middle is a mix of consumer and industrial uses. The number of industrial IoT devices and systems will likely outnumber the consumer market by several orders of magnitude.

Consumer devices are more visible and have already gained public interest. These devices include automobiles, home monitoring and control, personal and medical wearables, and life enhancement systems. Pretty much anything used by consumers who had electronics will move to IoT. Additionally, many new products will be created that were never thought of as "electronic" but now become IoT components (e.g., smart shoes, net clothing, smart food packaging, and others yet to be dreamed of).

"Smart" connected devices are already in personal, industry, and government use, including Wi-Fi routers and Internet hubs, smart TVs, home security and management, inventory control systems/sensors, shipping trackers, inventory monitoring systems, HVAC systems, connected security devices, health devices, appliances, and you name it. Many of these IoT devices are made and sold by the largest commerce companies and deployed by others in the retail, government, healthcare, manufacturing, transportation, IT, and telecom markets.

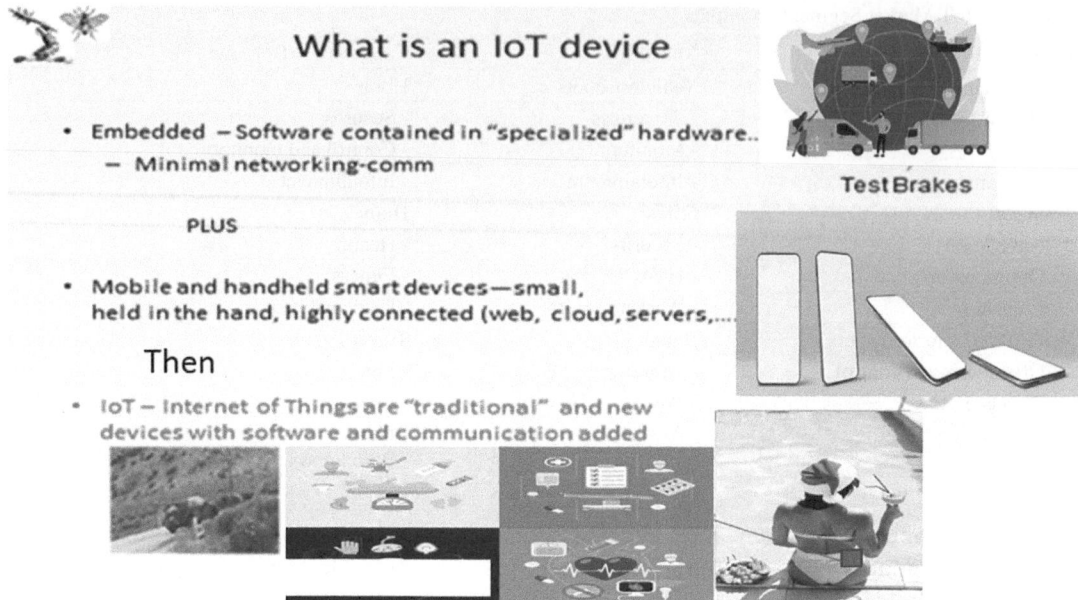

Figure 2-3. Embedded and mobile merge into IoT devices

Composite by Jon Hagar from preceding links

Many IoT systems will cross the bridge between consumers and industry to form the middle because they have communications to make the bridge easy. Industry and consumers will be users of these IoT devices. The division between the three is easily blurred. Consumer products now generate data that the data analytics community can use. Likewise, the industrial IoT systems will generate information and have uses that consumers will want. Companies working in the IoT segments may specialize in one area, but many companies will be in each segment.

From a testing perspective, there are differences in testing based on segment (as a starting point). For example, in Figure 2-3, a wearable device like a watch or monitor may have a "low" risk for bugs as an issue at first. Still, the more advanced versions of the watch might also monitor health data, making it a "higher"-risk device and, therefore, require more or additional testing. The lady in the swimsuit is a walking IoT system. She has a device at her hip that is monitoring her blood sugar. It communicates to her smartwatch (transmitting data), which in turn "talks" to her cell device (again transmitting data), which can, if needed, call her doctor's office. She is a walking, living system like the IoT devices in the car for control and handheld devices, and, therefore, these devices would require extensive testing to mitigate risks.

IoT impacts on testing may be hard to gauge because of these complexities. Consider the watch on the young lady, which at first was deemed "low risk" by the company when they first released it. Should it be retested when it is used differently? Experience shows us that many companies may say something like, "Well, the device has been used in the field. It is a watch; therefore, it is proven, no new testing is needed." However, now the watch has a communication link to health devices, a smartphone to link it to an external health system, and a link through a cellular network. Risks now include security, health privacy, death, false alarms, etc. What other risks can you think of given this scenario? We would recommend new testing of the system. However, many companies will slap on the watch a "user warning" that states "it is not the watch company's responsibility." Litigious societies would not allow that to pass as "normal."

A Sampling of IoT Challenges in Development and Testing

To contemporary software people, a few things seem as clunky and impractical as the old IBM punch cards. To computer programmers, punch cards were common and indispensable for working on a computer. In the 1960s, the cards were nothing short of magical to the general public. I can recall people who knew that I was a programmer bringing me stray cards they found lying around campus. Even when the cards were blank, these offerings were treated with great reverence when handed to me. Besides being just a piece of light cardboard in a standard size, the cards allowed a programmer to intervene on the sleeping computer and "make it dance." Of course, you had to wait for your turn in line to use the mainframe computer, pay for computer time by the second, and feed the card reader (Figure 2-4). Most significantly, it was a straightforward approach to instructing the computer with your wishes. Perhaps the computer operator might drop your cards, but otherwise, when you handed in your deck, you could expect a printed response on large sheets of paper called green bar (because it had green bars across it).

Output was the second part of the lifecycle of the older computers. When things did not go as expected, all the information you needed to fix the problem was right at your fingertips. What could be simpler than a large sheet of green bar paper with your results? Do some debugging, repunch the program cards, and then try again. All your observations were in one place and ready for annotation with a device that has been around since the sixteenth century – a pencil.

Finally, the last step was the results of your run. Data read from a printout populated reports and articles. During the 1960s, it was common in newspaper and magazine articles to read phrases like, "the CPU of the computer is the 'brain' which controls all the data processing" or "on large reels of magnetic tape, enormous amount of data is read and written." The public did not understand what went on with computers. Today, decision-makers are far more technically savvy. However, they may be as ignorant of IoT impacts as the decision-makers of 50 years ago were of computers at that time. How can they be expected to understand the risks when they do not understand the technology that makes up the lady's watch system? As a designer of tests, a reporter of test results, and an advocate of observations from intervening in the IoT system, you cannot neglect the challenge of explaining what is going on in IoT systems.

Flash forward to today and IoT development and test. For typical users, IoT will seem like magic for some time. "Do not look behind that curtain," said the great Wizard of Oz. Development teams will have integrated development environments (IDEs). We will rapidly pull in software reuse items with microservices, commercial off-the-shelf, and open source no-code/low-code software packages. Our projects will primarily be reused software and hardware elements that we leverage with language constructs and tweaks of tech and libraries. We will rely on third-party software to help with

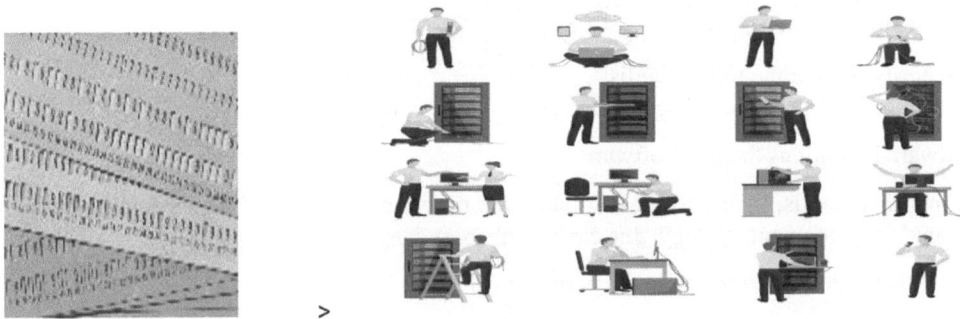

Figure 2-4. Punch cards ran old "big" computers

functionality via commercial off-the-shelf (COTS), open source software (e.g., libraries found in GIT), and other no-code/low-code development practices. These practices and not having security activities in place on your code can have risks (see the sidebar: Orion) [1].

ORION
"The routine (software) update, it turns out, is no longer so routine. Hackers believed to be directed by the Russian intelligence service, the SVR, used that routine software update to slip malicious code into Orion's software and then used it as a vehicle for a massive cyberattack against America. 'Eighteen thousand (customers) was our best estimate of who may have downloaded the code between March and June 2020,' Sudhakar Ramakrishna, SolarWinds president and CEO, told NPR. 'If you then take 18,000 and start sifting through it, the number of impacted customers is far less. We don't know the exact numbers. We are still conducting the investigation.' By design, the hack appeared to work only under particular circumstances. Its victims had to download the tainted update and then actually deploy it. That was the first condition. The second was that their compromised networks needed to be connected to the Internet, so the hackers could communicate with their servers." [1]

IoT devices, by their definition, will be connected to the Internet.

IoT functionality will span hardware, software, communications (comms), and operations (Ops) while generating large amounts of data, which we testers must also use. Hardware constraints and problems already abound, including

- Product standards
- Multitudes of device hardware configurations that change constantly
- Sensor input challenges
- Control output issues
- Battery and power concerns
- Memory usage restrictions
- Processor speed limitations
- Hardware lifecycles
- Just as a few examples

Software factors to address include

- Security and privacy
- Data processing
- Ubiquitous usability
- Third-party software impacts

  • Security risks at the top of the list
  • Off-the-shelf, no-code/low-code, and commercial components

- Hardware limitations fixed in software

  • An old saying is: "We will fix that hardware problem in software." If accepted into any company's development strategy, this carries inherent dangers or risks.

- Short lifecycles and interfaces to the hardware lifecycle

Communication challenges include

- Speed of the networks (local, nearby, and global)
- Drop and brownouts, even closed networks (e.g., China, Russia, Iran, and other countries)

- Multiple communication standards
- Device-to-device traffic
- Interoperability of data
- Timing (hold to download, stale data, latency, sequencing, throughput, etc.)
- Edge computing impacts

Operations will have various concerns, including

- Cultures of different organizations and users
- Addressing the preceding items on hardware, software, and communication
- Keeping users happy
- Drowning in massive data
- Staying alive to achieve any ROI
- Localization
- Globalization
- Security and privacy

Points of failure in all of the preceding areas will be unseen, unexpected, and have gremlins waiting. Organizations, especially testers, will ultimately have to address many of these using V&V/test to succeed.

Familiar test techniques will serve professional testers. Part 3 will cover essential test techniques. Quick tests apply to applications running on IoT devices. Testers will do domain tests, integration tests, and tests designed to explore risks. The science and philosophy of IoT testing will inherit industry test history. You have a good start for IoT if you know the art and engineering of existing testing systems, but it is only a start.

IoT testing, in practice, offers some formidable challenges. Starting to become aware of the challenges and understanding how to overcome them is the primary objective of this section of the book (see the sidebar: A Simple Heat Controller).

A SIMPLE HEAT CONTROLLER
Consider a typical IoT device – a heating control thermostat found in a smart home (Figure 2-5). Temperature controls adjust the furnace or air conditioning outputs. Rules might be as simple as reacting to ambient heat or cold or cooling things at night or only operating when the price of electricity is lowest. Suppose you were asked to test an IoT thermostat. What would your first considerations be to outline a testing strategy? What can you observe? You can see its display or on an app (running on a phone or in a laptop browser). If the thermostat connects to URLs on the network, can you see data coming and going? How can you observe and record the current state of the thermostat and its environment?

What inputs can be tested? You can click all the buttons and tickle every part of the user interface. How can you make the device think the air temperature is hotter or colder or spoof the thermostat into thinking it is such and such? How can you fool the thermostat into thinking it is now 2:00 a.m. when the time is just before lunch (11:00 a.m.), which is a much more convenient time for you to do the testing? Do you test or simulate what happens when the thermostat must be reprogrammed with codes, temperatures, and security settings when a new owner moves into the house? Do you do the reprogramming as part of a whole smart house environment? Can you overheat the house even with outside temperatures below 30F? Can you overcool the house even though outside temperatures are 30F? Can you hack the device using the outside Internet, cellular, or cable connection, and what things are allowed when the hack happens? Are user IDs and passwords available to you during the hack that has nothing to do with the thermostat device? Can you hack other IoT devices in the home using the thermostat as the "entry device" to the scheme?

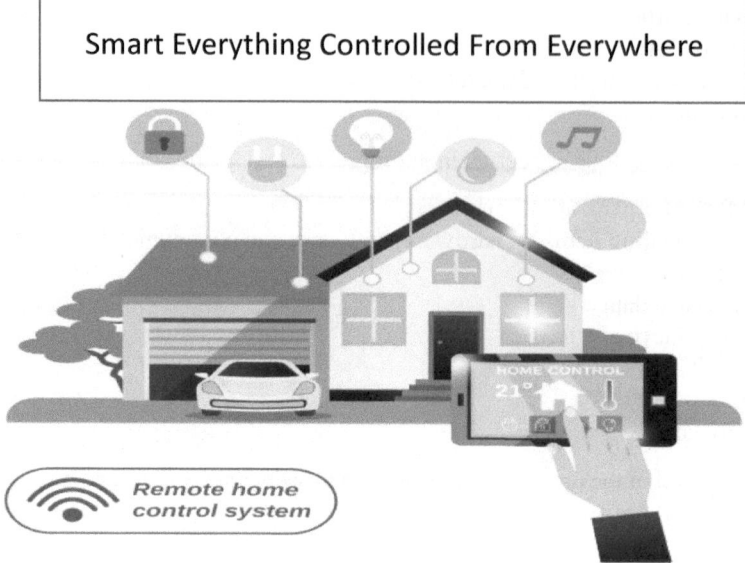

Figure 2-5. Smart houses will have many IoT devices (10s to 100s to 1000s), and do users want this and all that it implies?

What user stories need to be dreamed up for testing? Worse, do you do these tests for the first release, next, or last? You will design tests to collect information needed by stakeholders. As you collaborate with them, how do you educate them about the risks of IoT? Since there are always more tests designed to be built and run, how do you help stakeholders decide which tests are the most critical and why? How can you translate for simple stakeholders technical jargons like "Network latency and the unreliability of UDP messages to be dropped silently on embedded processors with no memory or resources to acknowledge as per the standard?"

What tests should be automated, and how? What exactly can we run tests against?

Suppose you get the idea that there is more to developing and testing IoT systems than just addressing the functionality of small devices. In that case, you begin to understand more of the total challenge of developing and testing IoT devices – or the world of IoT.

IoT Test Team Responses to the Development History of "Testing Is Dead"

Development in IoT will be fast-paced and often with a "startup" mentality (see Chapters 1 and 3). We will cover the importance of cooperation between IoT Dev, testing, security, and Ops in the coming sections of the book, but testers may face difficult everyday situations and have to think on many possible solutions. For years, we have heard "testing is dead," "we do not need testing, we are Agile," and "I do not make mistakes" from the development staff.

1. The hardware is not done, but the software needs to go into testing now, but we do not have time before shipping. What do we do?

- One possible answer: More than likely, this team did not think about testing until the hardware and software were completed. This case is old-school history for how to reach project failure quickly. Agile and most test efforts now advocate ongoing continuous integration (CI) and continuous test (CT) to reach continuous deployment (CD). Also, the tester must use planning, crisis, and general management.

  Note: There is more on these throughout the book.

2. IoT hardware and software are in flux, but testing needs to get started now and be done as soon as the Dev team is done. We must ship the device ASAP for prototype assessment.

   - One possible answer: This is a variation on case 1, and the answer starts with that answer, but we add that flux on hardware and software should be planned so that CI, CT, and CD are scheduled. Testing feeds into Dev constantly.

3. The IoT project does not budget for any testing because the effort is a startup, and Dev wants to stay alive after the first development iteration. How does the project do "something" to assess quality?

   - One possible answer: Agile testing is done with the Dev team doing some V&V assessment. They should also plan on testers coming onboard on the next development cycle.

4. IoT hardware is completed but has a problem, so the team decides to fix it at the last minute using the software because it is easy to change, and there is no time for retesting software. What do the testers do?

   - One possible answer: If I had a dollar for every time I heard this story, I would be rich. The risk item here is that the test team needs to have planned for this situation and have a fast and automated regression suite of tests ready to run, hopefully overnight. I have spent many late nights in the test lab.

5. The new IoT project uses off-the-shelf (OTS) hardware and software, so effort does not need much (any) testing because these things have all been in use.

   - One possible answer: We testers should remind or explain, and managers should know that previous use and testing did not address every use case of the *new* IoT device. Since new data and usage will happen, there likely will be new errors and functionality to show or prove the device is working. Also, there may be security issues that present at unexpected times. This answer may be a short test plan, but the project will have surprised users with no testing.

   - *Note: In this book, OTS software refers to commercial OTS, open source OTS, no-code and low-code OTS, microservices, operating systems, and others.*

6. The project has no users/stakeholders to talk to, so the Dev programmer will decide what is "needed" for the device testing.

   - One possible answer: This situation is that the development programmer has biased blindness. "They believe it will work simply because they created it." And while development should be doing testing, another independent, nonbiased tester viewpoint allows much less risk.

Examples of How the IoT Devices Can Impact Testers

As noted earlier, development impacts testers, and some of these scenarios are common to most organizations. But what in IoT may be "different" from classic IT testing? In this section, I present a few examples that have been encountered to date.

IoT testing includes all of the input and output hardware devices. First, testers will need to know and understand as well as test hardware. Many IT testers only deal with "hardware" issues in passing because the basic computer is generic. However, in IoT, each device will have unique hardware and physical world interactions with input and output subsystems. Yes, you may have the human user interface (UI) most of you are familiar with, but the input sensors will be gathering data often from nonhuman users (e.g., hot, cold, wet, slippery, numbers in ranges, etc.). Further, the IoT device will be controlling or getting feedback from something on the output channels (e.g., motors, on/off switches, actuators, etc.).

The next issue will be communication channels and standards. Here, you may need to test things such as Wi-Fi, cellular, Bluetooth, near-field communication (NFC), and/or any other communication standards. Worse, these communication channels "integrate" the IoT device under test with other systems. Does your testing stop at the interface to "the outside?" Who owns the system of systems testing (e.g., the home, the factory, the car, the city, the world)? Users will expect all of these things to work together no matter the configuration, and there may be many configurations, so testing everything will be difficult.

Governments are becoming more heavily engaged in creating "smart" cities and connecting the world. IoT devices control and monitor roads and traffic, energy usage, communications, and other physical systems that make cities work. Estimations are that money will be saved, and citizens will be happier. However, Ted Koppel's recent book *Lights Out* (`http://tedkoppellightsout.com/`) on the power grid threat indicates that many of our current cyber-physical systems are at risk of hacking, including power, water, and sewer, to name but a few. Many of these systems are adding IoT without the industry and government adequately accounting for their inherent risks. Some of us have been writing and speaking on these cyber-physical IoT security risks for years, but only now is our development and testing of these systems coming to the "visible" levels.

Okay, you say you are working on "low-risk" IoT devices and do not believe that testing is a big issue.

Was there much testing of the physical devices done? Possibly more testing is needed than you might think. Well, take, for example, a loudspeaker company that might have been in business for years making consumer speakers (Figure 2-6). The company historically had to test: electrical, fire hazards, UL standards, speaker life, long-duration test at max volume, different environments, etc. These companies learned that by product recalls and lost sales, testing must be done repeatedly. Now, take newer speakers from this company with added software to work with Wi-Fi or Bluetooth, or to play music, or allow a robot to talk in the house, or to allow other streaming services with many different smartphone connections. Does their hardware test team understand software testing of IoT devices? Not at first, but likely they will be learning.

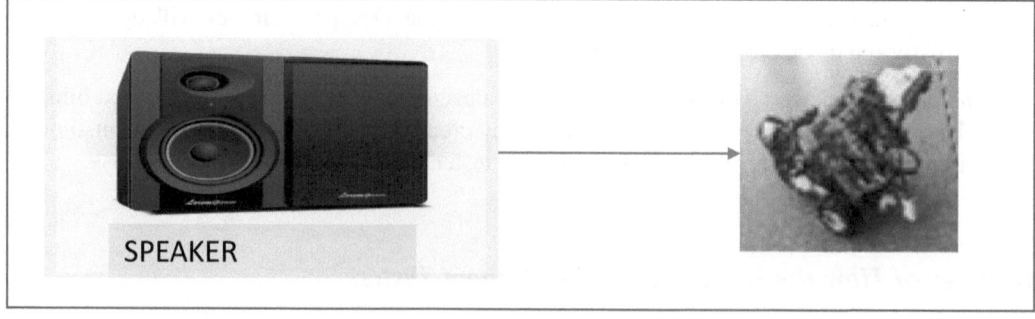

SPEAKER

Figure 2-6. Smart Wi-Fi speakers connected to a smartphone and house robot

Reference: Jon Hagar composite

A distorted or garbled song thumping from a speaker may cause you alarm that maybe hacking is afoot. However, what about when hackers are just collecting data on your listening behaviors? Would you even know if your playlist was sent to a marketing research team without your consent or compensation? Just like familiar "industry stamps of approval," that is, standards (CE, UL, EU security regulations, etc., covering protections for the consumer from a product that will set the house on fire or irritate the neighbors), are there similar assurances that companies can provide to consumers about the privacy of their information in a connected world?

Questions to ask yourself are: "What tests can you run to lower the risk of the loudspeaker not working with your favorite archive of music? Can I listen to music while the robot cleans my floor?" If you answer these sanity checks, some might call those smoke tests, and you may learn within a few minutes if the devices work together. These types of questions are the beginning of a test plan based on risk.

As a speaker company, you can post the results of simple tests on your product's support web page. Through a website, the update can be posted in real time. Incompatibilities tested and found between devices can also be published. Perhaps you can include an area where current and prospective customers can request tests to be run for in-use combinations.

Next, some medical device manufacturers see a 50 percent reduction to repair their connected devices in the healthcare field. Consumers can download new software to fix some problems with reported savings in customer service costs by $2000 for each problem resolved remotely. Great! However, consider the hacking of pacemakers in Figure 2-7 [2]. How much will that cost, and to what benefit?

Further, consider a tire maker using IoT to gain valuable insights about the performance of its products in near real time. The company uses an analytics platform to manage the vast amounts of data gathered directly from sensors embedded in the smart IoT tires of a smart car in Figure 2-8. The system allows the monitoring of the pressure, temperature, and mileage of each tire. By keeping these factors in acceptable ranges for industrial use, tire/care fleet managers can significantly impact fuel economy and safety, possibly saving over $1500 per vehicle per year. However, consider the impact

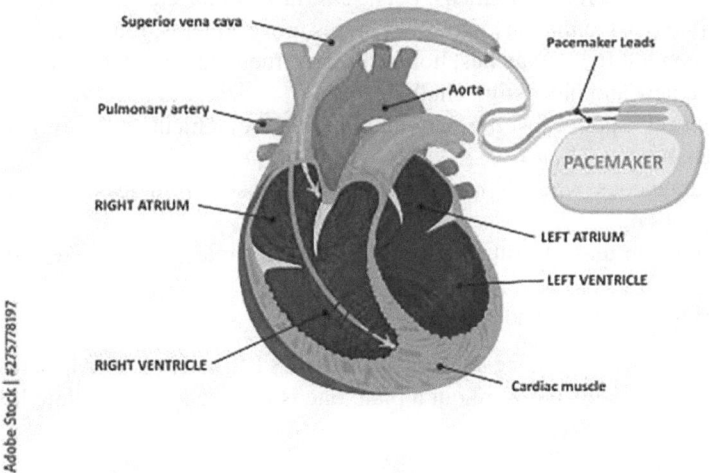

Figure 2-7. Smart health IoT device (pacemaker)

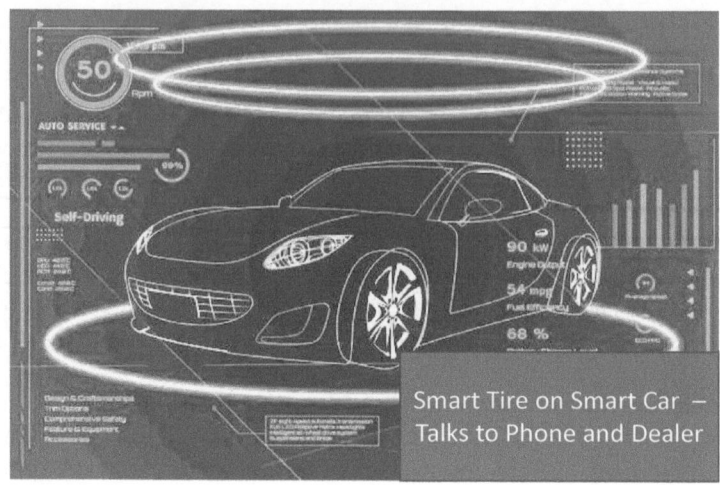

Figure 2-8. Smart tire

of a consumer being hacked the way a Jeep was [3]. Can the IoT tire system be hacked, and what would effectively make the car unsafe? Skilled testers should be able to answer this line of questions and provide the information to management and stakeholders.

Hopefully, given the last two situations, you have begun to think about another significant scary aspect of IoT security testing. If it seems like there will be a lot of learning, new testing, and new bugs making it into the IoT field, yes, that is 100 percent correct.

Summary

In this chapter, the key points we considered include

- The IoT world is vast – Homes, humans, pets, seas, mountains, etc.
- Industrial IoT (IIoT) – Companies, factories, stores, governments, roads, etc.
- Consumer and general IoT – Watches, home security, medical, entertainment, etc.
- IoT devices face new and old testing challenges.
- Where is the "sweet" spot of just the right amount of development and testing to be successful?
- Does success change over time?
- Testing is advantageous when it moves a decision-maker from less complete information to complete information.
- More information can make testing a success. (Less information can make testing and a product fail.)
- Who owns testing the system and system of systems?

In the next chapter, I will introduce and consider lessons learned in IoT project test planning, which is an important first step in testing. Without a plan, one is just wandering around lost.

References

1. www.npr.org/2021/04/16/985439655/a-worst-nightmare-cyberattack-the-untold-story-of-the-solarwinds-hack
2. https://threatpost.com/pacemaker-hacking-fears-rise-with-critical-research-report/120174/
3. www.wired.com/2016/08/jeep-hackers-return-high-speed-steering-acceleration-hacks/

Figure References

1. https://image.freepik.com/free-vector/illustration-satellite_53876-8504.jpg; https://image.freepik.com/free-vector/internet-store-goods-international-shipment_335657-2454.jpg
2. www.freepik.com/free-vector/vector-smartphone-with-blank-white-screen_7588609.htm#query=cellphone&position=36&from_view=search
3. https://image.freepik.com/free-photo/christmas-woman-relaxing-by-swimming-pool-funny-girl-celebrating-christmas-resort-with-cocktail-fruits_1157-49378.jpg; https://image.freepik.com/free-vector/surgery-medical-icons-design-concept_98292-450.jpg
4. https://image.freepik.com/free-photo/green-punched-card-programming_469558-8250.jpg; https://image.freepik.com/free-vector/system-administrator-flat-icons-set_1284-17354.jpg
5. www.freepik.com/free-vector/smart-home-with-smartphone-control_2458891.htm#query=smart%20homes&position=3&from_view=search
6. https://image.freepik.com/free-vector/loud-speakers-isolatedillustration_73621-990.jpg
7. https://stock.adobe.com/images/id/275778197?as_audience=srp&as_campaign=Freepik&get_facets=1&order=relevance&safe_search=1&as_content=api&k=pacemaker&as_camptype=test-sponsored-a&tduid=77396d47a6d0d41c1f50f29edb16bc2f&as_channel=affiliate&as_campclass=redirect&as_source=arvato
8. https://image.freepik.com/free-vector/blue-neon-sports-car-infographic_53876-99420.jpg

Chapter 3
Big Picture Lessons Learned in IoT Project Test Planning

In Chapter 2, I introduced IoT technology and the impacts it is having on development and test teams. In this chapter, I will dig deeper into IoT test planning, specific testing risks, and impacts on testers. Any new technology entering the technology curve encounters various problems in development. Some problems will be new to the IoT space, while others may be familiar from other software technologies. IoT companies and testers should start by understanding as many of these problems and solutions as possible because those that do not learn from history are doomed to repeat the historical lesson.

High-Level Issues at a Glance

This section considers likely high-level problems that may be encountered during the development and testing of IoT devices. It is where testers and planners need to start their risk analysis and test thinking. More IoT risks will be covered in later chapters.

#1: Creating and Releasing Software

One of the IoT situations that companies face, which have never developed software, is the challenge of creating software and releasing it in IoT products. Software is different from hardware. This is a common scenario. An example we have seen many times is companies taking existing hardware staff – qualified or not – and putting them in charge of software development and testing, only to have unhappy users when the IoT product is fielded. This is more commonly called "product failure" when, in fact, it is the management's and stakeholders' such as customers, who fail to understand the software requirements of people, products, and the success of both. Alternatively, we have seen hardware companies start adding processors and software by using outsourced subcontracts to solve their lack of software experience. Suddenly, new errors appear in their "new and improved" software-based IoT devices because the subcontractor did not understand testing. Finally, we see companies hiring new staff to be in charge of software development and testing, only to be surprised when the management of these new organizations does not understand the critical nature of software lifecycle efforts. These situations can be varied and mixed into other projects, but they form a fundamental set of IoT project risks.

© Jon Duncan Hagar 2022

J. D. Hagar, *IoT System Testing*, https://doi.org/10.1007/978-1-4842-8276-2_3

#2: Understanding the IoT Lifecycle

Another strategic managerial phenomenon is where a company thinks IoT is just like developing information technology (IT), web, or PC software. To be sure, in IoT systems, there will likely be IT, web, and PC components and/or some similarities. However, the IoT device is a mix of embedded systems, hardware, and mobile software environments. IT departments, developers, and testers will encounter situations familiar to the mobile, hardware, and embedded software situations, but much of it may be new or strange to some. For example, there are problems with the amount of available memory in embedded systems, limitations in battery usage, processor speeds, communication dip outs, and visibility into hardware states. Also, in mobile/smart devices, embedded testers detect bugs related to device interoperability and system integration issues with unique hardware.

Moreover, finally, as in traditional computing systems, one can have risks in basic functionality, nonfunctional quality features, and data concerns. All of these issues form another new set of risks for IoT systems that must be included in the development, verification, validation (V&V), and test planning. And these latter elements are part of the development lifecycle, which includes software but is not limited exclusively to it.

#3: Test Tools for IoT

The next surprise risk will be the development of test tools for IoT. The uniqueness of the IoT device will mean that a mix of traditional and specialized custom tools to create a development and test environment is needed. The staff should not expect sophisticated IDE debuggers and coverage analyzers for the code running on a resource-challenged IoT device, meaning a poorly planned device vs. a resource-constrained device with specific constraints or limitations. For more information, refer to Part 4.

#4: Avoid Impacts of Rushing in Competition

The goal of seeking fortune and a company's competitive edge, first and foremost, will lead to "fast" and incomplete planning, among other mistakes.

Most of the industry is or will be seeking a competitive advantage by having IoT devices. *Harvard Business Review* has written about this as in this reference [1]

Conducted in September 2014 on early IoT adopters, the survey shows that companies benefit from deploying IoT-based initiatives. Among the reasons they most frequently gave for adopting IoT were enhanced customer service (quoted by 51 percent), increased revenue from services and/or products (44 percent), improved use of assets in the field (38 percent), and acquiring more information to support big data/analytics efforts (35 percent).

Respondents said they have deployed or plan to use IoT in many areas, including asset tracking, security, fleet management, field force management, energy data management, and condition-based monitoring. Moreover, they give IoT high marks. For example:

- 62 percent say IoT somewhat increased or significantly increased their customer responsiveness

- 58 percent say it increased collaboration within the business

- 54 percent credit it with increasing market insight

- 54 percent believe it increased employee productivity.

As the quote indicates, many companies are entering and rushing into IoT since they believe there is quick money to be made. The money will come when the device sells and from selling services such as network information, data, and analytics. This combined product and service space will come with various new risks. Who owns the data? What are the privacy issues? Does the company have the ability and right to use the data as well as perform the data analytics? The answers to these questions are project dependent but may influence product development and testing.

#5: Advanced IoT Device Challenges

Advanced risks and challenges should be assessed for V&V/testing and development of IoT devices. These are listed here:

- "Physical environments where the device will be used need to be addressed by development and testing" – Hagar
- "IoT has unique aspects of privacy and compliance in regards to regulations" [1, 6]
- "Is security and privacy adequate outside of regulations (future-proofings)" [1, 6]
- "Machine-to-machine communication and interaction will be fast and frequent with IoT" – Hagar
- "Managing the numbers of devices and the data generated will be a challenge" [1, 6]
- "Interaction and usability will be a key to the success of IoT" – Jeff Yakmora (verbal communication)
- "No Code or vendor software and hardware reuse will spark development speed while still needing testing" – Hagar
- "Interoperability with different hardware, software, and networks will be a challenge for many years to come" – Hagar
- "Skills of the team to perform data analytics (and development) will be an issue" [1, 6]

While some of these items are for all software, IoT devices magnify them and deserve focus in test planning.

#6: Testing in the Complex World of IoT Systems and Systems of Systems

Finally, and maybe the hardest top challenge is the stopping point for testing. Does the job stop at IoT devices, the edge, the cloud, or the world? And who owns the testing past the stopping point? Again, answers will be project dependent.

The stopping point may be evident in IoT's planning development and testing. Consider Figure 3-1.

We start (at the bottom) with a device in the factory, which will get a single IoT test plan. Indeed, this device needs to be tested, and the development-operations (DevOps) team can draw a line (the black box) and say our ownership stops here. However, IoT devices can connect to the edge or fog computing. Who owns that? The next connection would be to the cloud. Who owns and tests that? And finally, everything connects to the world; see Figure 3-2.

The users of the whole world will have expectations that every element in the whole world scenario will work. Some IoT systems may have a clear definition of such a picture. Many do not. The word "user" is an overloaded term (used ambiguously) because other parts of the system may be nonhuman users. Bad actors or hackers are users that we don't want. Hardware can also be a user. The boundary endpoint may be unclear because of all the factors, users, devices, and teams.

Factors will change the way we create products and field IoT products. Users are sophisticated and will expect ease of use, simple solutions that provide more benefits than problems, as well as fun.

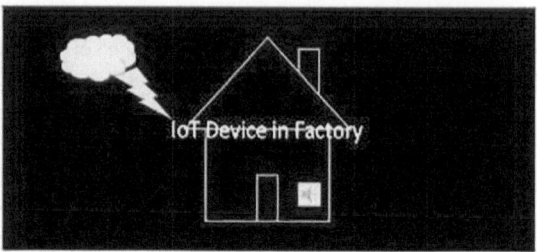

Figure 3-1. The factory black box moves into the IoT world

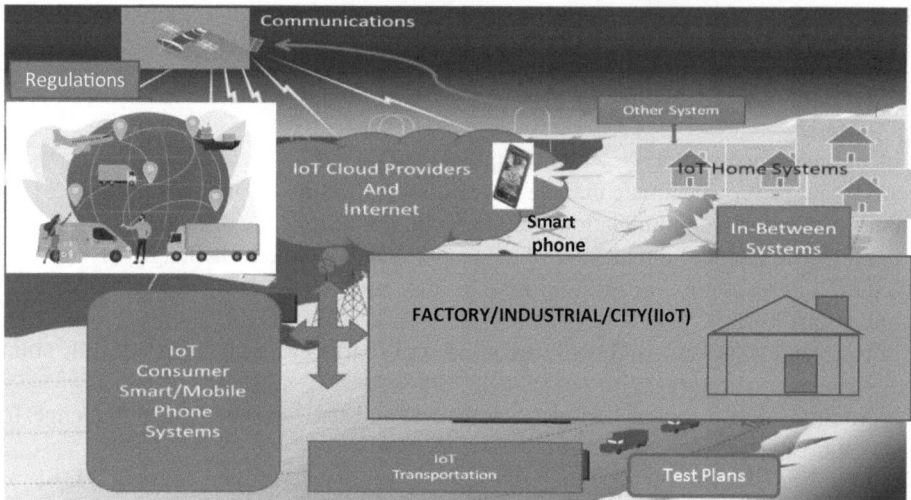

Figure 3-2. The IoT world

Reference: Jon Hagar composite

Many users do not want to know their devices are "smart" computers. They want the whole system to work, seem simple, and make their lives easier. If the system fails, they will complain, which could make the product a failure.

Because of these top-level risks, this book will present information based on three broad categories of IoT testing:

1. The device itself (single point)
2. Devices with a set of communication channels (e.g., edge, fog, cloud, to or from a system or system of systems)
3. The whole system or system of systems (or as big as it can get)

Project Test Risks: What Are They for IoT?

The concern with risks may be enough to make some companies reluctant to move into the "high-tech" IoT space from their more traditional endeavors, or they will accept the ownership of this "big picture" system. Some companies not making the transition to IoT may not remain viable businesses. We still have traditional (analog) Swiss watchmakers, but many Swiss watchmakers did not move to digital. They left the market to Asian companies; thus, Swiss companies lost their market share.

Figure 3-3. Market Darwinism from being upright to human to the smart brain to AI

Furthermore, the traditional watch market share has dropped worldwide as many people no longer have traditional watches. Instead, they use their smartphone and/or wearables to tell time. To stay competitive, companies will move traditional products into the IoT world, and many other new companies will conceive new products in the IoT space. It is likely that a majority of companies will move into IoT even if they do not have software system development and testing experience.

With this expansion, companies and users will look for the killer IoT apps and excellent IoT device configurations. There will be market volatility when customers expect "quality" (a value that someone is willing to pay for). Precisely what quality is may change over time with different products, but high-tech users have increasingly had higher expectations of functional and nonfunctional elements. For example, the early smartphones only partially looked like those that now dominate the market today. Users quickly posted bad reviews on social media regarding their opinions of systems, apps, and devices – whether qualified opinions or not.

Companies and people looking to start IoT projects should consider the impact of marketplace Darwinism, as depicted in Figure 3-3.

To survive the IoT market evolution, one needs a good idea developed to be just "good enough" to reach realistic sales estimations. Information provided by testing is part of how software companies determine if a software product is good enough. The software industry is littered with good ideas that were not good enough and projects that did not deliver on a good idea. Various industry statistics indicate that between 30 percent and 60 percent of new software projects have major delivery problems or fail to deliver a product. The exact number is not that important for companies looking to move into IoT. What is important is that companies and engineers in IoT understand that the risk of first-time failure in IoT is high, and this point should be given due consideration. That is not to say that all or any of the ideas in this section must be implemented. However, the lack of ongoing planning and risk analysis has resulted in many "not good enough" products.

Getting Started with IoT Test Plans and Strategy

One aspect that makes IoT different from traditional IT software–focused systems is the integrated communications that must take place between unique hardware, specialized software, networks, and operations. These are all significant areas, each of which could or do have dedicated test books,

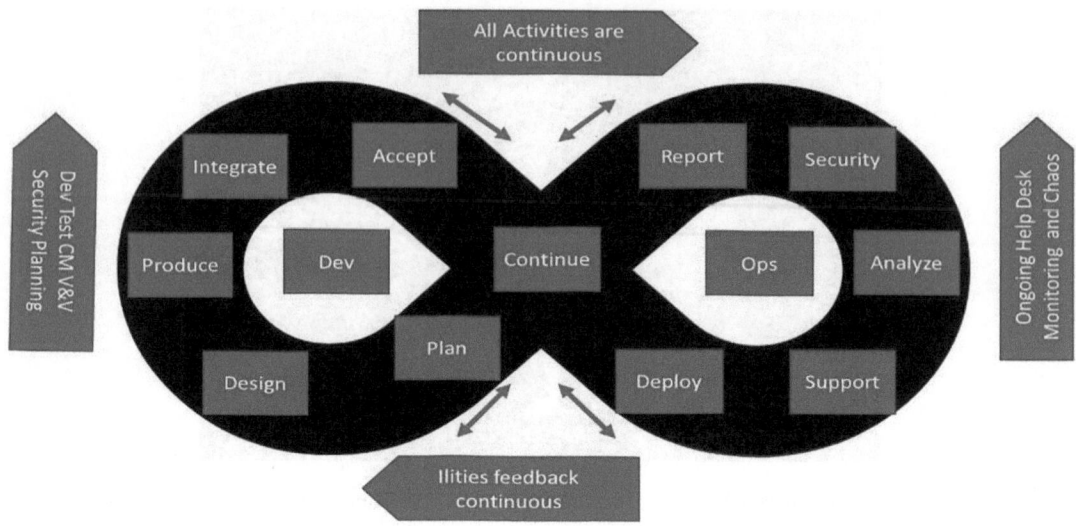

Figure 3-4. IoT development-test-security-operations infinity lifecycle

Reference: Jon Hagar

standards, and references. In Part 2, we will talk more about planning these environments and factors that teams must consider. This chapter provides a list to get you thinking about the areas and these factors. You can use the lists to look up more information on the Internet and other references, as well as read through Part 2.

The infinity lifecycle picture of Figure 3-4 connects to the ideas of traditional, Agile, DevOps, and other lifecycle depictions as an ongoing continuum that continues for the product lifecycle.

> **Note**
> All activities are continuous, repeated, and can be done/redone at any point.

The software lifecycle picture is often used and shown with directions such as a cycle, waterfall, V test model, etc. However, the double-headed arrows and "continue" box in the middle of Figure 3-4 depict activities and sub-activities on an IoT project that can be done at any point because lifecycles can be continuous and infinite, even though humans tend to think and write in a linear fashion.

The following list summarizes the IoT infinity lifecycle and adds examples:

- **Design** efforts can include the following areas:

  - Requirements
  - Modeling
  - Quality characteristics and goals
  - Interfaces and integration
  - Allocation to hardware, software, and/or Ops
  - Communications
  - Data allocations and analysis
  - V&V/test
  - Management
  - Support functions (e.g., configuration management (CM), quality assurance (QA), measurement, improvement, marketing, etc.)

- Hardware **design** engineering efforts include

  - IoT device requirements
  - Electronics
  - Mechanical
  - Packaging
  - Support functions (e.g., CM, QA, measurement, improvement, marketing, etc.)
  - Implementation and manufacturing
  - V&V/test
  - Data allocations and analysis

- **Produce** software (development) efforts include

  - Commercial off-the-shelf (COTS) software
  - Vendor-provided software

- **Support** functions (e.g., CM, QA, measurement, improvement, marketing, etc.)

  - Implementation programming
  - Data allocation and analytics
  - V&V/test at component and integration levels

- **Produce** hardware (Dev) with configuration management efforts include

  - Building the hardware unit
  - Building the hardware elements
  - Building the full hardware

- **Integrate** efforts include

  - Software integration
  - Hardware integration
  - Third-party element integration
  - Configuration management of hardware and software
  - Complete the full system

- **Acceptance** efforts include

  - Testing/V&V of the software
  - Testing/V&V of the hardware
  - Testing/V&V of the system
  - Network and communications testing and acceptance

- **Operations** include

  - **Deploy** the system with operations and maintenance in the field

- **Support** activities (as needed) such as

  - Help desk
  - Realtime user support
  - Rapid response time (security, faults, failures, etc.)

- **Security** of the system elements

  - Threat detection
  - Breach detection and cleanup
  - Vulnerabilities and testing (see Part 3)

- **Analyze** the operations data

  - Data analytics
  - Selling data
  - Product risk identification
  - Process and product improvement
  - Dependability, a.k.a. IEEE 982.1

- **Report**ing to/for management, engineering, and stakeholders

  - Data
  - Security
  - Legal issues
  - Outstanding issues
  - Support costs
  - Revenue, profit, loss

Finally, IoT support functions span all of engineering and management areas, including

- Engineering of hardware, software, and the system, each with their lifecycle
- CM and SCM
- QA/SQA
- Measurements/metrics
- Process/product improvement
- Testing/V&V
- Leadership and management
- Security
- Marketing
- Quality assessments and analytics

If you get the feeling that there are many things to do in the infinity lifecycle, you are likely still underestimating the job and all the possible subparts. The reference list supporting these engineering efforts has ties to standards and industry references measuring in the thousands of pages. If the previous concepts are unfamiliar to the reader, more reading on using these ideas is in order. The reference lists found at the end of this book are a place to start to fill in what this book only introduces. The lists and brief outlines are incomplete since this book's focus is on IoT testing. But, I feel it is essential to recognize these other activities.

Experience shows that companies and teams with expertise in one or two areas will underestimate and underappreciate many aspects of the IoT lifecycle. Each group and each company will be at different maturity levels and understanding in these IoT topics. A key for me is that groups must know their strengths and weaknesses.

It is quite risky to under- or overestimate knowledge and maturity. I do recommend that teams moving into IoT do some assessment of their weaker areas. Project risks should be considered areas of weakness (lack of knowledge, experience, or training) or overconfidence (demonstrating a know-it-all attitude). Further, ongoing process assessment and improvement are good ideas as any effort continues.

Startup companies will likely do their best within the limited budgets and schedules these groups typically have. Mature, large companies will have more resources to pull from and, hence, fewer weak areas, yet overestimation of abilities can be a risk in and of itself.

Note
Comparisons to standards, such as ISO, can provide a baseline for project test planning and thinking. I am not saying follow or use any standards or reference books verbatim as they exist. I believe that people should have some working knowledge of references, such as ISO and IEEE or even so-called "expert" books, and then be able to tailor project or product processes to meet the intent of such references – unless there is a contractual or safety requirement to do otherwise. Such documents are not meant as a "bible" but helpful information useful in triggering critical thought during your work. I have worked with places that only used 10 percent of an ideal while being very successful. I know groups that used all parts of a standard and then failed.

Introducing IoT Verification, Validation, and Testing Concepts and Standards

Many traditional software testers and projects are unfamiliar with V&V concepts as defined by standards such as IEEE 1012, ISO 15288, ISO 12207, and ISO 29119 [2, 3, 4, 5]. The lack of familiarity may be because many V&V concepts are related to system and hardware evaluation or because aspects of the software industry have focused on a broad definition of testing and quality assurance (QA), which includes the essence of V&V. However, in IoT, because I feel that product quality assessment activities need to "go beyond" traditional IT software testing, as defined by common practice or in standards, such as ISO 29119, I have outlined in this section the concepts of V&V and associated historic industry ideals and IEEE 1012. Readers following these concepts and wanting more details about V&V and industry integrity levels can obtain and reference IEEE 1012, other V&V reference books, or standards such as ISO 29119. Still, these standards do not address all aspects of the new "state of development," which is IoT.

V&V is part of building a system, including hardware, software, and operations. IEEE 1012 has decades of use on various critical large and complex software systems. V&V provides evaluations and assessments of development products (e.g., requirements, designs, documents, and the end deliverable user products). V&V is performed throughout the development lifecycle and allows test staff to help the organization deliver a "good enough" product. The definition of "good enough" evolves as a product matures (e.g., good enough for an early stakeholder prototype may not be as robust as the mass-produced device). Good enough is a consensus determined if the deliverables are correct, complete, accurate, consistent, and testable within the context of a delivery cycle.

The IoT project needs testing, V&V, DevOps, and support functions to support and address an IoT system completely. V&V is accomplished during IoT product development. The development organization supports the V&V, that is, you cannot test quality into a product once the product is done. V&V can be done within the development organization. However, independent V&V (IV&V usually performed by an independent subcontractor) uses a different engineering team that does the work and when more separation from the development team is deemed necessary. V&V/IV&V is included with testing in this book because IoT systems are more than just software.

V&V supports development, management, and other stakeholder interests. Stakeholders can include customers, third parties (e.g., regulators), and external engineering groups. V&V provides a source of information and data to these parties. Information can include functional and nonfunctional requirement qualities such as performance, reliability cost, schedule, etc. Test data can include models, input data, results, performance numbers, error reports, inspection notes, analytics, etc. Parallel development and V&V efforts allow the feedback of V&V data to development for quick product improvement (e.g., correct errors to avoid technical debt, understand performance issues, etc.).

Details of V&V are contained in IEEE 1012. IEEE 1012 provides V&V processes, integrity level criteria and usage, and V&V planning information. Readers in critical IoT product domains (e.g., health and safety) can benefit significantly from reading through IEEE 1012. However, most project use of any standard is NOT required except by a contractual specification.

V&V can be applied to devices, networks, systems, hardware, software, data, and a system of systems. V&V is addressed in IEEE 1012 V&V standard. IEEE 1012 does not provide specific information about V&V/test approaches or techniques. These can be found in references and standards such as ISO 29119 software test standard.

What Is IoT V&V in a Nutshell

Verification is a development activity that involves testing but also uses other assessment approaches. Verification uses a variety of activities, which can be done by the system, hardware, software, and test staff throughout the lifecycle. These are shown in Figure 3-5.

Verification takes the artifacts of one lifecycle stage and assesses if they satisfy the information from a previous step. For example, verification will assess if the software design and implementation (a.k.a. checking, as noted by James Bach) fully meet requirements. Verification assumes that source starting point information (e.g., an operational concept, requirement, design, etc.) is "correct and complete."

As shown in Figure 3-5, validation is a development support activity involving various concepts. It is often stated that validation is done at the end of a product's lifecycle, but this is an oversimplification and can have cost impacts. For use here, *validation should be accomplished on each product to be delivered and confirmed throughout the lifecycle as the product matures.* So, for example, it is good to conduct inspections, including a peer review with a customer on requirements to validate the requirements will meet the stakeholder's needs. This is why Agile encourages user involvement with concepts like stories, use cases, and even the produced code.

Further, validation may need to include modeling and simulation before moving to design for some critical requirements, such as the control laws for a self-driving car. Then, the modeling and simulation can continue throughout the car's lifecycle (from start to disposal). These uses directly support ongoing validation.

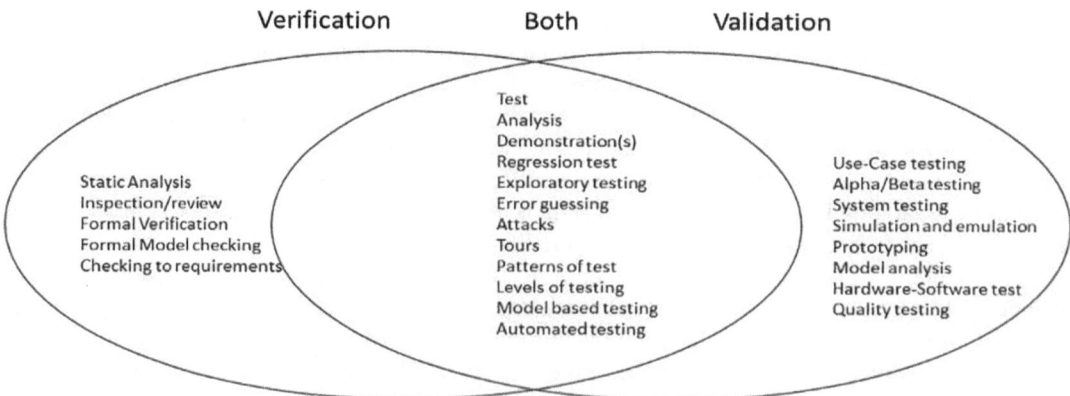

Figure 3-5. IoT device verification and validation – hardware, software, and system

Reference: Jon Hagar IoT Class 2016

Summary

IoT development, operations, test, and security are combined in an infinity lifecycle. Companies wanting to improve their competencies will jump into IoT and software without understanding the challenges, risks, planning, and strategies needed to succeed with this type of a project. Risk-based testing is a crucial mitigation concept applied by internal testing and, when justified, external V&V teams. Finally, the highest priority risk to consider, after the obvious one of errors in critical functionality, is security threats. Even experienced test teams may need to adjust their thinking and test concepts when security threats grow.

In the next chapter, I will explain the factors driving different levels of IoT testing and V&V. This will aid readers in the test planning and justification of test activities.

References

1. https://hbr.org/2016/02/to-predict-the-trajectory-of-the-internet-of-things-look-to-the-software-industry
2. "IEEE Standard for System, Software, and Hardware Verification and Validation," in *IEEE Std 1012-2016 (Revision of IEEE Std 1012-2012/Incorporates IEEE Std 1012-2016/Cor1-2017)*, vol., no., pp. 1–260, 29 Sept. 2017, DOI: 10.1109/IEEESTD.2017.8055462
3. ISO Standard 15288 Systems and software engineering – System life cycle processes
4. ISO Standard 12207 Systems and software engineering – Software life cycle processes
5. "IEEE/ISO/IEC 29119, ISO/IEC/IEEE International Standard – Software and systems engineering – Software testing" – Parts 1 to 5 series
6. https://hbsp.harvard.edu/product/BH685-PDF-ENG

Figure Reference

1. https://cdn.pixabay.com/photo/2018/12/20/06/54/evolution-3885331_960_720.jpg

Chapter 4
Factors Driving IoT Testing/V&V Selection and Planning

The previous chapter introduced lessons learned in test planning and V&V, so this chapter defines the detailed factors that a project working on IoT should consider in their planning efforts. As shown in Figure 4-1, four factors combine to create a decision matrix for planning purposes. This chapter explains each factor quantitatively and defines IoT test planning levels. One way to start planning is to determine a factor score based on integrity levels, product maturity, organizational (Org) development (Dev) ability, project size, and the other factors defined in this chapter. This scoring can be used numerically as outlined in this chapter or as guidance using each factor taken together.

The chapter now steps into each of the four factors to use in scoring.

Factor 1: Using Integrity Levels to Drive V&V/Test Planning and Strategy

Of consideration in IoT is that not every device and associated software will have the same importance to the users. Each device will have varying degrees of critical use. Therefore, the amount and type of V&V/testing will need to vary. Integrity levels range from minor user annoyance, when something fails, to people dying and lawyers getting involved. IEEE 1012 defines different integrity levels for

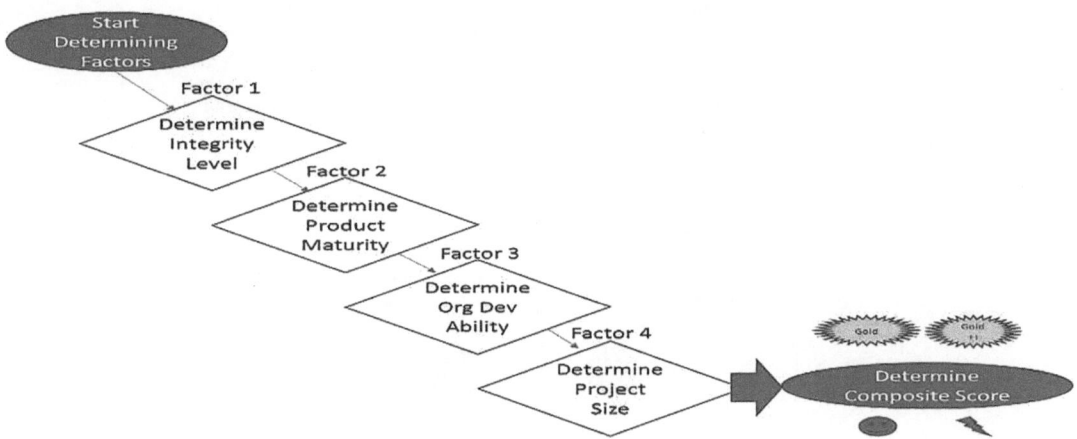

Figure 4-1. Determination of IoT composite factor score

Reference: Jon Hagar IoT Class 2016

J. D. Hagar, *IoT System Testing*, https://doi.org/10.1007/978-1-4842-8276-2_4

V&V. Here, I offer a summary and some basic ideas from IEEE 1012. Still, for any project needing more details and specific considerations, I highly recommend obtaining IEEE 1012 or another document on V&V that has much more detail on integrity levels, the impacts of their use, and how to classify different types of critical software.

Integrity levels provide a numeric ranking system to help in determining the amounts, tasks, rigor, activities, criticality, and approaches for V&V. The integrity determination is based on rankings of complexity, criticality, safety level, security level, desired performance, reliability, or other unique characteristics of software and/or hardware as shown in Figure 4-2. The producer, users, and other stakeholders concur on determining an integrity level. In IEEE 1012, extensive classification tables are provided to aid in the determination and associated activities of V&V based on lifecycle stages.

The four IoT integrity levels are shown in Figure 4-2 and summarized in the following list (V&V/test level classification and examples). The colored icons of Figure 4-2 are introduced in this chapter to be used for each level and factor classification as a visual identification aid. The icons are also used later in the book when I talk about IoT test activities related to levels and factors. These icons in other chapters and sections can be traced back to Figure 4-2. While the four divisions of Figure 4-2 are based on IEEE 1012, my classification is different, tailored, and simplified for IoT. These classifications can be a starting point guide for test planning but should be tailored to an IoT project context.

V&V/test level classification and examples

- Nature of integrity level = 1

  - Less critical device.
  - Device failure minimally degrades functionality or nonfunctional criteria.
  - Hardware is simple.
  - Software is simple.
  - Communication interfaces are well defined and based on industry standard.
  - Operations and data analysis are none to minimal.

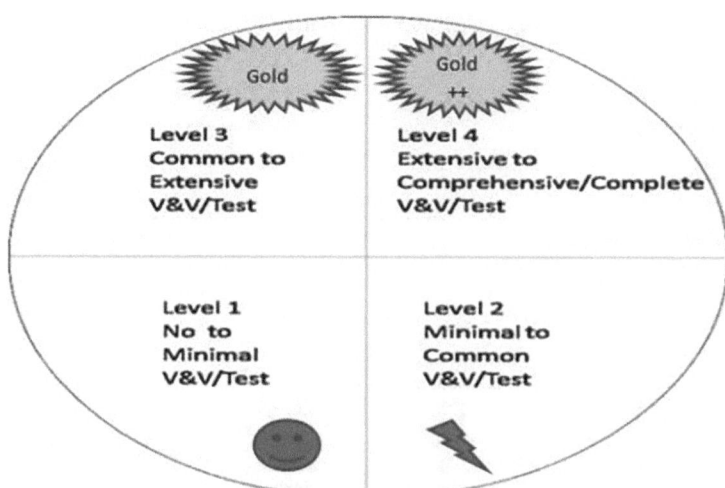

Figure 4-2. Example division of the component integrity determination

Reference: Jon Hagar IoT Class 2016

Examples

New or prototype devices used for demo.
For an IoT component that has a backup device, the primary component must be tested first.
Consumers "play" with a toy, such as a model racing car.
Wearable sports band (nonmedical).

- Nature of integrity level = 2

  - More critical usage.
  - Failure has a low impact or minimally degrades functionality and nonfunctional criteria (1–3 features).
  - Hardware is moderately complex [1].
  - Software is moderately complex and interacts with other users.
  - Communications are moderately complex and/or may use several media approaches.
  - Operations and data analysis are moderately complex and can impact business success.

  Examples

  Loss of money is possible (1000s to 10,000s or more).
  Losing part of a device's mission success or operational model uses.
  Impact business base and future sales (moderate).
  A home device that has security implications.

- Nature of integrity level = 3

  - Major criticality.
  - Failure has impacts or degrades functionality or nonfunctional criteria (50 percent of features).
  - Hardware is complex.
  - Software is complex.
  - Communications are essential to success.
  - Operations and data analysis are complex and critical to business success.

  Examples

  Major loss of money
  Failure(s) for a large part (or parts) of the required mission (or operations)
  Impacts on significant business base or sales
  Nature of integrity level = 4

- Life (someone dies) or severe business impacts (i.e., going out of business).
- Failure has an impact or degrades functionality or nonfunctional criteria (100 percent of items).
- Hardware is extremely complex.
- Software is extremely complex.
- Communication approaches are critical for success.
- Operations and data analysis are complex, critical, and large (big data).

  Examples

  National news and loss of money
  Total loss or failure of a mission (operation)
  Impacts major business base (goes out of business)
  Self-driving car
  Pacemaker systems integrated with hospital

The following list (examples of recommended V&V/test activities) provides an example recommendation of V&V/test activities that might be associated with integrity levels.

Examples of recommended V&V/test activities

- Level integrity 1

  - Example 1: New prototype IoT device – No written test plan/developer testing/fast exploratory testing
  - Example 2: IoT toy car – One-page test plan/agile testing stories/some test automation with exploratory test

- Level integrity 2

  - Example 1: Home IoT thermostat – Written test plan with strategy/developer testing attacks/exploratory testing/security attacks
  - Example 2: Wearable watch – Written test plan/developer testing/network testing/exploratory testing/security attacks/functional attacks

- Level integrity 3

  - Example 1: Home-personal nonlife-critical medical device – Written test plan conforming to standards/developer attacks/test automation/scripted testing/conformance testing/functional and nonfunctional testing/security testing

- Level integrity 4

  - Example 1: Self-driving car – Full up V&V and test planning/V&V over full lifecycle/developer attacks/test automation/scripted testing/conformance testing/functional and nonfunctional testing/security testing/model analysis and testing/safety analysis
  - Example 2: IIoT lighting and traffic control system – Full up V&V and test planning/V&V over full lifecycle/developer attacks/test automation/scripted testing/conformance testing/functional and nonfunctional testing/security testing/chaos testing in field/network analysis and testing
  - Example 3: Factory IIoT control systems – Full up V&V and test planning/V&V over full lifecycle/developer attacks/test automation/scripted testing/conformance testing/functional and nonfunctional testing/security testing/

I basically follow IEEE 1012 with four levels since this is an industry reference point, but nothing prevents more levels or even fewer; however, less than two makes little sense. The preceding table and figure are versions modified for this book and IoT devices/systems. This integrity classification approach is very simplistic and can be used as an example or starting point. For a more rigorous approach, refer to IEEE 1012 standard. Figure 4-2 shows a schema to determine integrity levels. It can be modified and customized for the local context. The integrity level determined will associate directly with V&V/test efforts and activities.

Scoring (if you are doing a numeric score): Assign your level integrity 1–4, which is your score for this factor.

Factor 2: Risk-Based DevOps and Product Maturity Testing

Closely associated with integrity level is the idea of risk and product maturity. A mature product is in use in the real world for a while and possibly has lower risks. However, just because an IoT product is mature does not mean there are no risks. Mature products can have risks introduced by new software releases, changes in hardware, or even new Ops usages in the field. Lower integrity levels imply the

IoT product overall has a lower risk. However, every product comes with risk and V&V/testing deals with risk. I thus advise that testers conduct risk-based exercises and risk-based testing throughout the lifecycle with the project's maturity and integrity level.

Many books, classes, and standards treat risk analysis as a prominent subject [1]. Many authors and standards on project management, development, and testing talk about risk analysis. In this section, I want to give a quick start to readers on the beginning of risk analysis to support test planning, but it should not be an end to risk analysis and risk-based testing.

The SEVOCAB [3] defines risk as (reference: `https://pascal.computer.org/sev_ display/index.action`)

> *(1) an uncertain event or condition that, if it occurs, has a positive or negative effect on a project's objectives (A Guide to the Project Management Body of Knowledge (PMBOK(R) Guide) -- Fourth Edition) (2) combination of the probability of an abnormal event or failure and the consequence(s) of that event or failure to a system's components, operators, users, or environment. (IEEE 822008 IEEE Standard for Software and System Test Documentation, 3.1.30)*

Risks play into integrity levels. Risks are, in part, subjective but have an aspect of objectivity (probability and consequence). I recommend teams conduct and repeat a risk assessment to drive test planning, even for a level 1 integrity device. Assessments include management risks, development risks, and product risks. Many teams and most managers tend to focus on management and development process risks, which are those things that may impact cost and schedule. Often, managers and teams overlook product quality(s), which is a test concern, until the point that quality (poor quality) impacts cost or schedule, but by then, it may be "too late."

This rush is to be expected on many IoT projects. Indeed, in IoT, many startup efforts only care about getting a product to market (on a schedule) for as little cost as possible since they usually have minimal money. However, assuming the startup has just enough quality and functionality to clear the starting gate, many IoT products will quickly worry about quality risks, which leads to more testing or company failures.

There are risk tools and systems to help in formal risk analysis, but most of us keep it as simple as possible for the context. A tester starts risk analysis with information collection, which can be done using interviews, history assessment, experience, review of taxonomies, independent checks, workshops, checklists, organized brainstorming, and/or customer interviews. The knowledge coming from these efforts should be captured and recorded. Capture can be done using note cards, spreadsheets, tables, text files, risk tools, or pictures on whiteboards.

Since risk analysis in support of testing should be an ongoing effort, it will continue throughout the lifecycle, and as the IoT system matures, risks will evolve and change. Thus, the test planning will change.

Risk Analysis Process for Testers

A risk statement creates a written definition of the risk to aid understanding by stakeholders and possibly drive testing efforts. Classically, risk descriptions capture a single condition followed by details of the potential consequence (potential problem or risk). One way to do this is with a statement structure of

<if condition>, then <consequence(s)> + <time factor>

If condition = a single phrase citing a single key circumstance or situation that can cause the concern, doubt, case, or uncertainty. *The "if" should not have an "and" statement since this would indicate multiple risks.*

Consequence = a single phrase or sentence describing the key, an adverse outcome of a condition. *A consequence can have "and" statements.*

A time factor is a single phrase or sentence that captures a time factor or implication of a time factor of a risk that can or may occur. *Time factors are optional.*

Note on the above

For more detail and references, see *Software Test Attacks to Break Mobile and Embedded Devices* by Jon D. Hagar.

Testers should analyze the common areas of IoT risk provided in Table 4-1 during risk analysis.

Risks exist in all projects, from prototype concepts to mature ones and even during maintenance. Each type of project will have different kinds and numbers of risks. Testing/V&V data provides project teams with risk mitigation and closure information. Testers should use the list in Table 4-1 as a top-level checklist to trigger thinking, which can lead to risk statements and considerations for lower levels of risk.

Many projects in a larger historic organization will have a formal risk analysis process [1]. Testers should support and build on these risk events. My experience is that many items that testers deem as a risk will not make it to a management-level formal risk process. I have often kept and worked on "lower-level" development and test risk lists and analysis within the test/V&V teams. Such actions were done until a time when the lower-level risk is addressed or bubbles up to a formal management-level risk list.

VENDOR OF HARDWARE-SOFTWARE SUPPORT PROBLEM

In one effort, I was dealing with a hardware-software vendor to supply a chip-computer system, software, and communication interface protocols (CIPS). One of the testers had personal experience with the vendor and warned that hardware and software issues would likely be another historic project. Management said they trusted the vendor and had other risks that were higher. The project-level risk was assigned to hardware and software from the vendor. However, the development and test teams tracked this lower-level risk. The team sprint planning asked the vendor for a prototype drop of a partial system. The vendor committed to the delivery date. The actual delivery was five months late. The first warning bell was sounded.

When the IoT hardware and software were delivered, the development and test teams had an immediate sprint priority (no waiting) to receive, inspect, integrate, and communicate with the system. This happened in just a few hours:

1. The hardware plug was wired upside down and backward – the vendor said to ship it back.
2. The development and test teams did not ship the system back but wired a "test harness" to allow the integration to continue.
3. Upon startup, the test team found a bug in the software (actually a series of bugs).
4. Management was notified and put the vendor/product on the project risk list.
5. Testing continued with new drops and hardware and software.
6. Bugs continued.
7. By the time product rollout came, the vendor's product was the number one project risk and almost missed a product delivery date, but the development and test teams had worked with the vendor risk as a high priority, so the delivery date was met. Yeehaw.

Moral of the story: Test often, focuses on risk, communicate to management with data.

Table 4-1. IoT Potential Risks

1. Safety – When the well-being of humans is threatened
2. Security and privacy – Data or information can or may be exposed
3. Hazard – Damage to equipment or the environment is possible. Hazards can include hardware within and outside of the device
4. Communications – Loss of information or control caused by communication channels, interfaces, and protocols. Comm issues can include dropouts, bad data input/output, internal and external comm to the device, slow comm lines, etc.
5. Business impact – Bad computations generate wrong information and end up impacting profit
6. Regulations and legal – The product could result in harm, not compliant with standards, or be at odds with government regulations, leading to legal actions
7. External environment factors – Impacts from hardware inputs for devices and electronics, which are susceptible to influence (noise) either systematic or random, including outside communication lines and characteristics; the "real world" (weather or conditions in the real world, such as wet roads or rain); and even human operations
8. The impact of input and output noise – Input sensors or outputs to devices and electronics that are susceptible to noise influences
9. Complexity – The size of the system or some aspect of the system makes missed cases more likely
10. Compatibility and interoperability – The ability to integrate with other systems, since often IoT will be in a system of systems
11. Quality factors not met – There are many qualities of a device or system that may be a risk (see other sections and books for discussion of qualities)

Table 4-2. Example Risk Statement Table

Risk Statement for a Smart Diaper	Impact	Likelihood	Test/Note
If the hardware from the vendor does not have proven wet sensors working by sprint 10, a software design may be unknown	High	Medium	Conduct analog-to-digital (A2D) and digital-to-analog (D2A) wet sensor tests
If the Wi-Fi communication protocol cannot handle sensor data rates, data dropouts may occur (wet user)	Medium	Medium	Performance test comm

I recommend that groups work on risk identification early and often over the product lifecycle. Personnel with test and product knowledge should support risk efforts. The developers and testers should always consider risk inside a test group. Risk analysis can be done as part of agile team efforts if Agile is followed or as informal team communication. I recommend the team effort because one gets a comprehensive set of risks (giving the team a chance to learn), but the individual tester can test risk analysis independently. However, that may not be as effective.

Internally, development and test teams should focus on technical risk. As the story told, I can use any risk to focus and assign integrity levels of V&V/testing. As the project matures and more information becomes available, risks will change in priority. Testing will not address all risks since usually the number of risks and areas to test exceed the budget and schedule, but as a standard like [1] details, risk-based testing can drive the test planning process. The resulting risk statements can be captured in a table such as in Table 4-2.

Once risk statements have been defined, they should be prioritized and integrity levels assigned, which could be another column in the example of Table 4-2. Setting priority for cases and risks can be challenging. The literature talks about many methods, but maybe the easiest is to decide how many tests you have the budget and time for during this attack effort, and then sort the testing into buckets: "test ASAP," "test if there is time," and "don't test." Teams can balance the test plan's budget, schedule, tasks, and risk.

Once development and test teams start risk analysis and risk-based testing, the number of risks will grow. Teams should review materials with stakeholders. Periodic risk review is a good practice.

However, the stakeholders must know they cannot thoroughly test every risk within limited budgets and schedules. Using risks to plan and test design is a learned skill that will benefit IoT teams.

Scoring (if you are using a numeric score): Assign your risk.

Score 1 – Low

Score 2 – Medium-low

Score 3 – Medium-high

Score 4 – High

And this is your score for this factor.

Factor 3: Organizational Ability Impacts on IoT Test Planning

Different organizations and people are active in the IoT world. Teams range from startup organizations with a great idea and little experience to big "mega" corporations with personnel having lots of experience. This section outlines basic conceptual examples of the different kinds of organizations. In later parts of the book, I will expand test thinking for many of these organizations.

This breakout is given to help understand how the organization plans IoT development and testing since it will differ depending on the organization's abilities and the context of the IoT effort. I suggest readers consider this breakout of ability levels and find one that seems closest to your organization's ability while recognizing these classifications represent a continuum. Your organization will not precisely match one of these breakouts but will probably come somewhere in the continuum. The placement of your organization will impact your test factors.

Hopefully, you will find this information helpful.

"Newbie" Companies – Level 1

I do not pretend to be an expert in startup companies, so what I write here has been gathered from friends, conversations, and some Internet research. I expect that there will be people setting up new companies as in the days of the web and .com world, hoping to be the IoT version of Google or Facebook. There is certainly room in the market for many of these startups, yet only a few may be the next IoT "google."

Here are some of the items a Newbie company should consider doing more research and work on:

1) Money – Get funding via a crowd network or self-funded.
2) Stay alive on the first version – Get enough working product to stay alive, and this may mean no formal testing, but a lot of developer testing and maybe some fast exploratory attacks should be considered.
3) Hardware – Use as much generic, off-the-shelf hardware as possible and find a suitable manufacturing source.
4) Software – You can do it yourself, but going agile with low/no-code should be a starting point.
5) Production version – Define what "good enough" is for your customers and users but expect changes (possibly significant changes) once you go live.
6) Assume risk – Expect problems, so some testing can help prevent these risks from becoming "big ones," but assuming risk will be part of the game.
7) Schedule – Most every experimental project takes two or three (or more) times longer than you first expect, so again think and be agile.

Probably the most significant area is funding, followed by "staying alive." I cannot say what your level of assumed risk will be nor who your stakeholders will be. I know several people who have played the so-called "silicon-valley lottery" (moving from startup to startup hoping to hit it rich, always hoping to "hit it big"). I know a couple who did well, but it is a massive gamble with many implications.

Do not risk more time, money, or reputation than you can afford to lose. Also, keep in mind that "good enough" is always different for each situation and changes over time.

Good luck.

Companies with Experience Moving into IoT – Level 2

Many companies are moving into IoT and software that are not traditional software IT companies. While they may have an internal IT department, they have done little or no complex software development to deliver a product. Further, I know of great software companies with little hardware or operations experience. Most companies will be moving into IoT, and they will have different areas of expertise and areas to work on. I outline the areas here. It is noted that all of these companies will require more and different types of testing.

Companies with Hardware (Electronics) Experience – Level 3

Companies that understand hardware production and maybe even have had some classic embedded software experience in their product previously will find the following IoT differences:

1. Communication with devices and to the network.
2. Colossal data volume to do analytics on.
3. Ongoing operations and the ability to update the software on the fly.
4. Security and privacy.
5. More complex graphical user interface (GUI) that, at the same time, must be simple to work.
6. Many protocols and standards, which are fluid.
7. The lack of experienced software staff is pervasive across the software industry.

Companies with Software Experience – Level 3

Companies that have software experience but view hardware as a "generic" problem (e.g., PC hardware, which tends to be very similar compared to unique types of IoT hardware) will find the following differences:

1. Unique hardware with sensors and controllers that interact with the real world.
2. Realtime performance issues.
3. More and new security and privacy vulnerabilities.
4. Hardware lifecycles and upgrade issues.
5. Hardware retirement and disposal concerns.
6. Many software and hardware protocols and standards are fluid.
7. Lack of experienced hardware and systems development staff.

Company with Hardware-Software Experience – Level 4

Companies with both hardware and software development experience may be in the best position for IoT, but even they will find some new challenges, including

1. Environments that IoT will function in and that change rapidly.
2. Communications to devices and networks (unless they have been working in the mobile space).
3. Scales of hardware from small (smaller than coin size) to large (city size) require a different kind of systems thinking.
4. More challenging security and privacy issues.
5. Resource limitations such as batteries, memory size, performance, and many others.
6. Quality problems in both hardware and software that are "different" from traditional PC issues.

> **Note**
> The experienced hardware and software development companies will likely encounter these differences once they get over the shock of their initial learning curve about IoT and evolve into it.

Companies with Systems, Hardware, and Software Experience – Level 5

Companies that have created systems and used system engineering have the advantage of thinking about the "big picture," and often these companies can deal with both the hardware and software, yet differences for this type of organization can include

1. Having outsourced hardware or software rather than doing things in-house (can be a steep learning curve).
2. Integration, which may be a familiar concept but may be new in the consumer or industrial aspects of the device where they have not played before.
3. Agile may be new since many systems companies are "older," and so they follow the older waterfall development models.
4. Agile with no-code/low-code and the security/trust concerns that come with these.

Government Organization – No Level but a Special Case

IoT will be in cities, counties, states, people's bodies, transportation systems, and you name it. I assume that government regulation will continue to expand (eventually) to cover IoT. Many leading IoT industries (e.g., transportation and medical devices) are already highly regulated and have existing standards, which are likely to continue.

I have heard regulators say, "well, the regulations that we have work now but must be applied." I am not sure that this is true for IoT. Many regulated industries seem very slow in how things happen. Many regulations are based on older technologies, years or decades old. IoT is changing technology very fast. Regulation always takes time to be worked out, while, at the same time, safety and hazard issues keep occurring, followed by the public's demand to "fix" things. For example, security issues in software products are experienced daily, but many government officials seem to think security is "not a big deal and is business as usual."

I have consulted with government organizations. Some seem to be "progressive" in getting ahead of technologies. Others seemed mystified, as if they were "seeing magic" when I talked about the development and test technologies that had been in use for many years. I am worried that poor and restrictive regulations will do more harm than good in a rush to plug holes in safety, hazards, security, etc.

The world is changing. Societies and governments need to keep up. I hope that any government readers will think about what is in this book and feel free to contact me.

Experienced Companies with Only Consumer Product History – Level 2

Finally, some companies have long histories of developing consumer products. Here, think of things like clothes, food, shoes, soap, and products you find in a supermarket or any place in big box stores, as shown in Figure 4-3. Many of these companies will first think IoT does not apply to them, but many will need to think again as IoT products appear in their business space. For example, I found a water filter pitcher with a "replace filter sensor" on it. The next step for this sensor is to become connected to the Internet. Ta-da, it's IoT. So can anything become IoT? You should look at the story case study of diapers for an example.

> **CASE STUDY**
> In class, I submitted an idea to an IoT contest regarding diapers. How can diapers be IoT? Read on. While holding a tiny newborn baby, some were challenged by saying that "we do not see how IoT can be used outside the tech space."
>
> Question: What is one of the problems parents deal with in babies?
>
> Answer: Diapers, specifically, when to change them. How is this currently handled? (1) Baby cries; (2) the smell test; (3) the finger test, with items #2 and #3 being done after #1 happens.
>
> Would it be nice if the parent got a smell notice when the baby needs changing?
>
> Can IoT do this? The short answer is YES. Diapers can come with a wetness sensor and a notification to an app on your phone instead of the baby crying and waking you at 2 a.m. needing a diaper change.
>
> However, is this idea a viable business space? Are we restricting thinking? Again, the answer is likely YES, because when I have used this as an exercise in classes, most engineers miss a part of the market space – the consumer diaper market.
>
> Can you expand your thinking on this topic? How about this? A big market is now adult diapers. It may be even more critical for adults or caregivers to know they need changing because many people using these products are either unaware or do not notify anyone that a change is needed. This situation can lead to bedsores and urinary tract infections (UTIs, a possible deadly infection). Infections are much more expensive to fix than simply changing a diaper.

So, many consumer companies may ignore IoT since it does not fit their business model. Some of the companies will get surprised by the Newbie Level 1 startup. They will then buy the Newbies. They will then have problems learning about hardware and software.

Figure 4-3. Consumer products move to IoT or die trying (Internet of Everything)

Worse, at the risk of being "skunked" by some other company, many traditional consumer companies need to avoid what happened in the watch industry some years back, as defined in the story "Swiss Watch Part 2."

SWISS WATCH PART 2

Years ago, a Swiss watch company invented the digital watch. It did not fit their "mental model" of what a watch was or should be. They let the rights go to other companies and countries. Before the digital watch, a large percentage of the watch industry (intellectual property rights or IP) rested with the Swiss. Now decades later, most people use and have digital watches, and the Swiss have lost market share. The moral of the story: Consumer companies with significant market share need to be aware of new technologies even when they do not fit their mental models. IoT may be one of these technologies.

Highly Experienced Company Including an IoT History – Level 5+

A few, usually big, companies have all the bases of Level 2 covered.

You are probably aware of their names. They have moved and will move into IoT. I have worked with some of these types of companies. I recommend they look at the list of Level 2 and ask themselves, "where can we improve?" My experience is that every company can get better. They do this by growing (training) their people, forming relationships with other companies, or acquiring small companies. I expect that acquiring companies will go on forever like it always has, and I hope that people will continue to grow their skills. I hope this book offers some help on growth in the IoT technology domain.

Process and product improvement areas and people/skills improvement areas have many books and references. I recommend that people and organizations review the list of organization types I offered and create their improvement or career plans. I have always used plans to work on areas for process improvement. I continue this work today with IoT planning.

Scoring (if you are using a numeric score):
Level 1 – Score 1
Level 2 – Score 3
Level 3 – Score 3
Level 4 – Score 2
Level 5 – Score 1
Assign your score and this is your score for this factor.

Factor 4: IoT Project Size and Complexity Impacts on Testing

The last factor, and maybe the simplest, is the IoT device's size with many hardware, system, and software elements. If the cost is low and the schedule is short, the amount of testing makes sense to be smaller and shorter. For example, testing a thermostat controller is not as complex as testing an IoT smart city system of systems.

However, just because something is small in size does not mean it is simple. For example, today's cell phones have millions of lines of software, thousands of apps and games, many communication channels, many hardware configurations, etc. So, the size and complexity of IoT factors still need to be considered carefully.

In IoT testing, size does matter, so let's consider an IoT large complex software (LCS) system in this example. IoT large complex software systems can include

- IoT with AI
- IoT with analytic systems done at the edge, fog, or cloud
- IoT networks of devices (multiple)
- Corporate IoT network systems
- Government IoT network systems

A problem facing large IoT systems and systems of systems is complexity, which often accompanies size. ISO and IEEE standards provide some useful definitions for complexity.

complexity

(1) degree to which a system's design or code is challenging to understand because of numerous components or relationships among components [2]

(2) pertaining to any of a set of structure-based metrics that measure the attribute in [2]

(3) degree to which a system or component has a design or implementation that is difficult to understand and verify [2]

This book defines attributes of size and complexity in an IoT software system to include

- Number of IoT subsystems (count of nesting levels or component systems)
- Number of IoT elements (components, no-code, low-code, off-the-shelf, reused) in the software
- Amount of IoT and support software (e.g., measured in terms of lines of code)
- Amount of IoT data input, stored, accessed, manipulated, and refined (Is data variability minimal or large?)
- Number of IoT hardware elements providing inputs or accepting outputs (located locally up to highly distributed)
- Number of IoT connections and interdependencies between subsystems and other components (Is the system extremely distributed?)
- Number of IoT system use purposes and user perceptions of these purposes (Is the use highly static or constantly evolving?)

- Number of internal IoT routine processes and interactions of computation
- Number of emergent behaviors and properties from internal and external sources (Are these positive or negative emergent behaviors?)
- Number of stakeholders (customers, users, experience levels, developers, support staff, etc.)
- Amount of autonomy (ranging from interdependent to totally independent)

Other characteristics in IoT contribute to size and complexity, which are difficult to count. These characteristics can occur in large complex software IoT systems and can be "counted." When these characteristics occur to a sufficient degree, an IoT large complex software system may become difficult to test. In test planning and design, the tester should consider the preceding factors and even measure them, since management will want justification for why risks and testing are "harder." However, testers should consider these characteristics in testing, including

- Nondeterministic patterns from external sources or systems (e.g., AI, realtime control, always-on features, never-fail, in-use adaptive systems, etc.) introduce difficulty in determining expected IoT behavior testing (e.g., planning, inputs, processing, expected results).
- External systems with morphogenesis of states and functions during usage impact planning and testing, since inputs to a software system under test may not be predicted, and the outputs cannot be reproducible all of the time in testing.

  - External systems may organize to produce nontrivial patterns, which become inputs into the software system under test, and are not in the specification or identified as a risk (e.g., cascading failures across software systems and elements, unexpected system conditions, hacking).

- Increased or changing contracts, customer bases, or regulations placed on the IoT software system cause compliance impacts for governance and legal usage.
- Social-technical drivers of the IoT system's software

  - Reclassification of software use and qualities (e.g., life-critical, safety, security, hazard, etc.)
  - Stakeholder changes (customers, different users, added regulators, threats, etc.)
  - Complexity of the test planning problem from management introduced factors, such as management style and organizational structures, number of contracts/contractors, and communication (or lack of it)

- Regression and retest planning implies cost and schedule impacts over time.
- Integration test impacts

  - Note: Integration test plans are a significant factor in some IoT efforts and can include new support software, test environments, updated apps, services, hardware, data, communications, users, interconnectedness, etc.

- Number of configurations and versions of the IoT software system under test being addressed by test plans and used in the field (configuration management issues).
- Impacts caused by changing developer activities and features during the entire IoT lifecycle that emerge over time (e.g., new requirements, code designs, new off-the-shelf, development methodologies, changed development plans, etc.).

These characteristics, when encountered in an IoT system, are, by themselves, strictly different from complexity and size. However, these characteristics can complicate the testing. From a software testing perspective, large complex software IoT systems tend to have

- Increased numbers of test plans and require coordination between different plans
- More test design and automation
- More test cases

- More results
- More test environments (see Part 3)
- More software issues/bugs
- Increasing project test plan activities and costs

    Scoring (if you are doing a numeric score): Assign your score number

- Size small and simple – Score 1
- Size medium and complexity – Score 2
- Size large and simple – Score 3
- Size large and medium complex – Score 4
- Size large and very complex – Score 5
- And this is your score for this factor.

Composite Total Scoring Factors (If You Need to Do That)

Many teams and managers will want a total "score" of the factors of this chapter. However, scores are like grades in school in that they may not mean much, but the effort of doing the analysis and learning to support test planning is essential. This scoring factor could be done differently for your local project and context. So this section is a start.

A scoring system and Figure 4-5 are relatively simple. Start by taking the factors given earlier in Figure 4-1, consider them in your test planning, assign each score per factor, add the scores up, and place your composite total score for test planning as shown in Figure 4-4. This gives you a reference point for your conceptual test planning and maybe talking to management about what is needed given these factors for an IoT test project.

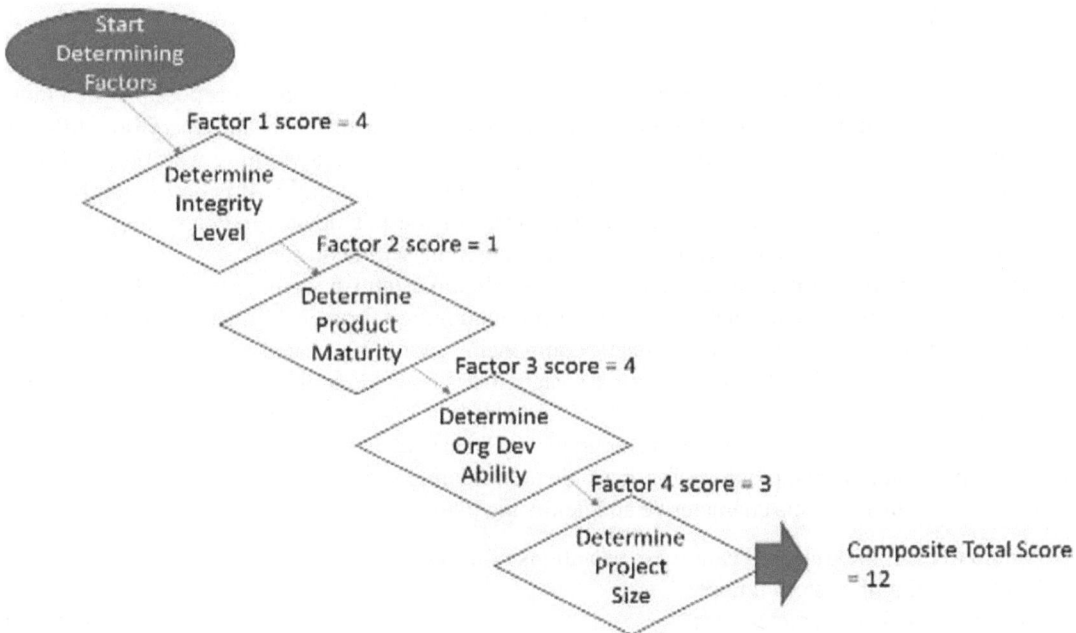

Figure 4-4. Composite total score

Reference: Jon Hagar

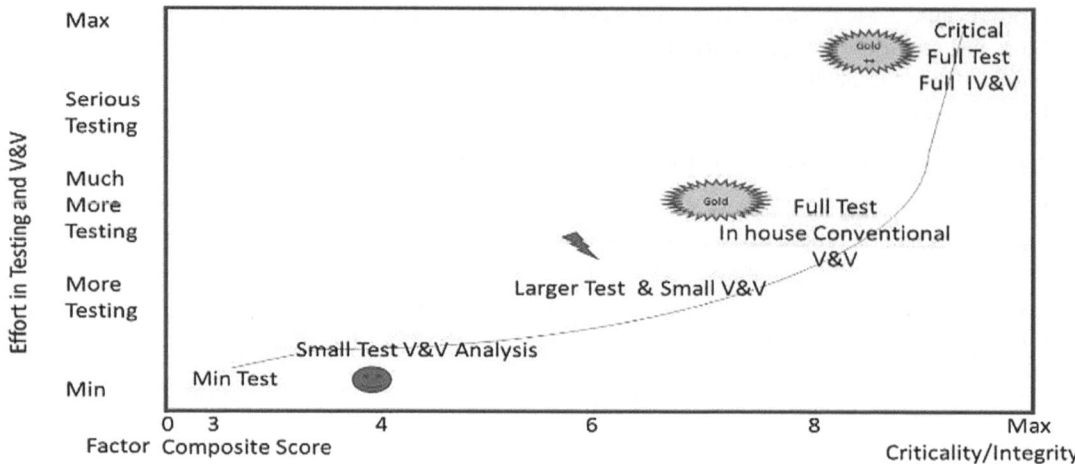

Figure 4-5. Determine composite IoT factor score and test/V&V levels

Note: The difference between V&V, iV&V, and IV&V is the degree of team independence from the developers. V&V has no independence. iV&V is done within the same company. IV&V is a totally separate company. The choice of independence between score 8 and Max is a project judgment call.

Reference: Jon Hagar

In this example, a composite total score of 12 makes the placement in Figure 4-5 near the top level of "critical full test and full V&V." The placement of icons in Figure 4-5 is also notional. Numbers, icons, and test levels must be adjusted to the local context. What should be clear is that there are many ways to plan, strategize, and implement testing for IoT.

How IoT Projects Should Mix and Match Factors

A situation some IoT efforts will have is a mixture of factor levels and scores. It is possible for a more extensive IoT system to have different system components and factor levels and, therefore, different V&V approaches and efforts. However, some care must be taken when working with project teams, subcontractors, or vendors to work the flow-down of factor levels carefully. It is best to apply factor analysis recursively to each element and subcomponent of the IoT device and/or system. Also, the factor levels are likely to change with time, so that replanning will be needed.

For systems with multiple elements, analysis tasks should establish factor levels for the associated subsystems and vendors. This analysis may impact planning, requirements, functions, hardware, and software. Software and hardware elements or components should be treated as a function of the IoT system. Thus, IEEE 1012 and factors require a system to assume at least the factor level of the highest-level component it contains. A lower-level component can assume a smaller factor level for the elements contained within a system if the recursive analysis reveals it is possible. When working with nondeveloped components, planning for these elements should use factor levels also.

The views of factor levels characterize and define the likelihood of system problems resulting from

- Failure to meet the trust and validity expectations of users
- Faults resulting in system failures
- Unverified quality characteristics

As factor levels increase, information control documentation and formality (e.g., tests and results) will grow in importance. High integrity factor levels often imply that legal and regulatory actions become possible. Records feed such a formal system and offer protection to the producing organization. Likewise, the involvement of vendors or contract items implies increased levels of documentation and formality. Added documentation does not say that agile ideals such as "working with the customer over contract adherence" or focus on code should not be considered. However, documentation as a backup for those involved in legal actions has proved decisive.

Scoring (if you are doing a numeric score): Assign your risk.

Combining IoT Factors for Better Test Planning

Factors can be determined following the flow of Figure 4-1. Increases in any of the factors may mean more V&V/test activities become necessary. The factors must be determined during project planning, starting at proposal time and continuing until a project is retired from use. The assessment of factors throughout a product's life is needed because the uses of an IoT device can change over time. *It is not enough to say, "well, the product has been used for x months/years, so it is proven."* For example, software and a device that is being used in a new environment and use case may encounter new, even unexpected, errors. Such cases have been seen when a car was put in a frigid environment and encountered use case errors that had never been tested.

Once factor levels are determined, actual V&V plans, processes, and activities can be tailored. There are many tasks, processes, techniques, and resulting documentation that are possible for IoT. These activities are discussed in other parts of this book (Part 2). Generally speaking, the higher the factor level, the more V&V/test activities to be considered in planning and implementation. The factor levels define activities, but these are not hard and fast formulas that give "the answer." Context and critical thinking are required.

Further, factor levels and risks should be evaluated and updated during the product's complete lifecycle and as plans are revised. For example, the story of wearable fitness devices that "morph" into being used in medical decision-making should be considered. Teams should expect (maybe even hope) that the market and usage of IoT devices will expand and change. Such movement can change factor levels, risks, and test plans.

Whether lowering or raising factor levels, the change should be agreed upon with stakeholders. The interplay of components, hardware, and/or software should be considered in changes since these are part of the "overall system." IoT systems are software- and communication-intensive systems. IoT interconnection and integration must be considered in establishing and changing factor levels.

Determining factor levels include products provided by vendor items. A new usage scenario may be tempted to assume a product is "good enough" and stable because it has been used under a specific set of scenarios. If the IoT device is going to be used in new usage scenarios, past vendor item performance may not be applicable.

CHASING A GHOST CAUSED BY BAD CONFIGURATION CONTROL
An outside vendor delivered a new hardware/firmware device for use in the test lab. The configuration management part number and revision history did not change on the device. Upon lab usage, new error messages were received in the test results. The testers assumed for a month that it was a problem in the software under test but later learned the vendor had changed communication hardware chips in the device. The vendor should have rolled the configuration identifier so that new testing of the changed device could be done in the IoT system as a whole, but they did not do this, and many months of labor effort were lost.

The actual selection of factor levels should be made by skilled staff with the support of stakeholders of both developed and vendor-supplied components. Low factor levels of simple IoT devices may be possible to have "no V&V/test," but the user may be very unforgiving if a device fails to function and "bad" things happen.

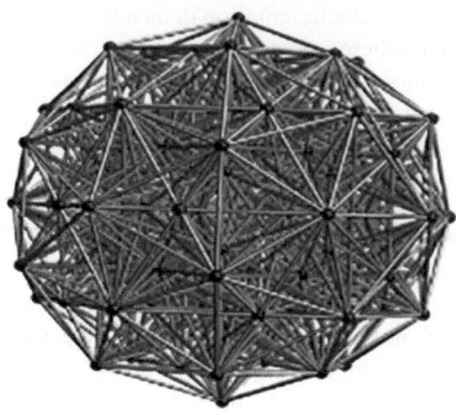

Figure 4-6. HyperCube of estimation factors

IoT Test/V&V Cost Estimation

The tesseract of Figure 4-6 is a projection of a four-dimensional cube. I would like to plot the four factors or more on a single figure, but like the tesseract picture, I will not try to draw and plot this in a book. I expect users to add the factors in the last section together to help calculate combined factors and cost estimation.

I get an early test planning question: "How do you estimate V&V/test costs?" Management always wants this. This is not a simple question to answer. However, this section provides a short starting point. Figure 4-7 starts with a simple estimation factor based on a percentage of development – the simpler the IoT, the lower the development and test costs. The estimation goes up to high critical integrity IoT systems of systems, where life or huge costs may be at risk. Figure 4-7 is used by finding your place on the curve, getting the percentage on the left side, getting the development cost number, and multiplying it by the percentage. Figure 4-7 builds on the earlier figures of this chapter.

The percentage approach is very general and quick but not very realistic or accurate. A better way to do estimates is to do bottom-up or function point estimation [3 or 4], costing task by task and test by test, but a rough guess estimate often provides a start.

To further use Figure 4-7, take the kind of IoT system you are testing and place it on the bottom scale between 1 and 5. Then travel up to the curve above your selection and read the percentage on the vertical axis. For example, a level "2" IoT system would need about 13 percent of the total development (Dev) costs, including hardware and software, while a level 5 life-critical full system of systems might need 50 percent of the total project budget, including testing, independent testing, and complete IV&V. Impossible, you say? Well, consider a new IoT-driven nuclear powerplant. Will it need much more testing/V&V since many lives and costs are at risk? The answer is, of course, it will need more testing.

I get the following question: "How are testing efforts distributed over the IoT lifecycle?" Figure 4-8 is a very rough estimating factor chart, but it is a way to get started. Here, I assume a complete end-to-end lifecycle where I start with an IoT prototype and end at the retirement of the system. I also assume test/V&V are part of the Ops and maintenance (O&M) efforts.

The way to use such a chart is that once your team has a budget for testing, say 100 percent, then in the early prototype, 5 percent would be used. Next, a first version of the product enters testing with 15 percent. Then, a functional but not complete version of the product is fielded to the test team, so another 20 percent of the budget is used. Finally, before the release of a product, a final fielding of the system is done with another 20 percent of the budget used. In this phase, 40 percent of the total budget may get used. Of course, this percentage number can vary depending on many factors. For example,

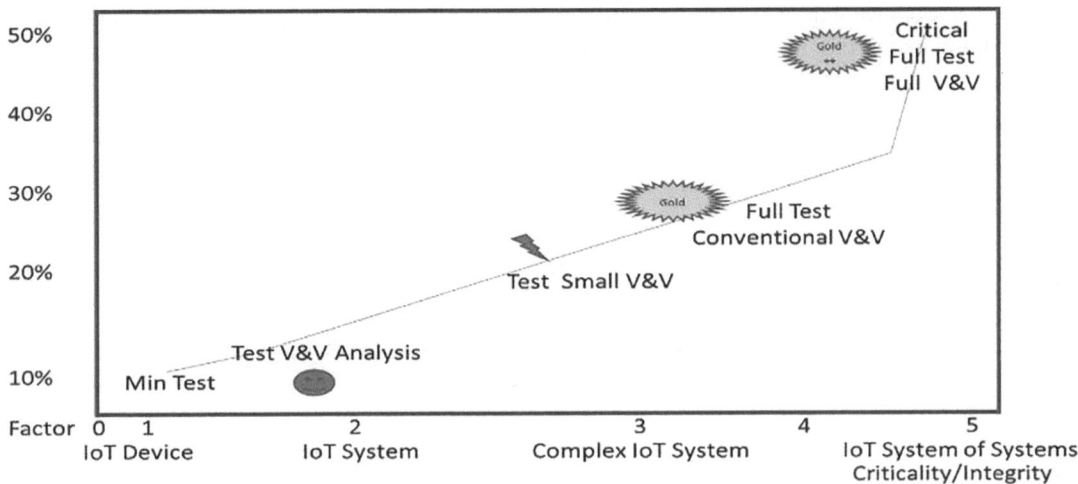

Figure 4-7. Test cost estimation as a percentage of development

Reference: Jon Hagar

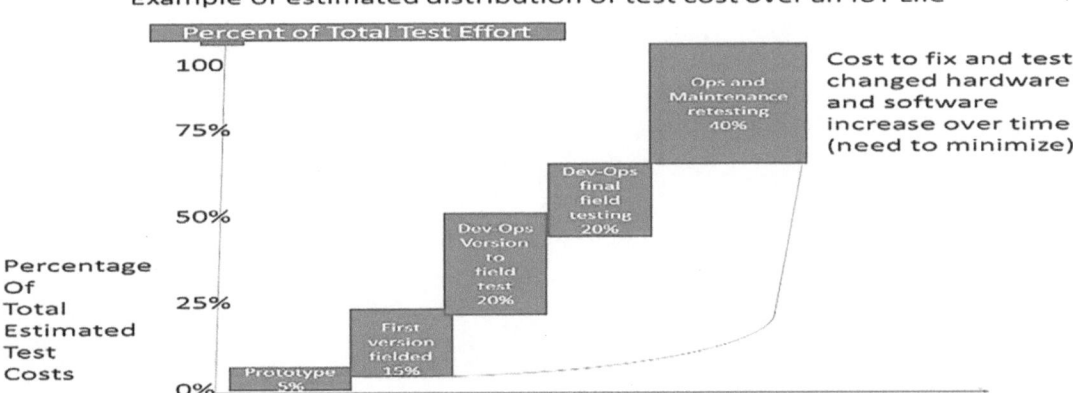

Figure 4-8. Distribution of V&V/test costs over a lifecycle

Reference: Jon Hagar

many teams will want to treat development separate from Ops. Okay, but make sure that some test budget is allocated to Ops, since new testing is likely to be needed for success. However, testers should not forget operations and maintenance, which may go on for years.

Summary

The key points of this chapter are

– You must have an agreement with stakeholders on V&V/test factors and plans.
– Good IoT testers think critically about risks, do lots of planning, and run scared.
– In your IoT test planning, you should consider four factors (project integrity, maturity, organization, and size) outlined in this chapter.

Now that we have explored the factors to consider in planning levels of IoT testing, in the next chapter, I will explain to the beginning test team or person the key ideas to get you started in IoT test planning.

References

1. "IEEE/ISO/IEC 29119, ISO/IEC/IEEE International Standard – Software and systems engineering – Software testing" – Parts 1 to 5 series
2. ISO/IEC/IEEE 24765:2017 Systems and software engineering – Vocabulary
3. https://en.wikipedia.org/wiki/Function_point – accessed winter 2022
4. https://en.wikipedia.org/wiki/Cost_estimate – accessed spring 2020

Figure References

1. https://pixabay.com/photos/grocery-store-market-supermarket-388302/
2. https://cdn.pixabay.com/photo/2022/01/11/17/40/polygons-6931148__340.png

Chapter 5
Beginner Keys for Starting IoT Test Planning

The previous chapter addressed the detailed factors that drive selecting IoT testing and V&V planning. This chapter offers some of the essential items that may make IoT device testing a success. Indeed, there are more keys to being successful with IoT, but I define six of the ones that I think are the first essential steps for success and to get you thinking.

IoT Key 1: Have a Ubiquitous User Interface (UI)

IoT could take lessons from experienced, professional web designers. For instance, the Google splash page is an excellent example of KISS (keep it simple, sir). There is underlying complexity that can be used, but the primary starting point is very minimalistic. Contrast this with some IoT apps I have seen, which have more options than can comfortably fit on any smartphone's small screen. Herein lies the rationale for a professional human factors consult (also known as industrial psychologists or human-computer specialists) for most, if not all, user interface design – providing that the IoT user is human. As for the UI being ubiquitous (meaning: found everywhere), nearly every person on planet Earth is connected in some way to the Internet. Along comes IoT, also connected to the Internet. Therefore, we must ask, "Should all UIs look alike or be simplified for IoT?" If that were true, the tester's and hacker's jobs would be much easier. However, context would be the overriding factor.

So, we need to consider what I mean by the user in IoT domains. Therefore, for IoT purposes, users can include

- Human interfaces (many different kinds and intellectual capabilities of humans are possible)
- Communication (comm) interfaces
- Hardware
- Security
- Other pieces of software
- Other systems
- Testers

There are two audiences to this crucial topic area. First is the development team, including local testers, who design the basic UI or graphical user interface (GUI). Testing of the UI should be early and frequent. I like the idea of "dogfooding," whereby the Dev/test teams "eat" or use their creations to prove to themselves they like the product. For example, if you know you had to eat dog food, wouldn't you make it out of delicious steak since dogs and humans love beef?

© Jon Duncan Hagar 2022
J. D. Hagar, *IoT System Testing*, https://doi.org/10.1007/978-1-4842-8276-2_5

The second audience is the team or person doing the UI validation. Here, it is best to separate this effort from the early work. The separation is to avoid bias blindness. The checklist of Appendix C will help in validating the UI/GUI. I suggest validation to ensure the device can be configured independently of the region, age, society, technical know-how, physical ability, etc. If the devices are ingenious (i.e., cleverly and originally devised and well suited for their purpose), they will accommodate these differences. Additionally, many users will want the IoT device to be plug and play (e.g., plug it in, and it works).

Testers should watch for statements such as "Users are going to configure the device, it is easy" or "no one would ever do that" on the Dev/test side. Most users cannot configure (i.e., set preferences or settings) on even a simple, smart device UI. Smart, to me, means the device foresees many of the things a user needs and wants and provides these without copious amounts of "easy user action." The device sets itself up, and its use will "hold the hand" of the user and make all actions simple, safe, and sane. (Don't hold your breath!) Also, on the "no one would ever do that" comment, this is what bad actors use, and hackers hope for. Developers should "think" like a hacker during development, and testers should think like a hacker during security testing. For example:

- The current infotainment systems of cars (circa 2016) have many IoT features. Many consumer reviews state that UIs/GUIs are not easy to set up, understand, or even find things. The car UI/GUI is not "intuitive." People either don't use them or often take them in for servicing. Other user interfaces in cars are a joy to use (or so it is reported). These reduce user anger and service calls. Companies want the latter case.
- The router in your house. How easy or hard is it to configure? Does it use system default passwords, which most people use? These cases may not work for IoT when you have 10s or 100s of devices in your house or usage area to configure. And the use of default anything, especially in security, should be a "no-no." For example, the Dark Web site on Shodan [1] lists default log-on passwords. That should be enough to scare you.

However, it gets worse since most of you, when you hear the terms "UI" or "GUI," may think human user interface. In IoT, users don't have to be human (e.g., interfaces to another computer, a piece of hardware, an AI system, etc.). IoT successful interfaces need to work with other machines in the IoT network through machine-to-machine interfaces or software application programming interfaces (APIs). It may need to seamlessly switch between different communication networks as their availability comes and goes. IoT nonhuman UI may need to work with sensors and actuators, maybe from many different manufacturers, without a human worrying about calibrations, environments, and other factors that will impact the hardware.

For instance, a semi-smart house has smart solar panels. The panels have a hard time talking and staying online with the routers. Why? Because the house does not have a constantly stable Internet connection, that is, no fixed broadband or cable connection. The communication (or call home ability) for the solar panels is handled through a cellular connection through an appropriate device. The solar panel company's IoT engineering did not consider or disallow a design for this communication configuration. The human user does not want to care about everything going on under the IoT hood. They want the device and systems to work. In this smart home automated configuration with an IoT system, hours are spent to get a partial configuration of speakers, smartphones, and TV, much less getting the solar panels to "call home." Many users will not want to do such time-consuming work since they want "plug and play" configuration ability.

The UIs of the IoT home "small" system need work, but it is cool when it does play. Moreover, even then, the total configuration of the system is a little clunky and not stable. For example, one has to push several buttons on one, two, or three remote controls to get speakers and TV to change sources. Further, Wi-Fi songs appear to drop out during play mode if two smartphone devices by different manufacturers try to control the speakers' volume.

Finally, our guess is an IoT UI will likely include voice input/output controls driven by adaptive AI to help the different levels and types of human users. These capabilities will take time to perfect. I foresee the beginning of a reasonable voice control UI with systems like Siri and Alexa, all legal and privacy implications aside. (Will we ever get to the technologies of HAL from *2001: A Space Odyssey*? Possibly.) Visualization will continue as an option, but users will start to interact with them the way we do with humans for many IoT devices. This may mean the "smartness" of these systems will need to understand the primary and nuanced aspects of human communication such as tone of voice, accents, body language, language, etc.

Plan carefully and thoughtfully for all interfaces. Due consideration should be made for the interface to actually operate it, whether it turns out to be graphical or activated through an API or another machine. Either way, interfaces of any kind can be tricky to test when you are not aware of what might be "triggering it."

IoT Key 2: Learning from Data Analytics

I have been talking about how data and big data will be generated from IoT in massive amounts. So far, the literature and interests seem to be from marketing people and business managers who see the data as a key to generating more sales and more profits. Many testers and development people seem less interested in any data analytics messaging, which is NOT GOOD.

Data analytics is a growth area in high tech. Many high-tech companies (e.g., IBM, Microsoft, Google, and others) invest big in data analytics and associated AI. The amounts of data available will be massive in IoT. Figure 5-1 illustrates the places where data and data processing will be happening for IoT systems. Data processing can happen at the IoT device itself, on the edge (a computing system with close connection to the IoT device such as a smartphone), in the fog (a computer system with more storage and processing ability connected to maybe a series of devices), or even in the cloud (very large storage and processing system).

So, it is curious that many Dev/test professionals seem to have no interest in this area of data processing. The interest of testers in the data and the IoT device seems obvious, since that is where we will begin our testing. However, why should testers limit themselves to testing just the IoT device? Well, to answer that question, let's take a step back and understand more about the devices and where all that data can be preprocessed or processed. Note: See information on data analytics (Part 4, Chapter 22).

The Internet of Things (IoT) and the Industrial Internet of Things (IIoT) allow connectivity to many millions of devices and networks to gather data. They share the data over architected and networked

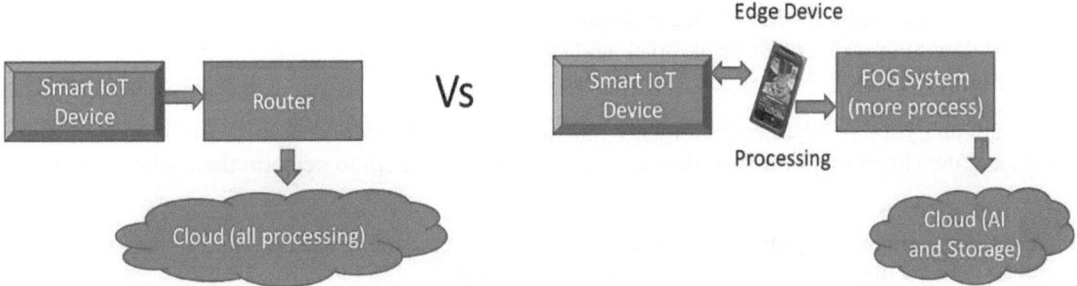

Figure 5-1. Processing at the IoT device, the edge device, or both

Reference: Jon Hagar Composite

systems, as well as over the Internet, hence the name. The Internet handles many connections to devices (routers, servers, PCs, etc.) plus networks. This creates so much data that it is unfathomable to think of "sizing it." Along comes OpenFog Consortium with architected solutions to standardize and promote fog computing (essentially distributing computing) worldwide.

We must understand the differences between IoT devices, the cloud, edge computing, and fog computing. Here are some definitions to help. IoT devices may be considered localized, on the ground systems generating Terabytes or even more data. Still, they may not process the data unless it is part of the device's control functions, requiring immediate outputs. This massive amount of data coming from many devices need to be analyzed. However, dumping that much data on a cloud for processing does not make sense. The cloud is (to use a simple analogy) a system (architected) of networked servers around the world that holds data from many sources but on which no "real" processing takes place, that is, "real" meaning taking direct inputs and performing IoT device processing to immediate outputs. And then comes fog and edge computing architecture to allow preprocessing or processing of data closer to "the edge" of its association where the data is created. "Many people use the terms fog and edge computing interchangeably because both involve bringing intelligence and processing closer to the edge where the data is created" [2].

My recommendation is that testers become data consumers, understanding where the data is coming from, what to do with the data (processing), and where the data needs to be stored, as illustrated in Figure 5-1. I have used data to improve my testing for years. I have data mind maps of errors. I have used real-world embedded device telemetry to verify my test plans and strategies. I have seen test labs that generated Terabytes and then Petabytes of data that testers and analysts could leverage [3].

Testers need new skills and specific analysis knowledge to become good analytic data consumers. Testers do not need to be statisticians. However, advanced testers will need to learn how to use analytic tools and be capable of asking questions of test technicians. This allows skill growth turning simple runners of tests into software engineers and data scientists. Some testers will transition to engineering during IoT, and some will remain testers. It would be best if testers decided for themselves their career path.

The history of computers, in part, is the evolution of data usage analytics. The nature of both computers and their data has changed over time. IoT is projected to generate vast amounts of data from what was once embedded devices, plus all the new IoT devices that will appear, generating Petabytes and beyond (are projected). This data will be generated during the operational use of these IoT device systems. Many interested parties may use the data. Testers should become involved in the operational use of IoT devices and the collection of operational data analytics. Benefits to testers may include

- Improved field-based error taxonomies reported automatically by devices
- Tasking some test activities to the user in the operational setting
- Mining help desk data to aid testing (e.g., use cases, problem reports, etc.)
- Analyzing fast, realtime data
- Mining social media for IoT device problems
- Using data to improve patterns of attack and test models
- Chaos engineering

Mining analytics will be needed to use such vast amounts of data, analytics, and big data. I hope testers and developers will think of data analytics as aides for them to perform their jobs better using

- Test taxonomies
- AI and self-organizing data analytics predefined for tester use to generate more interesting tests
- Smart data feeding into test data (old tests into new tests)
- Users as testers (generated data as part of DevOps)
- Causes of errors and prevention of future errors (process improvement and root cause analysis)

Tester data analytics and AI will be a career growth area. fog and edge computing are better suited to AI, given the amount of data that must be analyzed.

IoT Key 3: Unique and Specialized Hardware Working to Be a System with the Software

As mentioned previously, a difference in IoT is the unique and specialized hardware. Whether it is the components of a car, sensors, controllers in a house, or the elements that make a smart "city," we will see many everyday objects, maybe all things, becoming "smart" by applying the knowledge gained during testing and usage to improve them.

Testers of IT and computer software systems have always dealt with generic hardware, which usually worked. Sure, there were differences in computer platforms, peripherals, and communications, but IoT will be orders of magnitude different with specialized sensors, controllers, and attachments.

A tester who works with embedded devices will need to be more familiar with the unique and specialized hardware. However, embedded teams that move into IoT will get all the problems of embedded systems plus all of the problems of IT systems, plus higher levels of integration complexity than most of us have ever seen before.

Many of us have seen extensive systems and systems of systems pass lab tests, only to fail in the field. I expect the uniqueness of hardware and the system to grow in complexity and, hence, the likelihood of failures (Part 3).

Many have said we will live and work around such failures, but the idea of "good enough" will come into play here. The tester and the team will need to balance good enough with failures for software, unique hardware, and systems.

IoT Key 4: Level of V&V/Test Need for "Good Enough" IoT

A good enough IoT device [4] meets basic stakeholder requirements and expectations. The good enough device can ship and be sold but may have limitations or features implemented during future updates. It may not be perfect, but it can keep the project going for the next iteration. I think there will be a struggle for the right amount of V&V/testing for IoT to reach good enough. Also, the definition of "good enough" can and will change over time as an IoT device matures. The other parts of this book will explore this key more. There is no best or right level of V&V/testing; it all depends on context.

However, thinking testers are and will be needed and in demand. Standards may be a start for some teams, but critical thinking will still be needed. Experienced testers who already can think critically will evolve quickly for IoT teams. Testers who have minimal critical thinking skills may struggle.

The context will continue to be king for software, testing, and IoT systems. Management will always want to reduce testing and time to market. Some testers will still be viewed as "second-class" team members.

I always viewed my job as an educator for management and other stakeholders about what testing was about. I never felt like a second-class player. As I moved around, I found most engineering disciplines each felt they were "undervalued" too, so I valued everyone I worked with.

My interest in testing as an intellectual challenge meant that pay and recognition were less important to me than doing the best job I could, learning about testing, and communicating this to the other stakeholders. For IoT, the key to V&V/testing will be communications and perspectives given to the outside stakeholders.

IoT Key 5: Remaining Agile

The problem with writing technical books is that aspects of technology change almost daily. IoT, fog and edge computing, and AI are hot topics for 2022, and they are changing as fast as I am writing this book. Many timeless test concepts are included in this book. However, remember that something will change as the target systems (Dev-Test-Sec-Ops) evolve. I tell my students and readers in all domains that if they are not learning something new every day, the changing pace will leave them behind. Those of us in technology have spent *all of our lives learning*. Get used to it, be agile, and read everything with an eye to what stays true and what may be evolving. The more you keep learning while remaining agile, the better your chances of staying ahead of the learning curve.

IoT Key 6: Testing IoT, Systems, and Large/Complex Software (LCS)

Technology is now connecting everything. Where does the IoT device that a software tester may be testing stop? Who owns the V&V/test for the overall system? Often, the answer may be "system." This will stop when we get significant numbers of large/complex software failures. Big companies, governments, and even some individuals have already started realizing large/complex software V&V/testing must be addressed. Some of the concepts of this book can be applied to these IoT large/complex software systems. Some companies will learn the hard way after the devices have not been V&V/tested when they get hacked, crash, fail catastrophically, or even kill people. Understanding who owns the V&V/tests for the large/complex software will be a challenge. There will be many testers both on the development side and likely on the customer side. Here, I am not talking about the individual consumer customer. I am talking about larger organizations. A company that deploys IoT devices, governments using IoT, and other large entities will likely need large, possibly independent, software testers separate from the development team. This situation will occur because the development group often will not be responsible for the entire LCS. How can a smart traffic light that is networked with smart roads, that is talking to smart cars, and that is feeding a statewide control grid be tested except by a large-scale comprehensive test effort and overseeing organization? This will be where the large/complex software concept will come into play.

Summary

In large part, IoT will be all about the data that the device/system or system of systems provides, and you'll have to master and improve many keys to testing success. This chapter ends Part 1, which introduced IoT, testing, and planning. Part 2 (Chapters 6–9) gets into the nitty-gritty of IoT testing and planning.

References

1. www.shodan.io/ – accessed winter 2022
2. https://akira.ai/blog/difference-between-cloud-edge-and-fog-computing – accessed spring 2022
3. IEEE 982.1-2005 IEEE Standard Dictionary of Measures of the Software Aspects of Dependability, in revision as of 2022
4. https://en.wikipedia.org/wiki/Principle_of_good_enough – accessed winter 2022

Part 2
IoT Planning, Test, Strategy, and Architecture – Team Leadership

Part 1 introduced IoT development and test thinking. Part 2 of this book focuses on test planning. I will address not just the software test planning but, as the following image illustrates, the whole universe of IoT testing, which includes the IoT software, hardware, architectures, system development, planning, and operations.

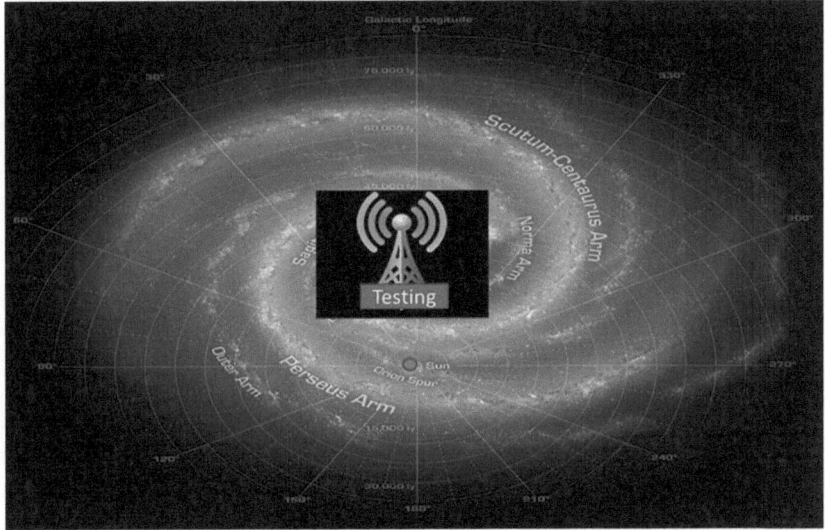

Testing is a significant key to the IoT universe in making products successful.

Reference: NASA image

While many new IoT projects will seek to minimize testing because it is seen as a "nonvalue-added" activity, in the long run, you will see this is a short-sighted perspective. However, I will explain how testing can be "ramped" up during the lifecycle of an IoT system to be a centerpiece of product success.

Figure Reference

1. https://cdn.pixabay.com/photo/2012/04/11/16/51/cellular-tower-28883__340.png – accessed winter 2022

Chapter 6
IoT Test Plan: Strategy and Architecture Introductions

This chapter introduces test planning with strategy for different types of IoT teams, which leads to test architectures. In many cases, the scope of an IoT device system is not limited to the particular IoT device itself. The chapter also considers that a test plan includes the key elements of a strategy and architecture which are the beginning of high-level IoT test concepts and designs. These introductions lead into the next chapter and the rest of the book.

Establishing the Mindset

Organizations must decide how much and what kind of V&V and testing is "right" for their context during the proposal, budgeting, and planning efforts. As teams start their work, it is optimal to have a test plan and strategy, which can range from the low end of "we will do no testing outside of what developers test and then let the users do all the testing" to high full end of testing which is "we will do full testing including independent V&V (IV&V) as defined by standards." The choices within such a range should be determined by the list found in Chapter 5, with regard to financial impacts, schedule resources, reputation, team skills, product maturity, essential quality, and others. Recognize that there is no one "right" or best technology and certainly no one answer for all of the challenges arising in IoT device testing. Further, remember that what is "right" in test planning is likely to change over time as the IoT product and its stakeholders evolve and mature.

IoT test planning is the first important step leading to a successful technology product. One of the hot topics on the technology maturity curve (link as follows) is IoT. Readers should understand the maturity of IoT technologies, devices, and systems. IoT, in general, is somewhat immature in many areas, as many companies are actively engaged in IoT development and deployment. There is likely to be much volatility and change in IoT, as well as a maturation of each significant IoT market segment. See Tech Growth Curve at `https://media.licdn.com/mpr/mpr/p/7/005/081/13a/317cb86.jpg` –accessed winter 2022.

As a technology grows and matures, projects and testers must know the fundamental historical principles.

The "ISTQB Foundation Level syllabus includes seven testing principles [1]":

1. Testing shows the presence of defects, not their absence

2. Exhaustive testing is impossible

3. Early testing saves time and money

4.   Defects cluster together

5.   Beware of the pesticide paradox

(Note: The paradox is that rerunning tests over and over does not provide new information.)

6.   Testing is context-dependent

7.   Absence-of-errors is a fallacy

I believe that test planning should be part of all project management efforts. It has been observed that a plan (a document or piece of paper) is not as important as the exercise of planning and thought efforts themselves. This should mean that in most successful human efforts, a plan exists, that is, some road map that takes the team from where they are to where they want to be, essentially to their end goal(s). Just aimlessly wandering around in the IoT wilderness is likely to be inefficient and ineffective and can consequently sacrifice company viability and profits, not to mention the careers of people working on the product(s). So, to improve the chance of IoT success, it is good to do the planning exercise and put forth a plan, such as a document, single page, whiteboard, mind map, etc. (see Part 3, Chapter 10). As a word of caution, too often, "the plan" becomes some "false truth" where people become afraid to change it and blindly follow it – even when it does not make sense or wastes resources. A plan should be more like the "Pirate's Code" (see the movie *Pirates of the Caribbean*), where the document is more of a "guideline" that can be tailored and changed as needs and context dictate.

Additional Considerations

The team must consider the IoT device, the edge, and cloud end users in complete software testing, as shown in Figure 6-1.

Starting at the bottom of Figure 6-1, many possible IoT devices and networked IoT systems exist. They have software and hardware that support operations (Ops). A development (Dev) team creates the hardware and software components. The hardware components may be off-the-shelf (OTS) items

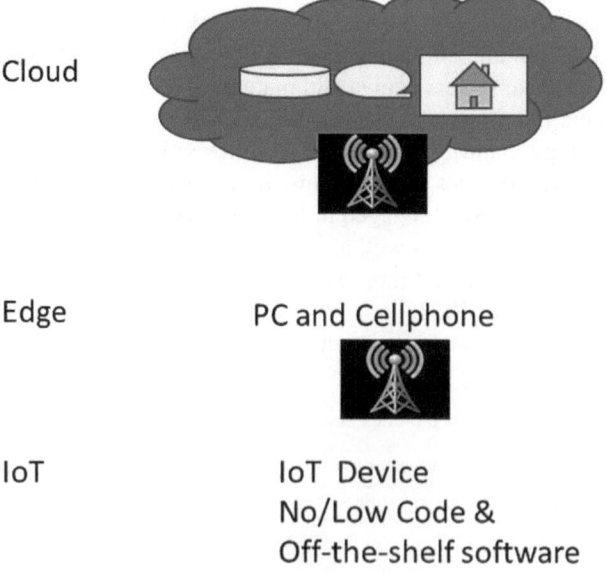

Figure 6-1. IoT system levels that testers may need to plan testing (device level to the cloud)

and/or custom manufactured hardware. Likewise, the software may be composed of OTS; microservices; "no-code," which is not written by the development team; "low-code," which has a mix of no and customized developed software; or newly developed code from an outside source (contractor). This means more and more components are OTS demanding a lot of "glue" to hook things together. This also means that in the IoT system, there can be a lot of hardware and software functionality and quality features that are not used by a particular IoT system. Throughout this book, we typically call such component configurations OTS. This book addresses later in test planning the implications of using off-the-shelf items (e.g., hardware, software, no-code/low-code, open source, etc.).

Next in the middle, the edge can be made up of computers, smart devices, smartphones, or other small systems which connect to the IoT system. Today, edge systems do much of the local processing associated with an IoT system. These help distribute the processing load. Finally, IoT information can be passed to a server cloud for more processing, storage, and analysis after the edge computations. As mentioned previously, the division of where to stop testing needs to be determined in the test plan by stakeholders based on the IoT system context.

Planning

Humans do the planning, and some do it pretty well, as illustrated in Figure 6-2. Many successful tech companies (Apple and Google come to mind) continue to exist due to careful, thoughtful, and strategic planning efforts.

We, humans, build and create things to a plan, a design, or a specification to hopefully make everyone's life easier, especially those who have to build something from it.

As stated earlier, the planning document is not as important as the planning exercise to organize efforts for the success and quality of an IoT device or system. So, let's begin planning at a high level in this chapter of the book with plans, strategies, and architectures, which will then lead into the following chapters as we drop down to the lower levels of plan and design. In the case of IoT testing, I hope that by doing the planning exercise on your project, you will help get a device that works, has the right qualities, and is created within a reasonable cost and schedule.

Projects go through different phases or periods. A cost-schedule period is when management takes a hard look at a forecast of cost and schedule. Costs and schedules are closely tied to development and testing efforts and workforce estimates for support staffing and resource requirements to deliver on any product(s). The PMBOK has more extensive knowledge on this critical subject [8].

Figure 6-2. Planning

Quality is a value that someone is willing to pay for (Part 3, Chapter 12). There are many classifications of quality (e.g., safety, security, functionality, performance, reliability, etc.). In IoT, as in many other products humans buy and sell globally, there are many qualities and levels of quality. Some qualities in IoT will be more critical than others based on the target application or users. Testers and development staff will need guidance from executives and managers to decide on the types and levels of qualities for the context they are working in and around. Every IoT device and system will be different, but project managers and leads must plan appropriately for the target level of quality.

While this section focuses on IoT specifics, having general testing background is essential to participate and add value to the planning exercises. I expect testers to have basic knowledge of test processes from the start (a proposal) to the end (retirement of the product). If you do not know about testing, start by investigating: ISO 29119 standard, books by Kaner, ISTQB, and others [1, 2, 3, 4, 7].

This book is intended to be used with other books on planning, budgeting, and strategies in testing (including the PMBOK [8]). As noted in the "Reference" section, I recommend that the readers have several references on hand because no single book can address all aspects of test planning for IoT. My tech reference libraries are significant and historic, having hundreds of works, which I refer to as needed. We need to learn from the new and old ideas in math, science, systems, and testing.

Since IoT devices range from consumer "fun" gaming toys to home systems, medical devices, industrial systems, and large-scale software systems of systems, the qualities and levels will also have a wide range. Gamers will want the device to be fun. Mothers will want their home IoT system to help them in some way. Doctors will want medical devices to keep people alive or provide data so that the doctor can help the patient. CEOS and CTOs of companies will want their IoT infrastructure to save money and be secure. Finally, the general public will want a smart city to be "smart" while not crashing under attacks, system failures, or poor performance. Given this wide range of stakeholders, no book, single plan, or strategy will be "best." Innovative teams and testers will require much test thinking for IoT devices to be "good enough."

Good Enough IoT Software and Devices

Developing organizations will want the IoT device to be just "good enough" at any time. Being "good enough" can be defined as making a "product that will make the company money, have a positive Return on Investment (ROI), and keep companies viable." But the nature of these factors and the definition of "good enough" will change over time. For example, the first web pages were raw but worked well enough for some brick-and-mortar companies to grow, yet now web pages are expected to be "great, flashy" with pictures and "easy to use on any device." What was once "good enough" to a PC-based web page user has changed over time. IoT devices and systems will go through similar growing pains. "Good enough" is a constantly changing balancing act that test planning must consider, which is another reason to have a plan and track things like "good enough," and it is a continuum, as shown in Figure 6-3. IoT teams must balance the effort and quality to find the optimal point

Figure 6-3. Good enough point between quality values and effort (cost and schedule) – a balancing act

Reference: Jon Hagar Class Production

for their effect. This balance will be reflected in plans, particularly the test/V&V efforts, as there is no correct answer for test planning.

Making money means that the stakeholder (user, customer, others) gets a good ROI with IoT investment. *ROI for testing considers two factors: the cost of testing vs. not testing.*

The cost of testing has a wide range – from almost zero (no testing) to as much effort as the development or more, and such costs change over time, with product usage and many other factors. Even teams who say "we do no testing" may mean we do not have a dedicated separate test/V&V team, but developers still have to review, compile, integrate, and run their software at some point to assess it as it goes into a system and once it is released. This effort is considered as ad hoc developer testing and can consume a third of their time plus or minus some percentage. Better developers do more testing, which, in part, is why they are "better" at their jobs. Such development teams may have test "specialists" embedded in them. Organizations where the cost of testing is more than the cost of development are few, but these are groups doing software for very high-risk systems (e.g., a control system for nuclear devices, which needs a lot of V&V/testing or lives are put at risk).

The cost of not testing can be real too. Consider if you must replace defective IoT devices because they were not "good enough." Automobile companies have had to recall *hundreds of millions* of vehicles due to faulty software. How much did that cost – not just in dollars but in losing credibility over many years? And how does a company regain that credibility?

COST OF FAILURE
Samsung recalled the Galaxy Note 7, which is thought to have cost the company $5.3bn and damaged the South Korean firm's reputation. Samsung stated that neither software nor hardware was at fault, only the batteries. But, batteries are part of the overall system and can cause the system to be "not good enough."
   `www.bbc.com/news/business-38714461` – accessed winter 2022
   So, consider the cost of bad press or loss of investors on a prototype IoT device that fails publicly. Then, consider stock impacts. See `http://money.cnn.com/2015/09/24/investing/volkswagen-vw-emissions-scandal-stock/` – accessed winter 2022.

Finally, consider the cost of going out of business. The high-tech world has far more companies that failed than succeeded because, often, their product was "just not good enough."

Most teams want a simple formula for how much V&V/testing vs. development will get them to "just good enough." No such "simple" formula exists.

If high tech were simple, everyone would be creating products and software while succeeding at every opportunity. The high-tech world can make money and has demand by many customers because *technology is not simple*. People want technology that works with a return on their time, money, or effort investment.

Many in the industry use the "good enough" concept of software and have some agile variations, as stated in ref `www.satisfice.com/articles/gooden2.pdf` – accessed winter 2022. This article is recommended reading. The concept is essential since most organizations use it in some form, though they do not always call it by this name. For me, "good enough" in test planning has the following characteristics:

1. Functions – Critical functions that a user or customer wants to work.
2. Works – The time between failures of functions is long enough for the user or customer to gain value.
3. Nonfunctional qualities – Valued quality items a stakeholder expects or pays for work (security, safety, fun, etc.).

A "good enough" product provides sufficient value, and the equation of value may change over time. Basically, "good enough" means the user or customer does not put it in the trash or ask for their money back. Users may complain, but if you ask if they want you to take the product away, they will refuse. Did you know that the first smartphones were heavy, dropped calls, limited coverage areas, variable voice tones, and other "features" that no one tolerates today? However, they were once "good enough," so now everyone has them even though what was defined as "good enough" has now changed. Now smartphone users expect them to be fast, on all the time, meet user expectations, and still be cost-effective.

My point on "good enough" is that the topic must be addressed as part of the planning process in IoT because implications and consequences can come back with a vengeance, which equates to the loss of ROI or failure of companies.

IoT Test Planning Basics

As Figure 6-4 depicts, teams should start test planning by defining "good enough" for their specific product.

Management will determine an acceptable level of product or project risk, which helps define test levels, a schedule, and a resulting budget. This "balance of cost and schedule to risks" helps set a good ROI for the company. But, eliminating the testing to save costs, time to work, and the costs of unhappy users go up, which impacts ROI and possibly the company's future viability.

From this point, the management or perhaps a test manager will determine test tasks, such as what to test, where to test, who will test, how to test, and when testing is done. In a cyclic-iterative effort, this information can determine the selection of strategy, which in turn refines the planning items. As the interactions continue, documentation should be produced to aid information retention and ensure everyone is operating to defined tasks and goals.

When I say documentation, I mean a range. It may mean agile planning games (software testing book by Crispin and Gregory [5]), or planning documentation can be something as simple as information

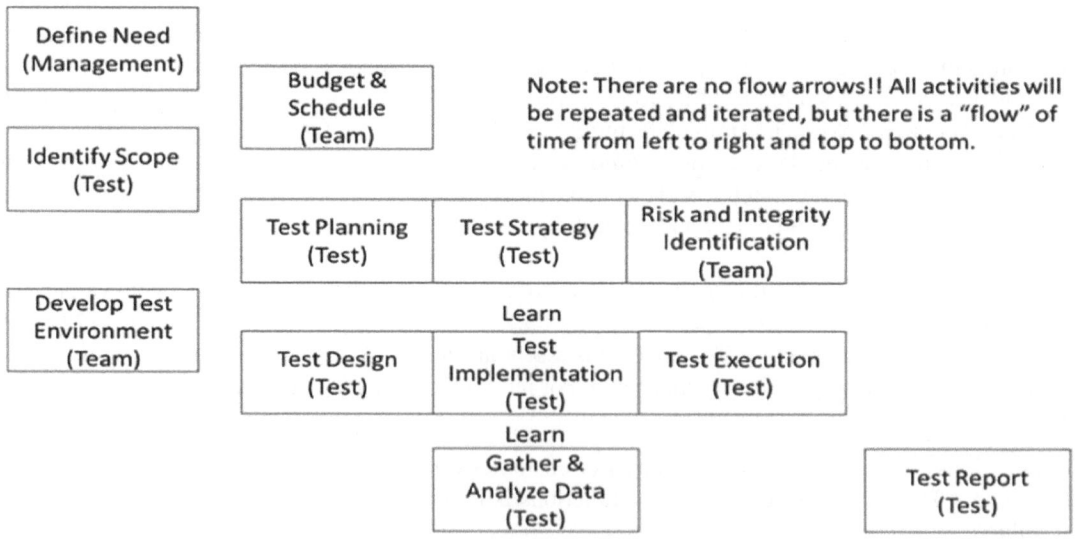

Figure 6-4. Test concept block logic chart

Reference: Jon Hagar Class Production

jotted on a whiteboard that a snapshot can be taken off and saved somewhere at work, perhaps some mind maps, or maybe a formal plan that includes all of these items (e.g., ISO 29119 parts 2 and 3 [2]). Again, there is no "best" or "right" way to document test plans. Context factors such as cost, schedule, regulations, legal compliance, organization practices, and other factors will determine the types and levels of documentation. I have experienced the whole gamut defined here in my test planning efforts.

A practiced and experienced tester or test manager should determine the nature of documentation and the formality of the document(s) reasonably quickly and possibly with some direction from other sources such as legal (lawyers), contract personnel, upper management, and company policies. Teams should not look to a "one-size-fits-all" answer. Moreover, just a quick note: if you are spending all your time planning, you are missing the point of planning since you need to do some actual test exploration and learning.

Further, like the product, test planning will reflect the nature of the organization that is producing it (see the Conway law [9]). People doing test planning should understand their organization, communication channels, and local team by knowing what the team believes is essential. Dogmatically demanding that testing "conform" to an ideal or standard will likely fail test planning when the organization does not support it.

In every test planning and resulting documentation type, one should expect the unexpected. I have seen "known unknowns" found in planning. I know that I do not have answers to unknowns, but they can be expressed as risks, provided you have a list of unknowns. However, you will also have the "unknown unknowns." These things will cost you dearly when they happen, and yet you may have no idea or knowledge of or about them until they appear. (The expression "hindsight is 20/20" comes to mind.) In test planning, I recommend agility, having a contingency option, and some resource reserves to handle the unknown unknowns. A project I ran had a 25 percent test management reserve allocated to test tasks, our reserve for the "unknown unknowns."

Further, a team should conduct another (more minor) planning cycle as soon as the unknowns become known. Agile teams are recommended to do this daily or at the start of the planning cycle [10].

In test planning, it is a good idea to have different levels of planning. In my experience, these will often be

1. Master test plan, which has the big picture and changes less. This plan organizes thinking and addresses the whole project. This effort may be made only once, though likely it will have updates as time and lifecycle stages pass.
2. Detailed test plan(s), which may be a series of plans addressing some specific phase or strategy of testing. Examples would include testing plans for a sprint, developer testing, integration efforts, a particular release to the public, a test cycle, etc. These plans would be subject to more changes than master plans.
3. Daily test planning could be daily stand-up meetings, email to-do lists, start work management check-ins, fire drills, etc.

Again, the formality of such plans can vary depending on the project context.

General Test Planning Outlined by Organization Classification

> **Note**
> The following are identifying "emojis" for use throughout this book.

In this section, I organize testing around general types of teams. Of course, there are variations and options within teams doing actual planning since there is no "best" plan for testing. I present these organization-based plan concepts as a quick start set of examples based on typical contexts I have seen. Learning will include examples.

At this point, the reader should make sure they understand test terms, including verification, validation, testing, checking, assessment, inspections, etc., which are used throughout the test plan outlines. These outlines are not intended to be a table of contents but a list of essential test ideas that should be considered for your test plan. Moreover, while the outlines look similar, they demonstrate how the team's work will differ because of context and the team's goals. Further, remember that these outlines should be viewed more as examples as a starting point for your specific context.

The Pure Startup, Single Device, and Small Team Who Are Trying to Stay Alive

Startup groups will do some IoT devices with only a few people (perhaps under ten persons). Such teams will likely be practicing Agile in development and testing.

Testing will likely be done by development and be "just enough" to assess if the device itself "works." Or, they will have a single person skilled and focused on testing within an agile group. This team has low expectations of "good enough" and is trying to "stay alive" with only the most critical features working.

A sample test plan outline for an IoT startup, single device, and small team:

1. Understand the IoT system and conduct a risk assessment exercise (listing what scares us or should).
2. Our hardware or another team's hardware checkout.
3. Any third-party software checkout.
4. Developer structural testing has test-driven development (TDD).
5. Integration and integrated test (make it work by brute-force if needed).
6. Small series of system checks against stories/requirements (acceptance test-driven development (ATDD)) and risks, performed in maybe less than a day's worth of effort.
7. Deliver (consider if the "customer/user" will be part of a DevOps test team).

These efforts will be early in the development cycle. The device needs to be "good enough" to make it to the next cycle. Testing is minimal and just "sanity checks." Nothing will be too formal. There is minimal or no documentation at this point. Exploratory rapid testing technique concepts [2, 3, 5] should be used here.

Mature Groups or Growing Teams Targeting Growth of Sales

So, the organization has made it through early efforts or has some history supporting development. The expectation is to deliver a device to marketing/sales and users. The team may have had several agile cycles behind it, where internally delivered increments are passed along to someone. "Good enough" here has a higher bar than earlier efforts. The risks and the need for quality are increasing now. The device needs work to the point where buyers will continue to buy it, and some good press (social media or tech reviewers) can be generated. The organization may be part of a larger company or a smaller organization. Still, the bottom line is the device must be "good enough" (e.g., better than first cycles or prototypes) so that people will continue to want it or buy it.

The IoT team has some hardware, software, and system skills. While the degrees of skill in each area may vary, the team can understand where the IoT system is stable or weak. Test planning must address weaknesses and leverage strengths during risk-based testing and strategy selection. The test planning can still be "compact," but the idea of no testing should be covered in discussions with management and customers so that the lack of testing does not become a program risk factor. More than likely, some formal testing will be expected.

A test plan outline example for a maturing group with a single device, and a growing system team might include IoT:

1. Risk assessment exercise and assignments to integrity levels
2. Verify the hardware
3. Check out any third-party software
4. Developer structural testing (TDD at higher coverage levels)
5. Integration and integrated tests (make it work)
6. Verify integration of hardware and software
7. Validate stories/requirements and the device
8. Small-scale system V&V
9. Deliver (consider if the user or customer will be part of a DevOps test team)

Optional strategies to consider for inclusion in an IoT test plan:

1. Expanded experience-based exploratory cycle, driven by risks and attacks
2. Math-based testing
3. Model-based testing based on development and test models (when models exist)
4. V&V of hardware, commercial third-party components, and system evaluation

Teams need to consider what "good enough" means to them in the project context. Good enough will vary from a startup to an existing large company. Questions to ask and answers for good enough at this point include IoT:

1. How many failures can be tolerated?
2. What costs can be dealt with and how?
3. How does schedule (time to market) affect the product, risk tolerance, and testing?
4. Who does the testing?
5. Who are my users and customers (i.e., stakeholders)?
6. Where do I do the testing? (Is an external lab or third-party testing required?)
7. When is testing done and being completed?
8. What are my team's strengths and weaknesses? (What do we have that I can positively leverage?)
9. Can something I do or suggest grow this product and improve sales?

IoT Test for a System with More Devices

The mature market is where many hope to be. The group has products on the market, is doing maintenance, has sales, and may want to expand with new or upgraded IoT products.

The range of IoT devices will cover consumer products, industrial products, government products, and everything in between. Hence, the risks and integrity levels (see IEEE 1012 V&V standard [6]) will be more significant and varied.

Test plans will need to cover the organization's interests, including IoT:

1. All the items listed in the "Mature Groups or Growing Teams Targeting Growth of Sales" section.
2. Does this product have increased risks or integrity levels?
3. Is this product the core business of the company (i.e., if the product fails, you are out of business)?
4. If the product is in maintenance, are any changes small or large (new test, smoke test, and regression testing impacts)?
5. If this is a new product, how can this impact the existing product lines?
6. What am I learning from user data (analytics) and competition?
7. What V&V/test improvement areas should be expanded as "good enough" changes?
8. Is it time to automate? What testing can be automated?

Test plan outline examples for this mature group, with many IoT devices in development, include

1. Risk and integrity level assessment
2. Verify the hardware
3. V&V any third-party software and hardware
4. Development structural testing (full TDD and analysis)
5. Integration and integrated test (i.e., make it work with other devices and system(s))
6. Verify the system
7. Test and verify software
8. Validate story/requirements and #9
9. Validate device via simulation, test environment, and field testing
10. Correct level of system V&V
11. Product quality V&V
12. Maintenance and regression V&V with any automation
13. Deliver products to stakeholders or customers
14. Ops and test with data analytics

Options to consider include IoT:

1. Expanded exploratory cycle driven by risks and attacks
2. Model-based testing based on Dev models (if any exist) with automation
3. Higher levels of test automation
4. Constant product and process improvement
5. Optimized regression and new feature testing
6. Team skill building, training, new and experienced team grooming
7. Formal data analytics and taxonomies

Teams in this category need to avoid "slipping" backward and losing market share (refer to Deming at [11]). Groups and companies once prosperous in this area can relax. Well, maybe the team should *not* relax. This situation opens the door to competitors wishing to take market share. Relaxing can be a bad thing.

Organizations in this category often learn that test/V&V are essential as these efforts provide the data needed to continue growth. This situation is not to say test checking and mechanical automated testing cannot be reduced or otherwise optimized, but such changes must be within process improvement and planning scope.

For discussion, the pesticide paradox and automation/regression testing problems should be considered (see the section "Wrap Up Test Planning: Regression Test Cases in IoT", Part 2, Chapter 7).

Figure 6-5 illustrates some common elements that a tester may include in a plan. It can be used as a checklist where not every element needs to be addressed. However, teams may want to consider the checklist items during their test planning activities.

This mind map can be used iteratively over the key areas. One can start with identifying what is "under test." Options here start with hardware, software, and the system. These can be refined downward into more detail. Next, the planner may want to consider lifecycle options, including DevOps, Agile, traditional, and mixed. From here, the tester can think about what resources may be needed, including cost (money), schedule (time), and the numbers and skills of test and support staff. Associated with resource definition is scope, where one should consider how much testing is done to address the testing of the system, hardware, and software with how much, if any, V&V will be included in these. And as these take shape, the planning should start to address processes, including risks, techniques and approaches to testing, and what documentation will be required. For this map, I chose to reference ISO 29119 [2] in these areas quickly, but other references on these subjects could also work. From these considerations, the lab testing architecture should be planned. Items here may include the software test architecture (STA), a hardware test lab, and possibly a system testing and integration lab. These architectures may be large or small, the choice of which is driven by the "levels of factors" (see Part 1). The planning map and checklist are applied repeatedly during initial planning and then revisited over the lifecycle.

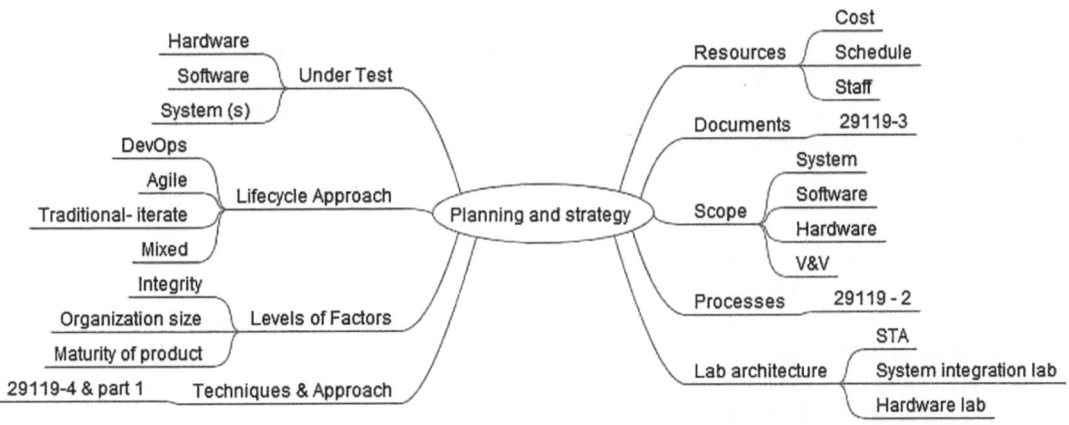

Figure 6-5. Planning and strategy mind map (note: full name ISO/IEC/IEEE 29119 [2] Software Testing Standard)

Reference: Jon Hagar Class Production

IoT Test Planning for Procuring Organizations: Governments, Large Corporations, and Others Gold

Some readers may be wondering why this section on test planning for procuring organizations (such as governments or consuming organizations) as a separate test planning category is necessary. The answer is because much of IoT will happen around names like smart cities, smart schools, smart factories, smart government, etc. The procuring organizations will be purchasing and using IoT with all the systems that surround IoT. These systems and systems of systems will be large or mega-large. Therefore, procuring organizations must possess knowledge before approving or spending limited dollars on IoT technologies or devices.

I believe that the world needs to develop "smart" IoT consumers. One of the biggest will be governments at all levels. However, large companies will come close, and even individual consumers may want to be "testing" IoT. Most large organizations, such as governments and companies, already have information technology (IT) departments. For years, many IT groups have had internal IT test planning efforts as part of the procurement and deployment efforts. I have seen companies with IT test labs devoted to assessing incoming hardware and software, including upgrades, before deploying to the rest of the company staff. This kind of testing is different from developer organization–based testing but follows many of the same ideas of test/V&V planning and strategy. While some in management might think what these people do is a waste of time, this kind of testing keeps most companies functioning and viable by mitigating many risks.

Companies and governments have learned in establishing and running internal "IT" departments that having testing/V&V efforts to do ongoing support to the organization becomes part of what they must do to succeed. Most of us as individuals conduct testing for the major procurements but maybe not with conscious intent. For example, would you buy a car (kick the tires) or a house (a viewing) without testing and evaluating it first? IoT systems should be considered in a similar vein.

What this means for IoT is a new procurement organizational paradigm must emerge. Procuring IoT organizations will need planning and product assessment for suitability and acceptance testing just as they do for IT efforts. Still, the efforts will have to expand to include more extensive hardware, systems, operations, and software domain.

Many procuring organizations may think that all they need to do is expand their IT department – and this can be a start, but what if they expect real "smarts" (e.g., AI, analytics, etc.)? If so, they will need more than just a traditional IT department skill set. Possible new skills that procurement specialists may need include the following IoT-related engineering skills:

- Operations understanding
- Data analytics staff
- Specialized hardware practices (not just for computers but perhaps for experienced telecommunications, network, or radio frequency analysis too)
- Human factors and IoT usage experts for the actual users of these IoT systems
- Psychology and AI experts
- Expanded and new software knowledge
- Testing teams with IoT understanding

The test plan outline for procuring organizations – governments, companies, and others – can include IoT:

- Risk assessment and integrity level exercises
- Verify the hardware within the new "smart" system of systems

- V&V integrated software elements in especially long-duration runs and chaos engineering (refer to Part 2, Chapter 7)
- V&V any needed software wrappers (containers that isolate a software element) locally created as part of Agile
- Integration and integrated test (make it work in system of systems)
- Verify the system and system of systems

  - Test and verify *all* software
  - Validate requirements and device
  - Correctly scale system V&V

- Product quality V&V
- Maintenance and regression V&V
- Ops and test with data analytics

  Options to consider for IoT procuring organizations: governments, companies, and others

- Expanded exploratory cycle driven by risks and attacks
- Model and simulation-based testing based on development models (if any exist)
- Higher levels of test automation
- Constant product improvement
- Team skill building, training, new and experienced team grooming in IoT
- Independent V&V (IV&V teams) with IoT specializations

The items listed here may seem onerous for the individual consumers, and likely most consumers will have ad hoc testing plans – like the ones they use for buying cars or houses. However, I feel that some consumers may want to follow some of the testing ideals since IoT at the consumer level will be more complex than current home "automation" systems and may be expensive and time-consuming. There may even be a market for consumer IV&V home automation consultants to help individuals build things (e.g., IoT home infotainment (smart house), small business IoT systems (small business IT), small community IoT (classroom, library, etc.)).

All of the previous lists are examples, and any variation can coexist and be helpful.

Impact of AI, Data, and Analytics on IoT Test Planning

IoT is projected to generate vast amounts of data from what was once embedded devices, plus all the new IoT devices that will be invented and marketed. Petabytes and beyond of data are projected. This data will be generated during the development and operational use of these IoT device systems. This section provides a brief overview of AI and data analytics for test plans to consider. These topics will be explored in more detail later in the book.

Many interested parties will use the IoT data, as shown in Figure 6-6. The first interested users tend to be IoT system marketing staff or management because they want to sell more and control the project. However, one of the interested parties should be the testing staff. Testers should become involved in developing and operational use of IoT devices and collecting analytic data from the operational data to use in testing.

Benefits to IoT test planning can include

- Improved field test planning based on error taxonomies reported automatically by devices
- Tasking some test activities to the user in a more realistic operational setting
- Mining help desk data to aid test planning (e.g., use cases, problem reports, etc.)
- Analyzing in real time data to improve ongoing test planning

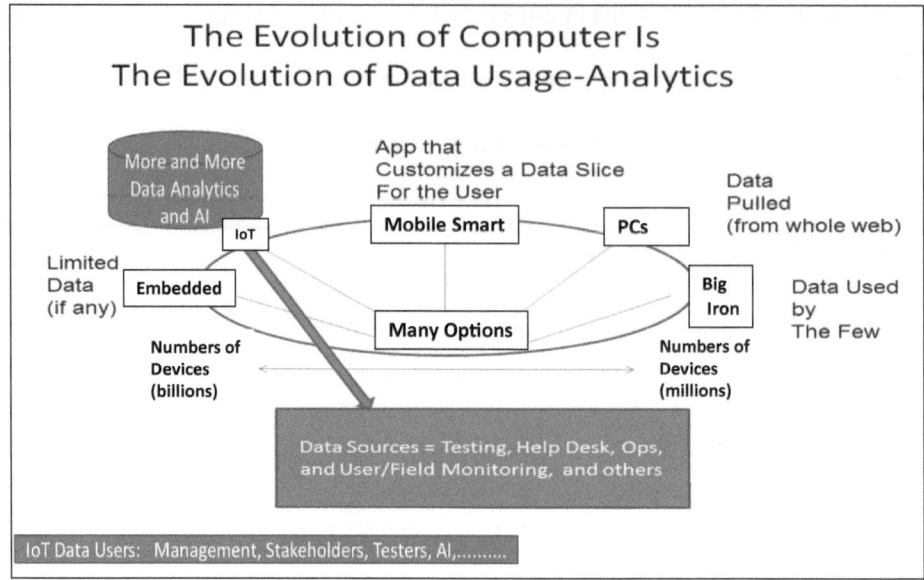

Figure 6-6. Data evolves to support IoT data users

Reference: Jon Hagar Class Productions

- Mining social media for IoT device problems
- Using analytic data to improve patterns of attack and test models
- Changing test plans over time with data-driven feedback

The test team will need analytics and big data mining to use such vast amounts of data. This will become an effort that must be included in master test planning over a complete project lifecycle. Planning will need to define the schedule and budget, skills, tools, and analysis types needed to accompany any IoT product. The tester will produce and use data-driven taxonomies (Part 4, Chapter 22).

Test management will need to advocate for this type of analytics. At first, I expect executives and management to see the testing data as less critical than consumer/sales data. They may see the operations and marketing people as having more valuable evidence. However, lessons learned from the embedded software systems industry showed that testers needed to look at the field usage data to see if it matched historical test case data. When a mismatch was found, it implied that new test use cases were needed and changes were rippling into their test models and test planning. Lessons were learned from data analytics as the software evolved. The new IoT system world must do the same.

NEW TEST MANAGER

I joined an existing project when a new device system was planned for the market. I came in with the proposal team to join a team already producing systems. The software management team was skeptical about me since they did not "hire" me, but I came onboard with the new system's team. However, my skills included data analytics, and the existing team had an extensive database of software issues. I gained access to the data and created an issue taxonomy (Part 4, Chapter 22). I asked management what their "big" risk areas were, and they answered "changing requirements." I then presented the taxonomy, which "should have shown" that requirements were an issue. However, they also had a high rate of software structural issues, which could be found by improved code coverage testing. The team introduced an improved development test process for the new system. So by doing this, the rates of escaped problems came down.

The organization and testers that master using data analytics to drive future test planning will likely be successful more often. In producing test patterns for the book *Software Test Attacks to Break Mobile and Embedded Devices*, I produced mind maps and taxonomies of historical errors that escaped testing but were seen in the field. In surveys of practicing testers, all of whom had error database logs, over 90 percent did not do *any* error taxonomy or data analytic analysis to improve test planning. The IoT test organizations or individuals who learn to use big data and analytics will benefit significantly from my story.

Product and Development Lifecycle Impacts on Test Planning – DevOps and Agile

So much of the world, including hardware, is evolving toward Agile, or at least groups say they want to be Agile. Agility can include many of the concepts from DevOps and vice versa. I suspect that IoT efforts will be confronted with DevOps and Agile, and sometimes aspects of traditional waterfall cycles, often from the hardware side. These will form "hybrid" efforts where the IoT goal will be to put together a product as soon as possible (ASAP). Teams will face time to market and try to keep funding sources happy. Agility will pressure development and testers to answer the following question: "What is or will be the minimum viable 'good enough' IoT product?"

For this book and IoT, the things I offer as possible answers to this question include

- The communications must work. (Bluetooth failed to work correctly or is just plain unreliable and has been viewed negatively in the industry and by many users.)
- The hardware must be functional and have reasonable reliability (quality).
- The software must work functionally plus have a few nonfunctional quality characteristics.
- The software can be conveniently updated to fix issues not met.
- Time to market must be met.
- Costs are minimized for developers or products and consumers.

An agile effort will stay flexible to meet these challenges, usually with small sprints, and watch the product unfold.

Some agile writers have said that "in Agile, we do not need test teams" (since testing is automated and owned by Dev). As authors such as Crispin and Gregory discussed in agile testing [5], this belief has faded.

So many IoT teams, particularly startups, have heard "no testing is needed." To complicate this, one of the industry's testing gurus, James Whittaker, has stated that "testing is dead." I believe he should have phrased it as "<traditional, end-of-life-cycle, scripted, unthinking> testing is dead." I agree with the revised statement.

I believe these statements could be accurate while advocating that a tester's role exists in many IoT efforts. The old ways of many testers should be long gone (e.g., late in the lifecycle; testing for days, weeks, and maybe longer; using long human-readable test scripts; making testers responsible for product quality; and manual brute-force testing), or at least for agile IoT, the testing needs significant changes. Additionally, for some early testing cycles (as noted previously) and for startups, "not testing much" may be the way to go.

Testing will integrate with development and during lifecycle operations, thus providing critical data (analytic information) to other efforts. Testers in IoT will need to be skilled in planning at a detailed level. The plans will need to address strategies (see the strategy section in Parts 1 and 2) and fit within IoT product costs and schedules. As I have been saying, there is no "best way" for testing, so that the need will be greater than ever for the test planning skills of managers and senior people.

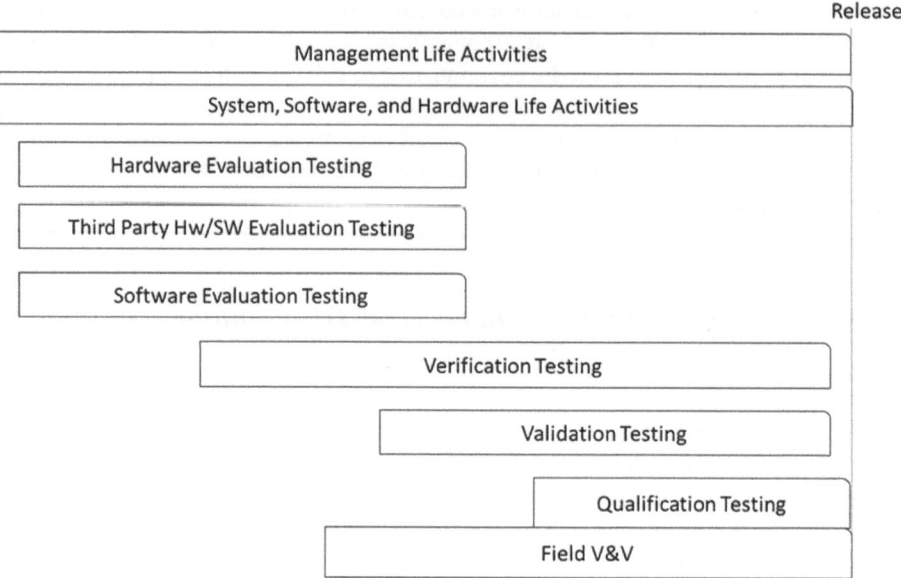

Figure 6-7. Example schedule for planning IoT testing (simplified)

The example schedule in Figure 6-7 is very generalized and very high level. It does not show dates or time lengths. However, it does show top-level schedule strategy tasks.

Management lifecycle activities span the entire time, as do hardware, software, and system engineering life activities. In early efforts, requirements, design, and implementation would be done. Later life activities would be maintenance and repair driven by data analytics. The test activities for the early efforts address hardware, software, and systems with an early focus on hardware and software as supplied by outside development. Midway through the IoT device's life, V&V is done before the IoT system is ready for acceptance testing. This example assumes product elements, hardware, and software are updated during the whole life of the IoT system, so ongoing V&V, qualification, and field V&V activities are shown in the bottom two task boxes. The scope of final qualification testing and V&V efforts depend on the IoT system's nature (e.g., small systems get minimal test tasks and extensive systems need more work).

Summary

This chapter introduced the ideals of IoT planning, test, strategy, and architecture. It is vital that a team of stakeholders determine what is "good enough" for each IoT device for their software and devices. Hence, test planning is outlined by organizational classification: beginner, experienced, agile, medium, large, and large, but new to IoT. Each of these will have team leadership and address IoT test planning basics, yet must tailor and customize the planning to their local context. The next chapter further addresses IoT test planning and strategy for hardware and software.

References

1. ISTQB website: www.istqb.org/ – accessed winter 2022
2. "IEEE/ISO/IEC 29119, ISO/IEC/IEEE International Standard – Software and systems engineering – Software testing" – Parts 1 to 5 series
3. *Lessons Learned in Software Testing* by Kaner, Bach, Pettichord, Publisher Robert Ipsen, 2002
4. *The Art of Software Testing* by Glenford Myers, A Wiley Series, 1979
5. *Agile Testing: A Practical Guide for Testers and Agile Teams* by Lisa Crispin and Janet Gregory, Addison-Wesley Professional, 2009
6. "IEEE Standard for System, Software, and Hardware Verification and Validation," in IEEE Std 1012-2016 (Revision of IEEE Std 1012-2012/Incorporates IEEE Std 1012-2016/Cor1-2017), pp. 1–260, 29 Sept. 2017, DOI 10.1109/IEEESTD.2017.8055462
7. *Software Test Attacks to Break Mobile and Embedded Devices* by Jon D. Hagar, CRC Press, 2013
8. PMBOK Guide: www.pmi.org/pmbok-guide-standards/foundational/PMBOK – accessed winter 2022
9. https://en.wikipedia.org/wiki/Conway's_law – accessed winter 2022
10. https://www.mountaingoatsoftware.com/agile/scrum/meetings – accessed winter 2022
11. https://en.wikipedia.org/wiki/W._Edwards_Deming – accessed winter 2022

Figure Reference

1. https://cdn.pixabay.com/photo/2018/03/01/05/46/business-3189797_960_720.png – accessed winter 2022

Chapter 7
IoT Test Planning and Strategy for Hardware and Software

The previous chapter introduced the fundamental ideals of IoT planning, test, strategy, and architecture. Different types of IoT teams will need test leadership to address IoT test planning that must be tailored and customized to their local context. This chapter further addresses IoT test planning and strategy considering both hardware and software. While planning IoT software testing, the hardware must be considered. In traditional software test planning, less attention is paid to the hardware on which the software executes because the hardware are common generic computers and associate elements. These elements are different from the more specialized IoT hardware. Verification and validation (V&V), as defined in standards such as IEEE 1012 [1], address aspects of hardware, system, and software test planning. In this chapter, IoT test planning of the hardware and software considers the architecture, product lifecycle, costs, schedules, and strategies. Finally, given the added complexity of specialized IoT hardware and software, the concepts of integration and testing become essential to plan. It should be noted this is not a book on testing hardware per se.

Traditional Testing Overview

As shown in Figure 7-1, I plan testing while development builds hardware or software in traditional testing. Development starts with requirements (needs), creates a design, and implements perhaps with some integration. During these efforts, the test team analyzes requirements and design, makes detailed test plans, and creates test designs and implementations (usually manual), and so after some implementation exists, they run tests. Everything is good. Of course, there can be different levels of testing (e.g., component, integration, and system) as they test from the bottom up. The lifecycle of Figure 7-1 worked well for things like hardware, and other times not so well for things like software, which can change quickly. The parts of Figure 7-1 are covered throughout this book.

Hardware, mechanical, physical, electronic, etc., did all right with the traditional model (often called waterfall or V model) when a test was added in. The traditional models worked well when teams understood that even the original name "waterfall" assumed iteration and cycles of change, and teams revised and modified as they built. Things were always done in repeated cycles. Problems arose with the waterfall lifecycle model when teams tried to create perfect products at each iteration and consumed too many resources at each step. However, teams produced working software this way. Yet the software industry has statistics showing that many (50 percent +/– 10 percent) software projects failed to one degree or another when they tried to follow a "pure" waterfall/V model.

© Jon Duncan Hagar 2022
J. D. Hagar, *IoT System Testing*, https://doi.org/10.1007/978-1-4842-8276-2_7

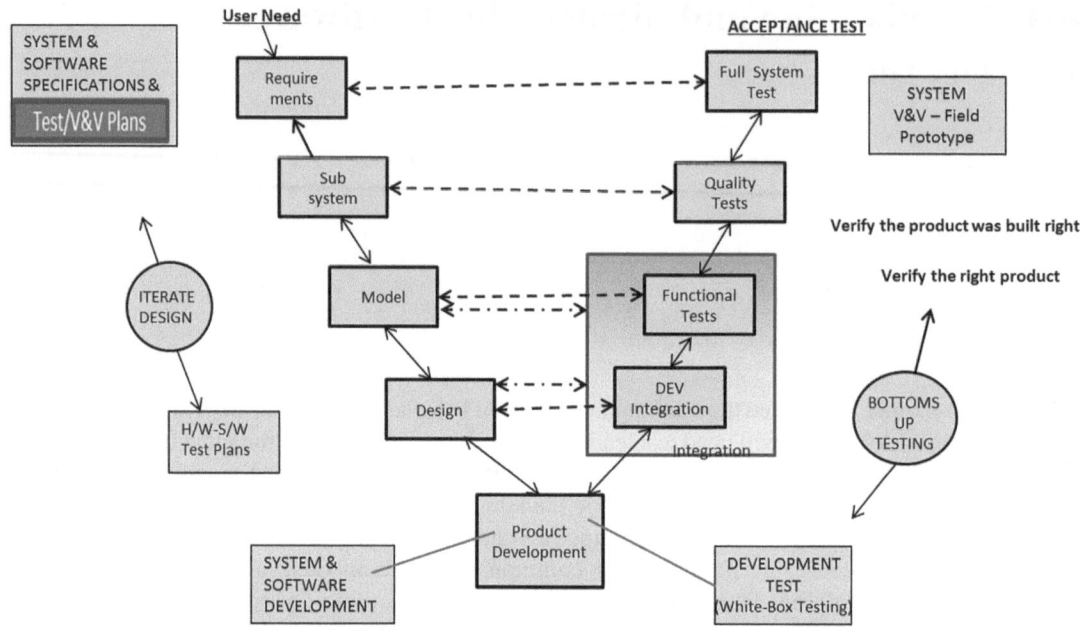

Figure 7-1. Historic V model for V&V/test efforts

Reference: Jon Hagar Class

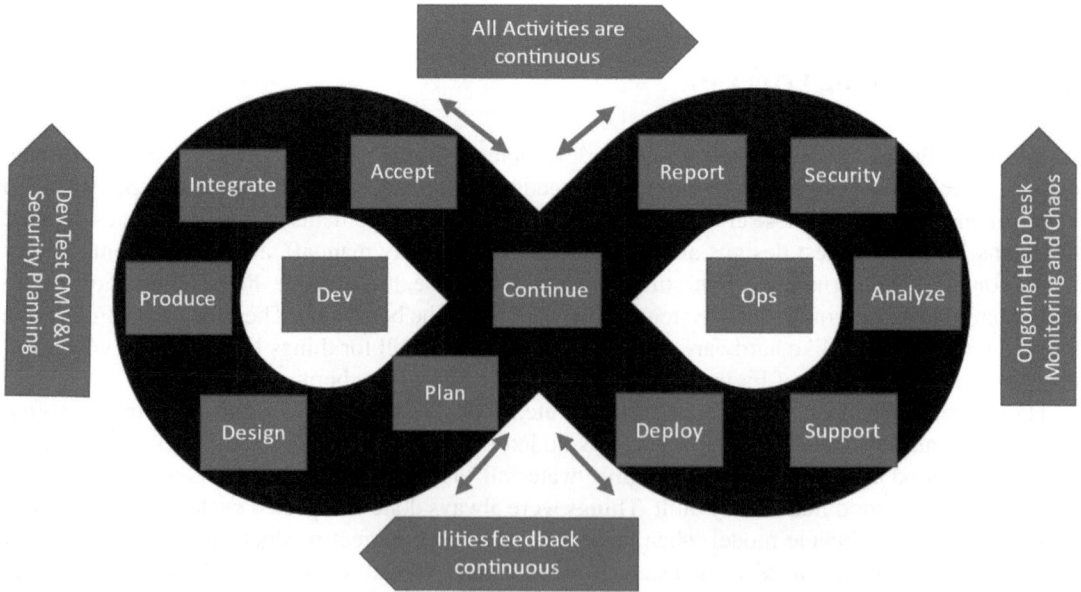

Figure 7-2. IoT test infinity model

Reference: Jon Hagar Class and [2]

Given that many hardware and system companies still follow a "traditional" approach, teams moving into IoT should expect and learn how to work with mixing and matching aspects of traditional cycles while adopting agile development. In traditional teams, the knowledge that iteration and cycles can still be possible with their traditional lifecycle is important. The mixing of traditional and agile models and techniques requires some work in scheduling and planning a lifecycle, as shown in Figure 7-2, which lasts as long as the project lasts.

IoT teams will likely not produce all of their own hardware or software. Teams will use third-party vendors by buying off-the-shelf (OTS) hardware, using no-code/low-code, microservices, and open source components. True, some IoT teams will create specialized sensors or components for their IoT device. Still, a vendor will openly provide most hardware and software (a.k.a. the third party). Outside parties can and will use different lifecycles than whatever your IoT project is using unless there is a contractual obligation to do otherwise. In this book's upcoming sections and chapters, I will address aspects of the Figure 7-2 model. Currently, many IoT projects will use agile and/or development-operations (DevOps) test lifecycle models, which the infinity test model directly supports.

Planning in the Agile/DevOps IoT Test Lifecycle

I expect most IoT efforts will follow or want to follow agile concepts or may be combined with DevOps, as the two are closely related. I recommend Gregory and Crispin's book on agile testing [2]. I have both books on my shelf and refer to them often, and I also talk to or listen to these authors every chance I get.

Most teams want to be "better, faster, cheaper" with a balance of these. The old joke was "pick two of the three," but *in IoT, teams will use balance and agility to adjust these three continually*. The balance on each of these will change over time and with product releases.

Readers should read and understand the Agile Manifesto [3]. It states:

We are uncovering better ways of developing software by doing it and helping others do it.
Through this work, we have come to value:
Individuals and interactions over processes and tools
Working software over comprehensive documentation
Customer collaboration over contract negotiation
Responding to change over following a plan
That is, while there is value in the items on the right, we value the items on the left more.

These beautiful and straightforward words took much time and thought to craft. It would be best if you took the time to read and appreciate what is put forward. I have found groups that claimed agile. For example, one team said they were agile, but, on quizzing, all they were doing was hacking the software and creating zero documentation. They missed the point. Reread the manifesto.

Next, I am picking several fundamental ideas from Agile that I think IoT teams should know and, within the proper context, practice in order to build successful system devices. These ideas are as follows:

- We are flexible in responding to changes in plans, products, contracts, etc.
- We use continuous integration at many levels (e.g., software to software, software to hardware, system, and systems of systems).
- Applying continuous testing, including

  - Verifying and validating by the team (e.g., developer testing, modeling, analysis, inspection, etc.)
  - Acceptance of test-driven development [2]
  - Acceptance/qualification cycle with short and automated efforts
  - Ops testing driven by users, testers, security, quality, and data analytics

Some critical IoT efforts might justify and need concepts such as outsourcing to testing as a service (TaaS) contractor or IV&V, but only if the integrity level [1] and factors (Chapter 6) justify the cost and time. Examples of where a high integrity level might be found are

1. Safety-related items.
2. Hazard items where the loss of property is possible.
3. Items where significant money can be lost. (Note: This may not be easy to spot.)
4. Security and privacy losses would or might be high.

For continuous test (CT), an agile IoT test plan outline might look like the following:

1. Test/V&V OTS as soon as they arrive.
2. Have an IoT team test/V&V person support any OTS customization issues.
3. Hardware tests.
4. Assist and evaluate test-driven development (TDD) using test techniques and metrics.
5. Develop acceptance test-driven development assessments and plans.
6. Conduct software testing using

   a. Acceptance tests
   b. Attacks in security and risk areas
   c. Exploratory testing using competent testers
   d. Testing tours of key areas

Continuous integration (CI) is conducting integration continuously within a test plan. CI includes

• Obtain OTS/third-party product(s) and determine what has been V&V/tested and what is at risk.
• Do integration cycles (bottom-up, top-down, iterative coil).
• Assess if wrappers are in use and whether the testing happen inside and outside of the wrappers.
• Build iteratively based on criticality, schedule, and test factors.
• Test interfaces.
• Build CI test features over time.
• Check CM/SCM components and work with various versions of the product as needed.
• Build data, stubs, and drivers as needed.
• Verify interface control documentation (ICDs).
• Validate to interface control documents.
• V&V to standards of products, protocols, communication, etc., as needed by the local context.

I take from the manifest and DevOps concepts the following points. In Agile, teams learn as they go, so they must respond to plans as the project unfolds. I communicate, which maybe should have been on the critical list, but I believe all engineering and business communication is as important as critical thinking. I agree that testing in Agile succeeds based on skill. Reading this book gives some, but not all, test knowledge. Skill is different from knowledge, so an agile team will have people who have practiced the craft of testing to build skills in software, testing, hardware, and likely even thinking about the system.

Notice I have not said in this section if you need a separate test team in Agile or DevOps. Again, there is the need for context and "it depends." Whether high performance or scrum based, agile teams need people with test knowledge and unbiased engineering, like those found in this and other books. Agile or traditional will not save a project if you do not have tester skills, continuous planning, communication with stakeholders, and critical thinking vs. just spouting ideas. Then, failure is likely. Some agile IoT teams will have a separate test group. Some projects will embed testers within a scrum type of team. Some will do a mix of organizational structures. The "it depends" requires careful thought as the context defines the paths to take during Agile, Dev, and Ops.

What Should Development and Test Do During Agile and DevOps?

AGILE AND OPERATIONS VS. TESTING

Consider the Tesla self-driving car (SAE level 3) story. In 2016, a person was killed while the car was in "auto-drive" mode. Tesla evaluated the data from the crash because maybe the system (car) was malfunctioned in the field. According to reports, the user did not take over control of the vehicle; instead, he viewed videos on his mobile phone. (Note: The data from several devices were used to construct this story.) Tesla has stated, "a user should not do what this user-driver was doing." However, users are human. Human users are likely to take similar actions on other IoT devices and systems, creating likely unplanned or unforeseen scenarios. Some of the human users will suffer the same consequences. Yet, like developers and testers, this raises the question, "Should we build the systems with this kind of likelihood in mind?" [4].

And the answer to this example is still up in the air. The US government's National Highway Traffic Safety Administration (NHSTA) is currently working on self-driving vehicle policies. However, testers in DevOps organizations cannot wait on the rules and standards to be created and then implemented because engineering and development are happening *now*. As the Tesla example indicates, as long as users understand the limitations, DevOps is the way to go. Unfortunately, many people do not take the time to learn about limitations.

I follow the agile testing approach (see [5]). Moreover, getting a product out using short cycles means many DevOps teams can significantly depend on test automation and manual experience-based approaches such as exploratory testing, attack-based testing, and testing tours. Test automation in DevOps testing could include

- Automate repeated or labor-intensive testing (e.g., regression, critical risk areas, etc.).
- Use data analytics to know where errors cluster and, therefore, where to automate more testing.
- Automation of CT, CI, and CD critical risk areas.
- Test outside of the UI (e.g., use API, unit, integration testing, etc.).
- Design the IoT software, hardware, and system to support testing automation.
- Plan and estimate testing to include automation (keeping in mind that automation requires resources).
- Measure automation to be able to report cost savings and identify improvements.
- Remember test automation is development and has repeated development cycles (be Agile).
- Leverage tools (e.g., OTS and commercial tools that fit your processes).
- Test in a cloud so resources (tools and computers) can be expanded without buying more software or hardware.

In DevOps, the team and testers need good configuration management (both CM and SCM) to function effectively with hardware, software, systems, and documentation, including

- Tools
- Development code
- Development hardware
- Development support documents
- Test environments/labs
- Third-party hardware, software, and systems that are in use
- Numerous device configurations

Finally, members on many Ops teams may not be titled "testers" but will need to think like testers. The Ops staff will need to watch for issues, find bugs, and define and document unhandled (untested) but repeatable use cases, especially those not seen before. These data should be reported to the Dev team through CM and SCM channels, and valid issues documented in the CM/SCM systems/tools by the Dev personnel.

Ops Team People Skilled in Agile Test Thinking

For example, some Ops teams will find data that is not a pure error. Still, if followed, a tester might quickly identify a trend that will find an issue. Testers will use data from exploration, ad hoc tours, attacks, and data analytics to find bugs. Testers develop mental models of software system "smells" and patterns, which, if followed, will yield valuable information or an issue that should be fixed sooner rather than later. Thus, I recommend the Ops staff have testers onboard who have advanced tester skills. For more, see [6].

Planning Using a Hybrid Agile Test Lifecycle

This planning is where the IoT test lifecycle should have a mix of traditional, Agile, DevOps, and iterative approaches covering hardware, software, systems, and operations. Smaller IoT devices may not face the hybrid lifecycle case or a more miniature mix. However, it is likely that if the IoT device is part of a more extensive system, or the team is dealing with a more extensive IoT system of systems, then the hybrid case is more likely.

Many IoT teams will want to follow a "keep it simple, sir" (KISS) approach and only deal with their device's small corner of the world. This approach is viable, but when done, I recommend the literature and information (user guide, contracts, customer specs, etc.) about the particulars of the IoT device define and state what is in bounds and what is out of bounds for the specific device or system. Most consumers will expect the IoT device and "system" to work out of the box when they just plug it in. They will not understand that the development and testing were done on a "smaller" view of the lifecycle and user operations space. The device may not function in their house with their router and other owner unique IoT devices. If teams can restrict their device to what was developed and tested, happier consumers will result. They will know where the boundaries are, and the next steps are needed. This results in "whole house" solutions from one vendor or consortium of vendors. For example, Apple does this already in some product lines.

However, most of us who have worked with systems and systems of systems know that during integration (of hardware, software, networks, IoT devices, and systems), really challenging problems appear, even after long periods of test or use. This situation is because, in software, coupling between components and elements can "hide" errors until the real-world inputs and device conditions are encountered over a more prolonged time.

SOLAR ARRAY HARDWARE CONFIGURATION EXPECTED
In one case, I was working with a solar array, which supported a connection to the Internet via a household router to report data generated back to a specific vendor (calling home, so to speak). The vendor expected that the router used with the solar panels was "hardwired" to the Internet. However, in this case, the router connected to a cellular hot spot (basically another type of router) that was *not* kept powered on 24x7. The connection configuration was shaky at best, given the terrain on which the solar array was installed. This integration issue ultimately was outside the vendor's control, or was it?

The hybrid system of Dev and testing is another topic, which cannot fully be addressed in a single book, but I do provide some starting points for IoT teams to consider. Many IoT devices and systems will be created in the hybrid development space of different lifecycles and approaches between hardware, software, and systems. In the hybrid case, learning from history to support test planning ensures the following happens:

- Test architecture is addressed.
- Test planning and strategy are considered.
- High-priority user needs (requirements, stories, models, etc.) are met.
- Design and implementation have V&V/test.
- Data analytics in development and Ops are considered.
- Test taxonomies and dependability of the system are outlined and worked.
- Ops and users are considered in testing (generated data as part of DevOps that is fed into new development).
- Integration, integration, and more integration in multiple phases and time points between hardware, software, and system.
- Security should be a high-priority topic and treated with a great deal of respect.

The IoT team working in a hybrid lifecycle will need to master and use the earlier engineering planning information and concepts listed previously.

Hybrid IoT systems include risks for the following:

- Environmental
  - Low signal quality, low power, lack of resources.
  - Input space is not thoroughly tested.
  - Physical environment factors.

- Design internal to the device
  - Functional and computational problems
  - Quality nonfunctional failures
  - Data and storage issues

- Output from the device is addressed:
  - Interoperability to other devices
  - Big data analytics and stale data (especially in notification messages)
  - D2A and environmental mismatch
  - Timing and performance issues

No system or system of systems can be 100 percent thoroughly tested, particularly as they get more prominent with more variations on configurations. Disclosure to stakeholders becomes a good option. Of course, the team needs to know what to disclose – boundaries, context, security, etc.

Another aspect of planning is the management side of budgeting, estimating, and scheduling, all critical areas for management but affect the entire project personnel.

IoT Test Plan Budgeting, Estimating, and Scheduling

A brief introduction to test budgeting, estimating, and scheduling is presented in the following to get IoT teams started. These are difficult topics that baffle even some MBA grads. For budgeting details, refer to information in the following sources:

- Test effort wiki [7]
- *Proposal Preparation* – Ann and Rodney Stewart [8]
- *The Tester's Pocketbook* – Paul Gerrard ([9] page 8 on the "equation of testing")

There are very few books strictly on test budgeting, estimating, and scheduling. I do not necessarily recommend nor follow precisely any of the preceding pointers, but they will give readers a reference point to continue learning. Most of us learned to budget, estimate, and schedule by doing and occasionally failing.

Estimating Schedules – A Brief Introduction to Support Cost Estimates

Any estimation is, in part, guesswork since we are dealing with the future. Estimates are only a forecast. There are many approaches for IoT teams to consider. See the following links to get started:

- `www.rationalplan.com/projectmanagementblog/traditional-and-agile-project-management-in-a-nutshell/` – accessed in winter 2022
- `www.mountaingoatsoftware.com/presentations/agile-planning-and-project-management` – accessed in winter 2022
- `www.agilenutshell.com/planning` – accessed in winter 2022
- `www.pmi.org/learning/library/combine-agile-traditional-project-management-approaches-7304` – accessed in winter 2022
- `https://en.wikipedia.org/wiki/Project_planning` –accessed in winter 2022
- `https://shop.acuityinstitute.com/collections/all` – accessed in winter 2022
- `https://en.wikipedia.org/wiki/Cost_estimate`

> **Note**
> These are only a few examples and certainly do not represent a complete and industry absolute truth.

Besides the information in Part 1 of this book, factors to consider for schedule estimation include Addressing the "big" picture as a first pass (activities and tasks that are known)

- Workforce – Get the right skills, small team or big, in-house or outsourced
- Materials needed
- Tools needed or developed
  - Prototype that may need to be created
  - Throw away items that are only needed in learning but still take time
  - First production run (one or more items)

YOU CAN MIX TRADITIONAL AND AGILE SOFTWARE LIFECYCLES

I was on a big project (millions of dollars), hardware (machines, buildings, many locations, computers, etc.), long-term (many years of development and decades of field use), and software (embedded, controlled critical functions, field and Ops processing, etc.). I did traditional formal scheduling for the "big picture" items. Schedules, plans, and progress were reviewed monthly with management. I managed and controlled the third-party contractors formally. However, I still stayed agile by responding to changes in the big picture with my test team.

Further, the day-to-day and week-to-week activities were very agile. I used daily stand-up meetings. Small scrum teams were standard. I had test automation. I worked directly with customers. The team evolved as the project unfolded. Finally, the first use of the project worked perfectly. The mix and match of approaches proved viable.

Estimation of Testing Size

Many managers and customers want the estimated test effort in test planning and contract setup. Size estimates typically can be done based on the following "factors" of testing:

– Number of estimated risks
– Number of test documents created and maintained
– Number of tests planned
– Number of meetings
– Size of the test environment, including production of tools, models, and support software
– Number of testers

Management likes "hard" counts of things. Other products and efforts (e.g., tasks) can be estimated. A sampling of common estimation methods is given in Figure 7-3.

Method	Project	Example	
Extrapolation Based On History	History exists Changes are documented Existing system costs understood Dev-test team has experience	Done before IoT system upgrade Past success Historic data can be used	
Analogy Based on Similarity	New system in similar domain New but known software Dev, test, & Ops have experience	Replacing old system Been there done this New but skilled team	Gold
Expert Consensus In the Team	New System Consult "experts" with knowledge Dev, test, & Ops have NO experience	Start up Experts provide advice Hire skilled team Risks are higher Take best guess multiple by 3	Gold ++

Figure 7-3. Sizing test efforts

The preceding methods are a beginning to get a tester started. Learning to make test estimates is based on the project context and takes practice, experience, and time. I have been working on my estimation abilities for 40 plus years, and I am still learning while often getting it wrong. I also have a secret rule of estimation method linked to me, but I learned it from other managers years ago. I take any estimate coming from Figure 7-3 and double it if I think it is likely to be right or triple it if I think there is a large margin of risk. Predicting the future is hard. For more information and articles on estimation, see [10].

Quality, Verification, Validation, and Testing

I put V&V/testing concepts after the money because I did not want the book to assume a waterfall appearance, and I wanted to put testers first. I understand most IoT startups and early projects will not think about testing or quality at first. Still, maybe they should have just enough thought about them to assure success as the IoT device evolves in the early first phase or two of development.

For details on V&V/test planning, design, execution, and environments, the reader should refer to Part 1 of this book.

Test and Quality

Some famous testing authors have announced testing is dead. This statement needs some context. Zombie testers should be dead (or are they already dead?). There is a project need for IoT information that V&V/testers can provide.

The "good enough" concept from software may mean the rev 0 phase of an IoT device has no or minimal formal V&V/test, but the development team must have some informal "checking." The team needs to determine whether the product is good enough to be released into the next phase. What "released" means can be very different from project to project. For more information, see the "good enough" concept by James Bach [11].

> **WHAT IS GOOD ENOUGH CHANGEs OVER TIME**
> Updates to software in IoT devices can change what "good enough" means. In the first version of a "cool" IoT device, which included a link to cell phone ability, users were happy with the phone and a simple IoT system. However, as the updates to the IoT device got more complex, the performance of the device and the ability to handle phone calls were impacted due to the performance of the overall IoT system. Users could not receive phone calls when using the IoT device. This made them unhappy. The users moved to other IoT systems, and the "cool" device died a slow death.

As the IoT product evolves and gains functionality, testing will also evolve. Expanded testing might use things such as

– Automation
– Analytics
– Exploratory testing

– Security checks
– Formal testing/V&V
– Verification
– Validation
– Reviews and inspections

The nature of the IoT product will vary over time and, thus, help determine the breadth of V&V/testing. Risks will drive the expanding list of qualities to test to get better and better products. In the following sections, the book considers V&V/Test planning.

Verification and Validation Activity

This section covers the basics V&V teams will need for planning. For more details on V&V/IV&V, readers should refer to IEEE 1012 [1]. V&V are higher-level activities that include testing and using other activities to improve and assess the engineering products. V&V, as defined in IEEE 1012, include general, hardware, software, and system V&V activities. V&V efforts span the entire IoT system lifecycle. V&V may be included in a test plan, but it can also be documented in the system, software, and hardware plans and their V&V plans. Where V&V is planned depends on the factors of the IoT system.

Verification tries to assess "did I build the product right." It needs some "truthful" reference information such as requirements, ConOps, or stories that I "check to." The primary inputs, actions, and results are shown in Figure 7-4.

Verification takes inputs of plans and resources along with risks, data, and system information. Along with these items, verification actions are performed, including planning, test design, and assigning an integrity level. Once the levels and plans are known, testing, analysis, and inspection of the identified verification products can be done. Verification support products include hardware, software, data, and system information.

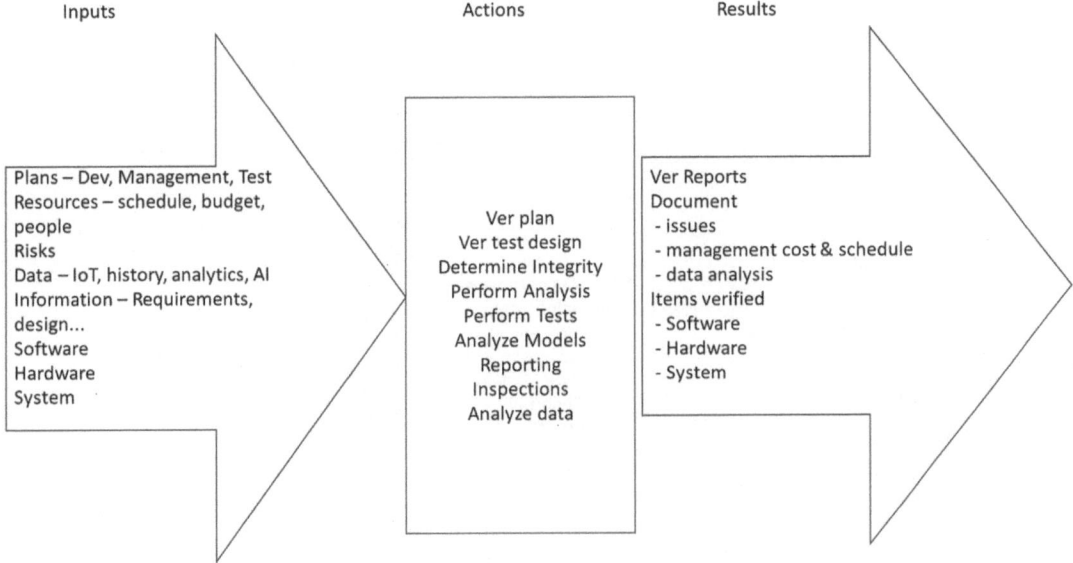

Figure 7-4. Verification activities

Reference: Jon Hagar Built figure for IoT Classes 2016

Verification approaches include

- Inspection – The tester visually looks at the item.
- Analysis – The tester uses tools such as modeling, math, simulation, similarity, or logic.
- Test – See early definitions of testing.
- Demonstration – The tester does a test with the actual system (unmodified) and no specialized equipment.
- Checking and test automation – The tester uses tools and automation to assess the software (minimal thinking).

Verification products include reports, identification of issues, results data, and verified products as shown in Figure 7-4.

Validation is focused on "am I building the right product" and is shown in Figure 7-5. In validation, I assume everything, including requirements or stories, can be wrong, but my stakeholder still wants some IoT system that meets their needs.

Validation is much more complicated than verification, but both are needed for a sound IoT system. Validation takes similar inputs to verification but includes consideration of the needs of the stakeholders. These needs may be unspoken and not written down, so the validation person needs an understanding of the IoT system "goals." Validation engineers usually have the same skills and use many of the same tools as the Dev engineering staff.

Validation focus includes

- A detailed review of requirements and design with an understanding of the "real" *needs* of stakeholders (needs may or may not be written requirements but are fundamental for a stakeholder).
- Analysis using modeling tools and simulations.
- Testing is typically done at an entire system and Ops level.
- Demonstration – The tester does a test with the actual system (unmodified) and no specialized equipment.
- High levels of test automation.

Validation products include reports, identification of issues, results data, and verified products as shown in Figure 7-5.

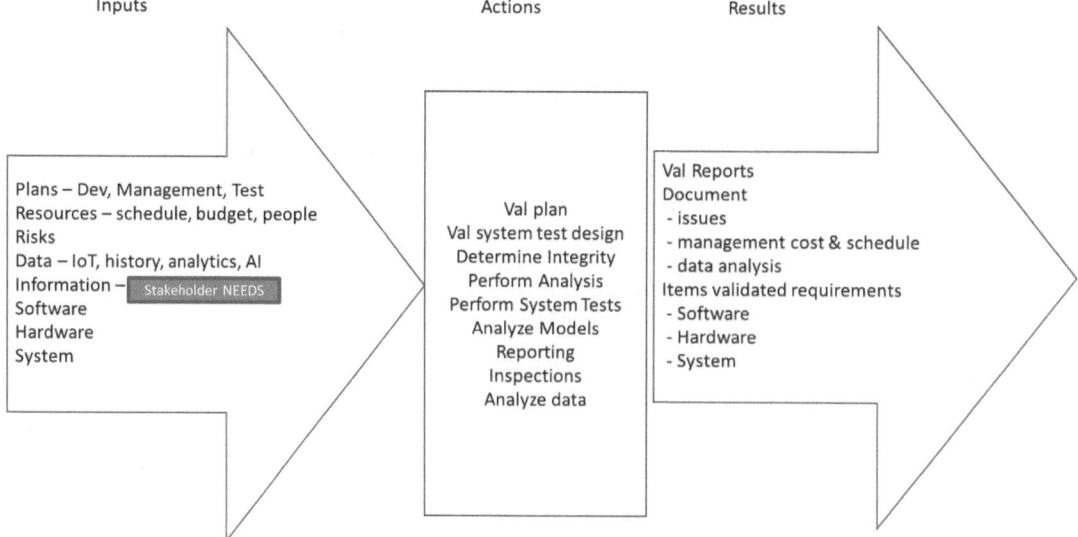

Figure 7-5. Validation activities

Reference: Jon Hagar Built figure for IoT Classes 2016

Wrap Up Test Planning: Regression Test Cases in IoT

Regression happens in software partially because of coupling [13] or lack of team understanding of design. Most software V&V/test teams are familiar with the idea of regression testing [12]. Teams understand that during development, as the software changes in areas that are already tested, regression testing is needed to answer the question "did we break something that was already working and tested."

Teams develop different strategies for regression testing. The steps include

1. Rerun all testing (time-consuming and expensive).
2. Rerun tests in the related function or code areas (complex to know if coverage is reached).
3. Rerun a standard regression test suite (takes planning ahead).
4. Rerun what the team thinks needs to be run to address items 2 and 3 (practical).

My personal experience on one project was that I did a mix of items 2, 3, and 4. I did risk-based and test planning. I changed the "old" test to avoid aspects of the pesticide paradox while still addressing regression risks. I made sure a test existed to assess the software's error or change. I did a lot more tests during the operations and maintenance (O&M) phase because I employed V&V/test automation.

Each of these strategies has plus and minus considerations. It is out of the scope of this book to define a complete regression strategy or approach. Indeed, there is a debate in the test community about the validity of regression testing [14].

The validity question comes from the pesticide paradox [15].

In the pesticide paradox, the ability of a test case to provide information diminishes the more times it is run without changes. Rerunning regression tests repeatedly is, therefore, of questionable benefit, even when one of the elements, the item under test, does change. However, most organizations employ some levels of regression testing/analysis.

On one O&M cycle, I had the following percentage allocations:

- Standard regression tests – 20 percent
- The expert team selected regression tests with 30 percent
- New tests (include developer tests) – 30 percent
- Tests traced to coupling – 10 percent
- Historical system tests not run in a while – 10 percent

Now my percentages are nothing magical. In fact, on other operations and maintenance (O&M) cycles, the percentages looked different. The key was that I applied critical thinking to planning with risk-based testing early in the O&M cycle. I got stakeholder acceptance of V&V/test plans, and I modified test plans as needed during the cycle. I was thinking, thinking, always critical thinking.

Also, the reader may notice the last category percentage, which may seem "new." The team added this case, which is different from "standard" regression concepts. I learned over the O&M years that it was a good idea to pull out historical tests that had not run for a while and rerun them with changes for the new O&M cycle. I did this during most O&M cycles, and slowly over the years, the team would cycle through most historical system tests. The project was looking for a regression that had escaped previous test cycles. If I had had the time and high levels of automation, I could have avoided this step, but I did not have the time and budget, so "cycling" over the historic test suite was a compromise. In one or two cases over many years, we did find regression errors this way, so the approach had some validity.

Now that I have stepped through a very short test lifecycle, I need to consider some of the unique aspects of IoT. Software IoT testing will need to account for hardware and software reuse in the planning.

IoT Test Planning: OTS Hardware and Software

In the DevOps test infinity model of constant testing, many IoT devices will be made using open source, low/no-code, and/or off-the-shelf (OTS) hardware and software. In this section, I introduce continuous test planning specifically for OTS. When it comes to OTS, teams should "trust but verify" (an old Ronald Reagan statement). To do this, teams should consider the IoT coil model in Figure 7-6. I iterate and repeat for OTS in the coil model as I do in the infinity model. The Figure 7-6 model has built-in interactions of development and testing supporting each other. The coil model can work for teams working with agile, DevOps, and traditional lifecycle production hybrids. Items A to F in the coil model are V&V/testing activities associated with the task of its column. These vary depending on the device, risks, and integrity level.

Management Activities of the Coil

Software or hardware management activities are classic concepts. Management starts with defining planning and controls. These can be formal and heavyweight (extensive color) or light and agile (color). The test team considers the Dev organization and teams, including Ops, support, etc. Closely associated with the organization are the supplier teams and management. And next, management, development (Dev), and test need the support efforts of configuration management (CM) and/or software configuration management specialists (SCM), process controls and evaluation using quality assurance (QA). Note, many teams consider test and QA the same, and they can be. However, international business and standards separate QA from test engineering. The contracts to implement suppliers, support efforts, and even engineering are ongoing lifecycle management focus areas over whatever model is used (e.g., Agile or traditional). Finally, historically, managers are trained to manage by the numbers, for example, data analytics in sales, marketing, processes, cost, schedules, etc. However, as this book advocates, testing using data analytics within a complete lifecycle set of processes is optimal.

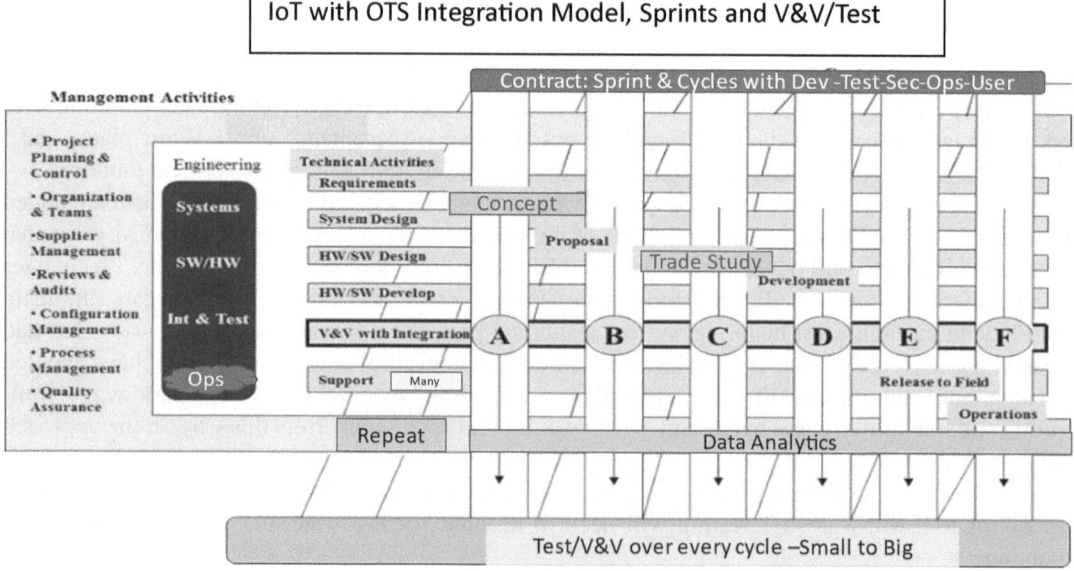

Figure 7-6. IoT OTS integration model

Reference: Jon Hagar Class Production

Engineering Using the Coil

The engineering teams for IoT coil model support systems, software (Sw), hardware (Hw), integration (Int), and test. These can be done in a single team (color) or up to many teams for each engineering area. Each engineering discipline will conduct technical activities repeatedly over the whole lifecycle. Each activity's nature, intensity, and depth will vary over each iteration for the coil spirals. For example, in the first cycle of the "concept," a simple prototype of the IoT device may be created for the teams to understand the idea before other engineering activities are performed in depth. Assuming the concept is proven and approved, a proposal is worked on by engineering with management to gain approval to proceed (including longer-term funding). In the proposal, requirements and possibly early designs will be expanded. Even in the proposal spiral stage, it is possible to have more prototype hardware and software, which is subject to V&V/test checks. Finally, as the iterations continue, the design, actual hardware, and production software are created, integrated, and tested. These later products will have been the subject of trade studies and selections over many iterations (Agile or DevOps). Final integration and testing must be completed before releasing the product to the field and operations. However, these activities and cycles get repeated over the IoT device's complete end-to-end lifecycle.

Software Dev-Test-Sec-Ops Team in the Coil

In the coil cycle model, the software development (Dev) team, with help from the testers, security (sec), and operations (Ops) staff, would do activities similar to those in this list:

- Concept – Define a need (story or requirements, this might be a concept team) and market.

1. Identify important product qualities.
2. Define risks for the project, hardware, software, system, and testing.
3. Prototypes of system, hardware, and software for go/no-go.

- Proposal – Define the detailed plans, requirements, and contractual elements (ISO 29119-2).
- Trade study

1. Conduct an OTS selection (e.g., trade study, competition, and decision) (Chapter 9)
2. In the process, testing should be done to verify vendor claims (vendors almost always overpromise).
3. Obtain the OTS items, test, and integrate (schedule problems are likely, so have prototypes).

- Development

1. The Dev team conducts development of local products (Parts 2 and 3).
2. Conduct development, integration, and testing as early as possible.

- V&V integration and test

1. The team does the integration, V&V/quality, and OTS acceptance testing.
2. Continue Dev test cycles, including regression planning.
3. Final integration and testing.
4. Final V&V/Testing.
5. Final system and software acceptance testing.
6. Release the product to the field.

- Operations – Repeat the preceding activities during the maintenance.

In the coil model of Dev testing, the team does testing as early as possible in the lifecycle, just like shift left advises [16]. The team should plan for new or updated OTS items as they arrive. Remember regression, smoke, and new testing are likely (see Part 4).

The following references contain the kinds of test information to put in the master test plan and strategy when dealing with the more traditional lifecycles with OTS, hardware, and software. These references cover a broad representation of views on testing, yet they do not necessarily all agree with each other. Again, testing is complicated, and there is no single right or "best" reference. For more details on planning, I point to the following references (in no particular order of preference):

- ISO 29119 Software Testing Parts 1 and 2
- *Systematic Software Testing*
- *Testing Computer Software*
- *Agile Testing*
- *Software Test Attacks to Break Mobile and Embedded Devices*
- *Testing Embedded Software*
- Software Testing
- ISO 29119 (as necessary)
- IEEE 1012 V&V (best V&V standard)
- Finally, do your own Google search on test planning, and you will find millions of references.

A word of warning here, in software test planning, *using OTS that has historic use does not mean that new issues and limitations will not occur in the new use of the software.* Since most IoT software will be heavily based on OTS software (e.g., OTS, interface protocols, hardware drivers, etc.), the test planning must address some OTS and third-party vendors of both hardware and software.

SOFTWARE THAT IS REUSED STILL NEEDS TO BE TESTED

A piece of software had been used for years. The reused software was installed without testing on a new system with other new software. The device accelerated (moved). The old software tracked speed in a numeric format (fixed point) with a set speed limit that the old system could never exceed. However, the new system hardware had much more acceleration. The fixed-point format flowed, and the computer system failed, which was lost on first use. The failure would have been avoided if only a little testing had been done on the OTS parts.

Does this mean that I must thoroughly test all the OTS or third-party hardware and software? Probably not, since this is not possible, but some checking or V&V of OTS is a good idea. *The OTS elements must be the subject of risk analysis and risk-based testing for high-priority areas.*

Can this be done in one or a few tests during planning? The answer is "yes," but the less test/V&V, the more risk is assumed within acceptable risks. Also, some of the testing may be done during decision selection analysis (Chapter 9) on an early cycle of the coil model.

IoT teams should understand how their lifecycle works when working with OTS vendors who follow traditional lifecycles. Late changes to hardware may be complex. Changes to some OTS elements may never be possible, ever.

This leads us to the idea of architecture, wrappers (Part 4), and solving OTS component limitations in implementations the IoT development team controls. Once we have the IoT software testing planned, as the coil model shows as an example, the critical quality factor testing needs to be considered.

IoT Test Planning: Security and Critical Quality Factors

Quality factors beyond basic functionality (Part 3, Chapter 12) often need to be tested for IoT. Factors such as reliability, dependability, safety, performance, usability, and security are essential for many IoT systems.

A basic test planning process is

1) Assess and list essential quality factors.
2) Determine risk(s) for each listed quality factor.
3) Assign a risk criticality value for each risk, for example, from one (low) to ten (high).
4) Identify a cost for each risk above a cutoff value (e.g., #5 next in this list).
5) Review each cost and risk factor with management and adjust as needed.
6) For high-value and cost risk factors, define the quality risk factor's mitigations and/or test activities.
7) For those high-value factors requiring testing, define test strategies and engineering to address and/or retire the risk.
8) Conduct testing/V&V.
9) Repeat this list over the lifecycle.

This general starting point for planning works for IoT qualities. However, planning for security risks is my number one concern, given recent history.

> **NO SMALL LEAKS**
> An IoT-based fitness app linked to a database with 61 million records was leaked to the outside world. While the impact may not be apparent right away, just the idea of so much information being leaked is and should be of great concern. See **Unsecured fitness app database leaks 61M records, highlights health app privacy risks** by Jessica Davis.

IoT systems will become a popular target for hackers because

- Ever increase new functionality in devices constantly create new vulnerabilities.
- Processing constraints on system, hardware, and software resources.
- Number of devices.
- Newness of devices.
- Many locations of devices.
- Connections and communications of devices.
- Amount of third-party software, including that there may be malware contained in any of the devices or various configurations.
- Large amounts of personal and critical data in the devices.
- Vulnerability of the user base of the devices.

To do system, software, functional, and nonfunctional quality testing, testers need a supportive architecture. This book calls this a software test architecture (STA). Because of these and other possibilities, this book addresses security tests and planning in detail (Parts 2 and 3).

IoT Test Planning: Introduction to Software Test Architecture (STA)

In system testing, I integrate and bring the IoT hardware, software, and maybe other systems into an integrated architecture of a system or system of systems. Test teams should also have an STA defined in the test plan.

I know that not everyone likes or uses standards, but I consider them a good reference (although they do have a cost to purchase). I suggest them as a good starting point and the earlier mentioned books by respected authors. Each of these standards/books contains a great wealth of knowledge and lessons well worth your time. In any case, standard or reference book, they must be tailored and used with care and thought. For STA, I use ISO 42010 architecture as a starting point.

The ideal system testing is that the lower levels and components, including integration testing, have been completed before I work the whole system. A small project may jump to system testing with fewer levels of testing, but the more complex an IoT device system is, skipping too many early levels is not advised. I advise this because if a lower-level problem has been missed and appears at a system level, it can be hard to tell where it is coming from, and it may be more expensive to fix.

I have seen the team chase "unverified failures" (faults that could not be repeated on every test) because software faults interact with hardware to make it look like the hardware had a problem and vice versa.

I will present a planning list, which assumes earlier and lower levels of testing have been done or understood. System test planning includes but is not limited to

1. Normal usage tours
2. Off-normal usage tours
3. Stress testing (to the breaking point and beyond)
4. System quality verification (against requirements)

    a. Repeat selected hardware and software tests

5. System validation with users/customers or surrogates for them in the loop
6. Simulations within the system testing
7. Testing in the field (real world)
8. Demonstration (a system as it is to be delivered)
9. Alpha/beta testing (Part 3)
10. Crowd testing (Part 3)
11. Quality(ies) testing
12. Usability testing

Item 6 may be needed for many IoT devices because the testing in the real-world systems where an IoT device is intended to be used may not be available or may be challenging to implement. For example, hundreds (millions) of home systems might be in use. I cannot do system testing with all of them efficiently in the lab or the real world. Thus, testing the following numbers, 7 and 8, becomes problematic. Using system testing, items 9 and 10 may help, but you run into combinatorial problems (Part 3, Chapters 10 and 11). System IoT testing in many IoT projects will be taking a "best guess" on some of these.

However, some system testers will only be doing system testing (e.g., a so-called smart city testbed). There are organizations doing assessments and tests of procured OTS and IoT systems. Groups doing this testing use names including IV&V (IEEE 1012), testbed assessment team, IT agencies, etc. These groups will not be doing the lower levels of testing except in audits or assessments. The test plans at each level of these teams will look very different from each other and the development test plans.

Development groups doing system testing will need to think long and hard about test planning, including STA. STA is defined in more detail in Part 4. The STA and test strategy are part of a complete test plan.

IoT Test Planning: IoT Strategy

My definition of the strategy for this book starts with a dictionary:

Strategy – The art of devising or employing plans or stratagems toward an overall goal [1].
Test strategy – Part of the test plan that describes the approach to testing for a specific test project or test subprocess or subprocesses.
Test basis – Bodies of knowledge used to drive test planning and design, including techniques and processes, which lead to detailed test activities.

As testers, we know that testing is about providing information about the product's qualities to stakeholders. Many people have observed that quality is value for which someone is willing to pay. Most people think that any product's first quality is functional (verify that it works to specifically defined requirements). However, this view is too narrow. Products have many qualities, also known as nonfunctional requirements and quality characteristics (ISO/IEC 9126 quality standard [21] and IEEE 982.1), that people should investigate during testing. Nonfunctional qualities are crucial and include performance, safety, security, usability, acceptable risks, operability, reliability, dependability, etc. In IoT, some of these qualities become very important for any strategy to address. While some characteristics may be formally specified in requirement statements, others can be implied or inferred by the development and test team. In any case, these should be captured in the test planning process as part of the test strategy and agreed upon with stakeholders.

There is one quality that most IoT stakeholders want to avoid, that being the quality of having errors or bugs (bad). Some past authors on testing advocate the only good test finds an error or bug. This thinking is also simplistic. Many tests that do not find a bug still provide helpful information about a device. Testers would indeed like to find errors when they exist, but so should planning, designing, and modifying tests to optimize error detection while also providing other information about the IoT device. I will address V&V/testing's "bug hunting" aspect throughout this book (Part 3). Many of the test techniques, patterns, attacks, tours, etc., are based on the ability to find errors, but keep in mind there may be other kinds of valuable information V&V/testing should provide. Test strategy as part of planning should support many kinds of information gathering.

Many of us separated system, hardware, and software V&V/testing in the past, but these divisions must blur with IoT. Given these viewpoints, V&V/testers in IoT should understand that testing just the software will not be enough. I advocate testers will also need to take on system testing, including assessing the unique aspects of the hardware. This will make a better qualified IoT tester by improving their hardware skills.

The subsections of this chapter provide an overview of test strategy concepts. Experienced testers can skip these sections, but readers not familiar with strategic concepts can begin learning from the following sections.

Strategy – The Basics

Do plans include strategies, or do strategies drive plans? Well, yes, both.

Project test strategies are the starting point of details in the test plans and later test designs. Test strategies and plans can be governed by organizational test strategies and policies (i.e., regulations and standards). Test strategies are the high level (referred to as the 10,000-foot level by some), leading to the master test plans, then more specific test plans (the 1000-foot level), and even daily test plans (ground level). ISO 29119 part 1 defines that strategies include the following: "the test practices used; the test subprocesses to be implemented; the retesting and regression testing to be employed; the test design techniques and corresponding test completion criteria to be used; test data; test environment and testing tool requirements; and expectations for test deliverables." Test strategies are typically documented as part of the overall top-level (or master) test plan.

IoT Test Strategy

I find one or two common but often unspoken test strategies in use in IoT test groups. These strategies may be used in combination with each other or alone. Testers use requirements-based and risks with expert-based exploration, but there are others. Many of these ideas are shown in Figure 7-7. Testers who use only one or two test strategies repeatedly, as if they were the only way to do testing, will limit their test thinking, resulting in missed errors, expensive testing, and testing that takes longer than in combinations in the test project. The test plans should document the new and complementary strategies that are to be used for IoT testing. Einstein's definition of insanity comes to mind: "doing the same thing repeatedly, expecting a different result each time" [17]. Testers need to use a variety of test strategies in combinations in the test project. The test plans should document the new and complementary strategies that are to be used for IoT testing.

Reference: Jon Hagar Class As depicted in Figure 7-7, test strategy and test basis form the supporting spokes for improving test plans, leading to improved test designs. There are and can be other strategies; some readers may have different names for them or feel something else should be addressed strategically, but for this book, this is how I am defining strategies to help in IoT master test planning. After a master test plan takes shape on an IoT project, more detailed test plans, based on the master, can be created for each test phase or cycle. These plans can be updated periodically and even daily for agile projects.

Here are some typical IoT test strategies with the names as shown in the strategy wheel that you may wish to consider for your test planning:

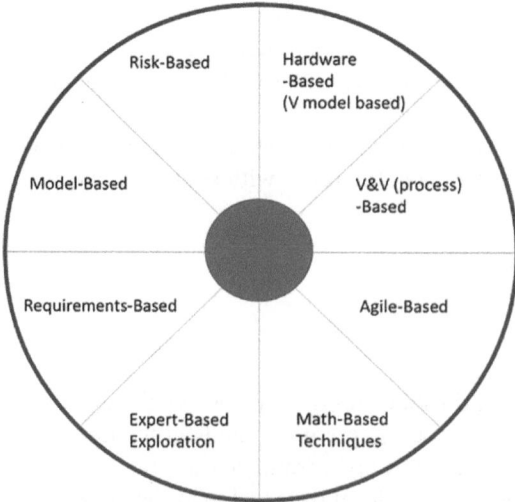

Figure 7-7. Test strategies and basis of support for test plans 2.0

1. Risk-based testing – In which the management, selection, prioritization, and use of testing activities and resources are consciously based on corresponding types and levels of analyzed risk (e.g., see ISO 29119) and integrity levels (e.g., see IEEE 1012). IoT critical risks include failures in function, security, safety, reliability, and other qualities that stakeholders value.

2. Model-based testing – In which models (not mental models), such as Unified Modeling Language (UML), UML testing profile (UTP), and others are used to drive one or more testing activities, such as to manage, design, implement, execute-automate, and report testing [ref: TR ISO 29119-8].

3. Agile-based testing – Software development methods, including testing, in which requirements and implementations evolve, being done by cross-disciplined, self-organized, and collaborative teams, including testers. These can be done for a whole product cycle (a written plan), a sprint iteration (recorded on a whiteboard), and even daily (verbally in a stand-up meeting). In Agile, testing is typically practiced by the whole team throughout the lifecycle.

4. Hardware-based testing (V model based) – Is a sequential development and test process in which activities progress steadily through phases, such as proposal, initiation, requirements analysis, design, implementation, testing, production, operations, and maintenance. In traditional efforts, a specialized independent group often does the testing toward the end of the lifecycle. V model–based testing is common and likely to continue for hardware developers and teams. Many software teams have used V model–based testing, though many teams "struggled" with it (Part 1, Chapter 3).

5. Requirements-based verification checking and documentation testing – Testing in which the requirements or other software document artifacts show that the code satisfies requirements and stakeholder artifacts (e.g., standards, contracts, etc.). These "checking" tests are typically performed and documented in written (scripted) test procedures and reports for legal reasons. There is often a legally compelling reason to show that the "shall" statements of a set of requirements are met and demonstrate compliance with standards or regulatory specifications.

Note
Some would consider this the same or a close cousin to V model based, but I separate the strategy since verification checking or documentation can be done in any testing.

6. Expert-based exploratory testing – Uses the tester's knowledge, skill, and historical practice to plan, design, implement, learn, and report the testing. Many (possibly most) testers employ some aspect of experience-based testing, although some use it more extensively. Supporting concepts include error guessing, ad hoc testing, and exploratory testing. It is common and often very productive testing.

   a. Break it/attack based with patterns and tours – Testing in which the tester attempts to find errors or address risks by using patterns (attacks) or meta-patterns (tours) to assess the software. Some testers document this by trying to break the software using record/capture tools. This strategy is closely related to experience-based testing and often includes aspects of risk-based testing.

7. Math-based testing – Is planned, designed, implemented, and analyzed based on mathematical concepts and techniques. The mathematical concepts/techniques include statistical, design of experiments (DOE), formal proofs, combinatorial, random, fuzzing, and domain (set theory, such as equivalence classes and boundary value analysis). Math-based testing ranges from very formal (proofs) and, therefore, uncommon from using techniques such as combinatorial, equivalence classes, and boundary value analysis, which are more common and easier to practice.

8. Verification and validation (V&V) based – Testing in which testers try to show that the development efforts have created the "product right" (i.e., meets requirements, design, standards, etc.) and "right product" (i.e., meets what the users and customers want). Products undergoing V&V would

include operational concepts, requirements, design, models, and implementations since any of these can have errors. Because of my "systems" view, V&V should be a common strategy for IoT systems (IoT device, edge, cloud, etc.). When a team does V&V, they usually use several earlier strategies on this list.

Most comprehensive project test plans will use several of these strategic methods in combination, where one strategy may be chosen as "primary," and then others used as needed. There is no "best" strategy or single one to use. You can refine your strategy based on your local context. Also, you can find and define other strategies, which you may wish to add to this list. Finally, since some projects follow a maturing model defined earlier, a beginning strategy may vary minimally, yet more complex strategies emerge as the IoT device matures.

What Is a Checklist for Strategy Selection?

Since there is no secret, single, "best" strategy except to consider a project's context at all points in time, a tester or test team must use critical thinking to select a good mix of strategies to go into any test plan. Further, testers should expect changes to strategies and plans once a strategy and plan have been defined, although detailed lower-level plans will likely change more than master test plan strategies. Here is an outline of a checklist to consider and use as a starting point for test strategy selection:

1. Focus on project context. The context will include factors of cost, schedule, product nature of size and complexity, product maturity, organizational policy, regulations and standards, customer or stakeholder expectations, and test team skills.
2. Include the risk and factor level of the IoT device and maybe even the system or systems it will function with or in. The more risk or higher the integrity level, the more planning and strategies are needed.
3. Consider how strategies can be mixed and matched for an optimal mix given context.
4. Which strategies will offer the most effective testing (i.e., how can you optimize your testing, meaning to "kill several birds with one stone")? Some strategies may be hard to implement but offer very effective testing, in which case you may want to consider the extra work to set them up. Other strategies may be straightforward but not produce many results. So are they worth the effort?
5. Include V&V of hardware, software, and system aspects of the IoT device, system, edge, cloud, and beyond. IoT V&V/testing goes beyond and above classic software IT testing.
6. Each strategy requires skilled and even expert testers, so ensure a complete test team with known and proven experience and skills.
7. Finally, how does the product's lifecycle impact the strategy? For example, a new startup product will have a different strategy mix than one that is historic and only running IoT maintenance testing on very critical but long-lived systems.

Approach an IoT test effort by critically thinking about your strategy and test basis. You do not want to be trapped by thinking there is a single "best" strategy (have I said that enough?) or that the strategies you are most familiar with should be what you select just because you are familiar with them. These traps are common for many testers, and yet they complain that something needs to change to make testing better (remember Einstein's quote on the definition of insanity).

Good testers can balance and optimize the test strategy and basis for IoT test planning. The test strategy will flow into the master test plans and then into the test design implementation. When teams incorrectly limit strategy basis, they limit test results, and IoT product success may suffer.

A BEGINNING TEST STRATEGY

An IoT project I know had a strategy of verification and validation based on the risks of the system with specific plans to use models, test attacks, math, requirements checking, inspection, and analysis during a traditional lifecycle development model. It was costly, but then the system involved millions of dollars and even loss of life. The IoT team cared deeply about testing, standards, policies, strategies, plans, and doing the actual work.

The team had strategies of V&V and experience-based testing. In the past, they had problems with devices that were left on for long periods and then had hardware "meltdowns." So, their test plans would always include an extended duration test of the hardware performed at a max level of usage (a stress test). However, when it came to a new IoT system, the long-duration hardware test did not stress test software, so they were not doing an extended software stress test. After going live, they got reports of the software failing after being left on for very long periods (thousands of hours), with the software receiving stress usage cases. They ended up having to fix the software.

They updated their test strategies to include some more of the earlier strategy on IoT systems with plans to include more software stress testing using long-duration practical usage. So, the test group expanded the test basis by having a risk focus, multipronged approaches, verification, and validation test strategies. These approaches had success in the long run as they continued to evolve over many years of the product's use and maintenance.

What Is a Strategy for the Individual IoT Tester?

I hope that testers in even the lowest level of IoT will practice more strategic thinking about testing before just jumping into detailed test plans, design, and running tests. You should keep in mind the "bigger picture" at the 10,000-foot level. When appropriate, add strategies over requirements verification scripts such as math- or experience-based software testing. Testers should work on mastering the application of other strategic approaches and bases. Enhancements can be started even in the daily detailed test design and planning activities on agile projects. Applying enhancements prepares testers for more advanced planning in their IoT test career.

IoT System V&V Planning: Start with a Combination of Test Strategies

A strong start in V&V/test planning covering the hardware, software, and system could include a list such as the following example:

1) Risk-based test planning to refine the following selections on an ongoing basis.
2) Requirement verification checking allocated during sprints and automated at the "end."
3) Analysis of components during development.
4) Review and inspect component products as they are created and evolve.
5) Developer or supporting testers doing attacks during development using lower-level test techniques.
6) Experience-based exploratory testing early in life to "learn" the software and find bugs.
7) Math-based testing as the components integrate and mature.
8) Experienced, attack-based testing as other strategies and components are completed.

9) IoT device testing at end sensor, edge, comm gateways (fog and cloud) to the system end.
10) Cloud-based IoT platform testing.
11) End-to-end testing of crucial quality characteristics, particularly functional, reliability, performance, and security.
12) Realistic field trials by third parties, IV&V, or government, if these make sense.

Depending on personal skill and the organization's maturity, I might add other strategies or reduce efforts on the preceding list. Still, I have run projects with a plan outline supported by the preceding strategies.

Now, let's refine this starting point based on a project's focus areas (e.g., hardware, software, etc.).

Hardware V&V Strategy (Chapters 8, 12, and 19)

In this planning, a test team has more of a focus on the hardware. This is not to say that software can be ignored, but teams sometimes focus on a component while leaving the software to other parts of the strategy. I will start with the list from the "system combinations" and refine it with the following:

- More focus on experience-based testing using inspections.
- Have experts analyze the hardware (e.g., electrical, mechanical, form, fit, function) before the "final" hardware is created:
  - Prototyping and hardware "models"
- Off-the-shelf product selection analysis and decisions (make, buy, reuse, etc.).
- Hardware production facility selection and analysis decisions (in-house or outsourced).
- First production line run test and inspection (prototype):
  - Risk-based checks
  - Math-based (everyday use, prolonged use runs, test to failure, test to stress)
  - Requirement verification checks
- Quality control checks of randomly selected samples once full production line runs start.

Software V&V Strategy

Again, I would start with the "system combinations" strategy but refine it for a software-focused team. The team cannot ignore the hardware during testing because, at some point, the software and hardware must integrate before proceeding with the system testing. I have seen cases where a software bug "burned up" expensive hardware after integration on the actual hardware happened. However, I am interested in making the software "good enough" to move onto the hardware and into system efforts.

BURNING UP PRODUCTION HARDWARE BECAUSE OF SOFTWARE PROBLEMS
I was testing a system with a limited budget, hence limited time for testing, so management decided to use the system's actual hardware as part of the test environment for software. This strategy reduced the budget because complete, specialized software test environments did not need to be created. So, during one test cycle, the test team installed the software, powered up the system, and started the test. Immediately a piece of production hardware had a motor actuator turn on full power, run up against the hardware stop, and stayed there at full power (think if your car was against a wall with the accelerator fully on). Before the testers could stop the test and power the system down, the motor on the hardware burned up. The costs were (1) $100,000 for a replacement motor system, (2) the test manager lost her job, and (3) it was a learning experience for me.

Here are my refinements for another system strategy. These have a more software focus, including the following, from [4]:

1. Static code analysis attacks as the code is developed
2. Software structural test, attack-based testing including logic, data, and missing cases
3. Integration attacks using emulation, simulation, or prototype hardware before going to real hardware
4. Test attacks after integrating the hardware in either the lab or the real world
5. System software tours and math-based testing, including combinatorial, IoT app testing, and cloud end
6. Regression and retest issues planned

If the organization and skill base are mature enough and have the budget to support them, I would consider test automation, model-based testing, and other concepts. However, automation needs care [18].

The preceding list is an example to use as a starting point. I would expect most real IoT efforts to be different.

Ops V&V Strategy

In this book, I talk about DevOps in various places, and I hope you have read that part so that this short section makes sense.

In Ops V&V, I expect the V&V/test staff to be in play (i.e., actively using Ops information for more cycles and product changes). I would also expect short and frequent development cycles. Yet, while building on early test planning and strategies, the test strategy will differ from new development. I believe that test/V&V strategies will need to address the following considerations during Ops, including

- Data analytics
- Rapid response-correction teams
- Security and privacy
- Safety or other "qualities" deemed by the stakeholders to be "critical" to the business
- Product lifecycle stage (new, mature, near retirement)
- Risks and integrity of the product
- Regression and retest

- Users as testers (feedback from the actual usage in the field via data and data analytics)
- Financial, etc.

Since IoT spans devices from industrial to consumer use, the selection of strategies during Ops will vary. The preceding strategy list may be expanded into project use as teams understand more about IoT Ops and mature. You and your stakeholders should create your factor ranking (Part 1, Chapter 4) and a version of this type of list.

Now that we have the software test planning introduced, the tester of IoT devices need to move past testing just the software and consider impacts from the unique hardware and system of the device.

Hardware Test/V&V Planning (IEEE 1012)

This section presents a fast overview of IoT to expose software testers to the hardware side. Teams should review the references if more detail in planning is needed.

The following outlines some example tests included in an IoT hardware-focused test plan. These detailed selections may be better suited for a lower-level hardware test plan than a master test plan.

Hardware basis IoT tests can include

- Functionality tests
- Form, fit, and function tests
- Test to failure
- Stress test
- Environmental tests (heat, humid, salt air, cold, etc.)
- Battery test
- Power test
- Material test (strength, color, feel, etc.)
- Noise and vibration testing
- Long-duration test
- Destructive testing
- Production-manufacturing variance (6 Sigma) sample following quality control sampling
- Quality verification checks (see ISO quality characteristics that follow Figure 7-8)

Hardware verification and validation testing have a long history in the industry. Industrial hardware engineers and designers are well schooled in it, but the system and software testing become necessary to include "smarts" in devices.

Figure 7-8 shows a general listing of external and internal quality characteristics based on international standards, such as ISO/IEC 9126.

The questions to answer in hardware test planning include

- Should the tests be done in a specialized organization's lab?

  - What environments with hardware, software, and system tooling need to be in place?

- Should an independent third party do the test?
- Should the vendor's tests be done or repeated, and how do you trust/verify/witness such?
- Should the test be done using simulations/simulators?
- Should the test be done in a controlled environment (e.g., test track)?
- Should the test be done in the field (and how do you get the resulting data)?
- Should testing be redone on new revisions of the hardware?
- Which areas of hardware need to be tested, including mechanical, electrical, packaging, form-fit-function, integration, etc.?

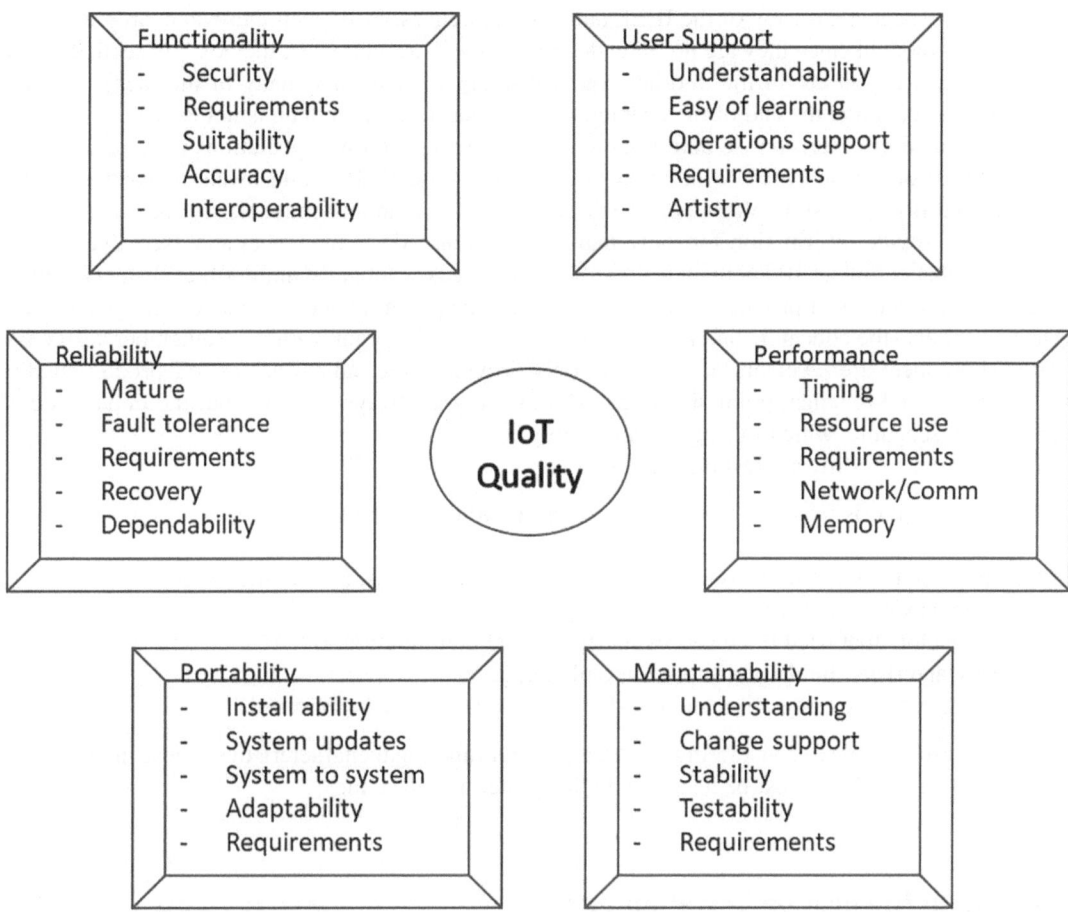

Figure 7-8. IoT quality block based on ISO-IEC-9126

Suppose you get the feeling that hardware testing is complex. IoT teams moving from software development into IoT hardware software system will need to do a bit of study and learning about hardware testing/V&V. If this is the case, you may need to get some new, skilled hardware test people and references. Teams with system and hardware test experience should be familiar with the hardware side and may need more learning on the software side. This is not a book on hardware testing, but the hardware influences software system testing, which I consider next.

Test Planning for Software Systems

Software test planning has been extensively written about in the literature. There are whole books and standards on the subject. I will not try to cover all of that information. I have provided reference books (see the "References" section) to help you in your studies.

Since IoT testing can range from zero to infinite in size, these are examples and starting points. Each will need serious tailoring and customization for your IoT planning. If you are starting near zero, follow the agile idea of doing what is most important first and work until time or money runs out. If the product you are working on is more critical and your processes more formal on a large complex

IoT system, I suggest a review of the IEEE and ISO references. Still, even these standards will need some careful thought since they are not intended to be used "straight out of the box"; instead, they are just guidelines. If you are in the middle, you will likely want to mix many of the ideas from the "References" section at the end of the chapter. (Remember, I said high tech is not easy.)

Teams should not overpromise in test planning. You should set low expectations with management teams if there are "near zero" levels of testing or even so-called middle levels of the test effort. Further, even if more testing/V&V is possible, teams in the planning must clarify that all testing is sampling to provide information for those making decisions. Many managers and executives do not understand that complete 100 percent testing of everything is practically impossible. Testers are messengers and journalists of product quality information. The planning needs to make these points clear as testers are also the educators and information providers to the organizations. Remember, test/V&V *cannot* show there are *no* errors or issues but only provide some indications of IoT product quality under specific and practical point situations. More testing is always possible, but the impact is cost and delivery schedule, which managers care deeply about.

Critical points in software test planning include

1. Test at many levels from the lowest to the highest (structural and functional).
2. Planning needs to address strategies.
3. Planning needs to fit with resources (skill, budget, people, tools, environments, etc.).
4. Planning needs to be flexible.
5. Quality factors that need to be addressed (functional, nonfunctional).
6. Integrity and risks should be factored into planning.
7. Hardware-software integration and interface efforts need to be highlighted.

System testing leads to considering the testing of nonfunctional characteristics. These include critical quality factors which must be tested in the integrated IoT system.

IoT System Integration Test Planning

Most traditional IT and software development teams do the integration and perhaps then find issues. Teams should be aware of integration approaches and strategies such as bottom-up, top-down, continuous (Agile), mixed, etc. [19].

Many pure software teams are less familiar with hardware component integration and set-based design [20] simply because they have traditionally been software focused, and the element is just a generic computer, which always worked as needed. For IoT systems, the software, hardware, hardware-software integration, quality, and dependability will be important [21, 22].

The integration of hardware elements and software should have schedules and plans and be part of, if not THE critical path on a top-level project master schedule and plan. The more complex the system, the more integration planning is needed, and the more likely plans will change, affecting budgets and schedules. The team did weekly and daily integration schedules and planning at places I have worked. Things changed rapidly, and I always looked for workarounds to encounter integration issues, requiring creative thinking and problem-solving. For example, if real hardware was not available for integration, I might need simulators or mock hardware (Part 4, Chapter 20).

Here are some general challenges teams may face during integration testing:

- The hardware will be late.
- The hardware will have problems, which are often fixed by software changes causing rapid coding and minimal testing/retesting.

- The test team will get "crunched" (meaning hardware schedules will move to the right, but the end ship date will not move), and software testing schedules will be impacted. Heads up! That can mean overtime for the test team.
- Integration testing will find problems in hardware, software, and/or system.
- System testing will find more problems.
- Regression and retest will have more cycles than planned (e.g., I always planned for three cycles but expected six in scheduling).
- Customers, managers, and stakeholders will *always* want better, faster, and cheaper.

The more complex, unique, and extensive a system is, the more integration and integration testing of the system will need to be planned with a focused and specialized team. Teams will need to select V&V/test integration strategies and plans based on the local context, including factoring in challenges such as these. I do not give an example strategy here beyond these considerations in integration strategy selection, since I view them as out of scope of this software-focused book. However, I would certainly start with the "system integration strategy combination" list as my starting point.

IoT System and System of Systems V&V/Test Planning – A Conceptual Overview

My bet is an IoT system, and IoT system of systems testing will be ignored or lacking for some systems once integrated. Development organizations will focus on individual IoT devices and not the integrated system. There will be industry "standards" for product interfaces and protocols, but currently those are in a state of flux. IoT products will be promoted as "compatible" with some system interface, but minimal real testing and assessment will have been done. In part, this is because the question of IoT integration ownership may be undefined.

Who owns the IoT system or system of systems? Which owner is responsible for the IoT devices playing nicely together with the possible interfaces and protocols? Does the city government own the "smarts" of the city? Many will say, "yes, the owner owns the problem," but currently many "owners" do not have the skill and knowledge to test systems or systems of systems or, in some cases, do not even know they face this kind of a challenge. For example, many people now own a smart IoT house. This lack of "smart" situation can be seen with homeowners who use default router passwords and configurations for their current modems and computers. Is this really smart? Alternatively, governments who can barely handle existing IT challenges and systems will take time to expand their IT departments to be IoT/IT departments. These stakeholders lack much skill and knowledge in the new IoT frontier. For these reasons, I advocate better IoT devices, ubiquitous UIs, and more IoT device testing by the producers and end users to consider employing IV&V integration teams.

I would like to see expanded systems and system of systems IoT testing. In this short section, I introduce my current thinking, but many of us are actively researching and supporting this area of systems IoT.

In systems IoT, I recommend integration test planning, noted previously, to be a significant effort. This would be done after hardware and software IoT testing has been planned and achieved.

I believe IoT systems will be built from the bottom up on the hardware side, with many integration cycles. On the software side, CI and CT will be good practices. The more complex a system or system of systems is, the more test cycles, steps, and planning must happen, driven by integration experts.

In complex systems, dedicated IoT integration and system test/V&V teams will likely appear. Such teams should consider

- "Good enough" definitions of many qualities with degrees of V&V/testing.
- Active and evolved integration test planning.
- Note: Integration and test planning should be activities that different teams often do on more complex systems.
- Integration test team involvement with hardware, system, and software architectural efforts.
- Integration-system risk analysis formally and informally on an ongoing basis.
- Having early users be unknowing informal "testers" with monitors, logs, and data analytics.
- Agile testing.
- Classic old-school V&V/IV&V/testing as needed for hardware, software, and systems.
- Advanced V&V/testing such as math-based, model-based, and experience-based approaches.
- Support to the "external" organizations, which are likely to include vendors, third-party providers, standards groups, IV&V, and customers.
- Communication to external organizations (e.g., standards groups such as ISO, IEEE, security, Underwriters Laboratories (UL – www.UL.org), etc.).
- See system test planning and strategy sections (Part 2).

Large systems and systems of systems I have been involved with often have problems that appear once integration happens. We have all heard the stories of airplanes that fail during early test flights around the world, cars whose infotainment systems are less than expected, and smart systems that get discarded:

- See www.computerworld.com/article/2515483/enterprise-applications/epic-failures--11-infamous-software-bugs.html – accessed in winter 2022.
- See www.scientificamerican.com/article/pogue-5-most-embarrassing-software-bugs-in-history/ – accessed in winter 2022.

Finally, system test teams may find some specific V&V/test ideas/techniques in the following list (use the Internet or Part 3 to find specifics on these ideas):

- Field testing
- Simulation analysis
- Modeling
- Prototype testing
- Destructive testing
- Electrical analysis, including sneak circuit analysis
- Burn-in testing (long and hard usage)
- Shake and bake V&V (vibration and heat)
- Stress (physical) testing
- Dog food test (when you eat/use what you create, so make sure it is delicious)
- Quality control in production

Summary

Key points of this chapter include

- Test/V&V planning is an essential early step.
- These early steps are often overlooked or not done until later in development.
- The IoT infinity and OTS lifecycle test models should be continuous and repeated for software, hardware, and the IoT system.
- Testing/V&V planning with strategies and architectures address functionality, qualities, and risk assessment.

The next chapter gets deep into the more detailed planning of the IoT software test architecture and environment details.

References

1. "IEEE Standard for System, Software, and Hardware Verification and Validation," in *IEEE Std 1012-2016 (Revision of IEEE Std 1012-2012/Incorporates IEEE Std 1012-2016/Cor1-2017)*, vol., no., pp. 1–260, 29 Sept. 2017, DOI: 10.1109/IEEESTD.2017.8055462
2. *Agile Testing: A Practical Guide for Testers and Agile Teams* by Lisa Crispin and Janet Gregory, Addison-Wesley Professional, 2009
3. https://agilemanifesto.org/ – accessed spring of 2022
4. www.theguardian.com/technology/2016/jun/30/tesla-autopilot-death-self-driving-car-elon-musk – accessed in winter 2022
5. www.ibm.com/developerworks/library/a-devops9/ – accessed in winter 2022
6. https://en.wikipedia.org/wiki/DevOps – accessed in winter 2022
7. https://en.wikipedia.org/wiki/Test_effort – accessed in winter 2022
8. *Proposal Preparation* by Stewart, Rodney D., Stewart, Ann L. ISBN 10: 0471552690/ISBN 13: 9780471552697, Published by Wiley-Interscience, 2022
9. *The Tester's Pocketbook* – Paul Gerrard, ISBN 13 9780956196200 (page 8 on the "equation of testing"), 2015
10. www.stickyminds.com/users/robert-sabourin – accessed in winter 2022
11. https://en.wikipedia.org/wiki/Principle_of_good_enough – accessed in winter 2022 and work by James Bach
12. https://en.wikipedia.org/wiki/Regression_testing – accessed in winter 2022
13. https://en.wikipedia.org/wiki/Coupling – accessed in winter 2022
14. http://kaner.com/pdfs/gui_regression_automation.pdf – accessed in winter 2022
15. https://en.wikipedia.org/wiki/Paradox_of_the_pesticide – accessed in winter 2022
16. https://en.wikipedia.org/wiki/Shift-left_testing – accessed spring 2022
17. www.brainyquote.com/quotes/quotes/a/alberteins133991.html – accessed in winter 2022
18. www.satisfice.com/articles/cdt-automation.pdf – accessed in winter 2022
19. https://en.wikipedia.org/wiki/Integration_testing – accessed in winter 2022
20. www.doerry.org/norbert/papers/SBDFinal.pdf – accessed in winter 2022
21. ISO/IEC 9126-1:2001 Software Engineering – Product Quality – Part 1: Quality Model
22. IEEE 982.1-2005 IEEE Standard Dictionary of Measures of the Software Aspects of Dependability

Figure Reference

1. https://en.wikipedia.org/wiki/ISO/IEC_9126 – accessed in winter 2022

Chapter 8
Planning for the IoT Tester on Environments and Testing Details

Part 2 of this book has focused on test planning. I have found many teams that do not do enough planning and then waste time redoing things, particularly the environments, tools, and detailed/daily test activities. I advocate that planning is the first step in testing that each tester must consider during the complete product lifecycle. The previous chapter looked at IoT test planning and strategy for IoT hardware and software since test teams deal with "things" driven by software. The focus was aimed at managers. In this chapter, I continue the IoT test planning by moving to more detailed test plans and the environment (labs), which individual testers should be aware of in addition to the management plans. This chapter ends by addressing testing during the completion of development, operations (Ops), and the end of an IoT product's lifecycle.

Assuring the Test Environment

The book *Software Test Attacks to Break Mobile and Embedded Devices* [1] has extensive information on setting up a test environment for embedded and mobile devices. The planning concepts from that book are a good starting point for IoT planning and will not be repeated here. However, I have extracted a summary outline of the critical IoT test planning points here, which includes the following:

- Developer-level testing environment and tools

  - Static code analysis tooling
  - Structural unit test tooling
  - Unit code data generation tooling
  - Error or backlog tracking facility tooling
  - Connection to development environment tools (optional)
  - Interface with model-based test tooling
  - Data visualization and analytic support tooling

- Integration environment lab and tooling

  - Hardware, internal and external connections
  - Software, internal and external connections
  - System/systems of systems connections
  - Timing and performance analysis tooling

© Jon Duncan Hagar 2022
J. D. Hagar, *IoT System Testing*, https://doi.org/10.1007/978-1-4842-8276-2_8

- Communication systems (Wi-Fi, Bluetooth, cellular, etc.)
- Simulation and emulation tooling

- Functional and quality test labs

  - Hardware support
  - Software support
  - Internal visibility support
  - System support
  - Simulation and modeling support
  - Security sandbox support
  - Quality supporting areas including reliability, safety, hazards, performance, etc.
  - Cloud support (virtual labs, devices, app testing, simulation, emulation, etc.)

- Real-world test labs and beds built-in

  - Smart cities
  - Smart homes
  - Smart buildings
  - Smart companies
  - Smart offices
  - Data analytics with real users
  - Test in the real world with controls in place to keep testing information available

Building on this outline and getting started during test planning involves selecting the test environment, tools, and data analysis elements that support the software test architecture and significant parts of the test plan and strategy.

Selecting the Right Test Environment

The more significant and complex an IoT device or system of devices is, the more complex test lab environments will need to be. As a start for your planning, here is a list of example labs/environments that I have already seen being planned for IoT:

1. "Iron bird," where the avionics systems of an airplane can be "flown" without ever leaving the ground through extensive simulations where every environmental factor can be induced as inputs.
2. "Flight test bird 1," which was an airplane, version 1, that could fly but whose only mission was to test the integrated system, and not to carry passengers or fight wars.
3. "Test car 1," which drove around a "test lot" within a company's facilities where traffic patterns and roads were actual.
4. "Mannequin 1," a "test dummy" that simulated heart and lung conditions so that things like heart attacks could be simulated.
5. "SIL" (system integration lab), where the components of devices were all present but not in a form for consumers so that new hardware and software could be "plugged in" and tested quickly.
6. "Mount Evans Test Bed" (link: `www.mountevans.com/MountEvansCom/Mount-Evans-Things-HighAltitudeAutoTestLab.HTML` – accessed winter 2022), which was a test lab in a test building, where cars from multiple manufacturers could be tested in the real world of the Colorado Rocky Mountain areas. Mt. Evans offers many extreme environments to stress cars (e.g., snow, cold, altitude, heat, dirt or unpaved roads, etc.).
7. "The fun lab" was where IoT toys were "played" with by workers and their kids.

8. "The Barcelona test city" (see `www.iotsworldcongress.com/the-iots-world-congress-will-showcase-the-potential-of-industrial-iot-through-10-testbeds/` – accessed winter 2022), which is a whole "smart city" test bench. (I expect to see more of these in each country.)
9. Chaos testing was done on production systems in the real world (see Chapter 16). I expect a lot of this kind of testing in IoT systems.

The nature of a lab depends on the IoT device system. The examples and labs/environments listed are starting points for planning specific IoT test facilities. There is no one size fits all.

Cost and schedule are driving factors in test labs, tools, and environments. I have spent thousands to millions of dollars setting up test and simulation environments. Such large ranges are driven by IoT device systems' nature, context, and regulations. Teams will need budget, schedule, development, support functions, and testing of all test labs. All of these will require management's understanding and approval.

Startups and smaller projects may have labs for the IoT device, an interface computer, power, Internet, and a table setting in some tester's work area. These setups may be enough for smaller projects.

More significant IoT efforts will need more extensive environments, including going into whole buildings and crowd or chaos testing. Many companies and testers may need to be involved.

The use of cloud-based testing will support assessment environments in IoT. This trend follows what I have seen in the mobile smartphone world. I expect the cloud to be used for hardware and software support, so smaller IoT organizations do not have to have their configurations. However, some care must be taken for quality concerns, such as security and privacy. Some cloud and device configuration services are already being advertised. Vendors will be ready to market these test services, tools, and automation for your IoT system teams.

Planning for Automation in IoT Tool Environments

One of the first things managers want to see when "improving" testing is the idea of test automation. It seems a logical way for management to improve testing and reduce costs. For decades, companies and industries have introduced test automation in computers, software tools, and even robots. These updates have resulted in productivity increases and money savings. If you are strictly a manual tester, you need to be working on your skills because automation can do your job.

An ideal automation area is for tasks that involve a concept where logic or an algorithm can be introduced. Here again, testing has these also. Conclusion: There are complex activities in testing that happen very quickly or involve other tasks where humans can fail, but computers can shine. Test planning activities need to leverage test automation whenever possible.

Communication Traffic Monitors and Network Sniffers are applications used to monitor the traffic in the interface, source/destination host addresses, etc. These can be over the air (traffic monitors) and/or on the wire (network sniffers). The list of generic tools that follows introduces classes of tools to help in test planning:

– Security probe tools – Software, such as disassemblers, that allow binary code to be "read" looking for security holes and bugs
– Fuzz tools – Security tools that aid in fuzz testing

- Combinatorial test tools – Tools that implement combinatorial test attack data selection
- Test execution automation – Tools that aid test execution (see the following note)
- Spoofing tools that support security spoofing attacks

> **Note**
> Many software lifecycle tools address areas such as management, planning, reporting, configuration management, error reporting, and general aspects of testing, but these are out of scope for this book, and information on them can be found in many general testing reference sources.

Hardware support test tools include

- Hardware in-circuit controller is similar to a software debugger, which allows control and variable step-by-step inputs to the hardware and software. These systems can be used in monitor mode (record CPU) or input mode (inject information into hardware under test). Many of these systems also support performance/timing analysis.
- Oscilloscopes and electronic probes check and record electronic/hardware events with timestamps, including power supply and signals.
- Software-driven radio is used to emulate radiofrequency systems' receiver and transmitter functions for a range of wireless gateways.
- Control and routing panels (also called patch panels) are usually custom-built electronic systems that allow the patching via cables to different hardware and software configurations.
- System tools
- Field support tools – Recording and tracking systems, which instrument the IoT system to collect data but do not interfere with its operation.
- Data analysis tools – IoT system will generate many data streams, which finally are stored in the cloud. Analyzing data using tools and AI will be necessary for the project, particularly for many testers.

> **Note**
> For more information, see Part 3.

Data Analytics with Tools

Given the volume of data IoT systems are likely to produce, tools will be needed to analyze and reduce the data streams into information humans can understand. There are many tools under this category, and new tools are coming online constantly. A good source of information on test tools is www.stickyminds.com and Part 3, Chapter 16. This site has a list of primarily vendor-maintained tools, so it tends to be up to date.

The planning of test tools, environments, analytics, and labs starts at a high level in the master plan. Still, details for larger project test planning typically are put into a detailed test plan and of direct interest to the individual tester.

IoT Detailed Project Test Planning (After the Master Test Plans)

Beyond the master test plan are lower levels of test planning details. These levels may be associated with a test cycle, a product release date, a significant build release, or even a sprint (if Agile). These plans outline specific tests and are often called test specifications. The plans are detailed and subject to changes as the development test efforts unfold.

Projects test products at many levels, and these tests are often included as part of development cycles, quality control (QC) in hardware manufacturing, and quality assurance (QA). Products are also tested as they are mass-produced, and this checking is often called QC. I do not address factory hardware QC production in this book. It is a large subject area but necessary for mass production runs of IoT devices on the hardware product side.

This section addresses some concepts and examples IoT teams may want to consider for detailed test plans.

Hardware Planning a Tester Should Know

Detailed hardware test/V&V patterns to consider at this low level include

– Circuit analysis
– Sneak circuit analysis
– Electric analysis
– Battery demonstration
– Mechanical system physical tests
– Power tests
– Radio signal tests
– Color tests
– Physical feel and packaging tests
– Acoustic tests (drop, shock, and vibration)
– Temperature and thermal cycling tests (bake, cold, hot, normal, stress, cycles)
– Hardware usability
– Long-term wear tests doing vibration and heat cycling (shake and bake)
– Quality checks (see ISO 9000 series [2])
– Integrity V&V (see IEEE 1012 [3])
– Integration and interface
– Reviews and inspections

Note
The details of these patterns and techniques are not defined in the book but can be found in hardware or electronic books and standards.

Detailed Software Testing

Detailed software attack test/V&V patterns to consider at this low level include

– Attacks found in the book *Software Test Attacks to Break Mobile and Embedded Devices* [1]
– Software test techniques
– Software integrity checks of IEEE 1012
– IoT test attacks, tours, patterns, and frameworks
– Reviews and inspections

> **Note**
> These are detailed in Part 3.

System Test Plan Patterns a Tester Should Know

All of the lists that follow are at the system (or system of systems) level. Many of these features should have been tested during other testing, so reuse of tests may be possible, but here all the parts of the IoT device/system are in "play" together, possibly for the first time. It is tempting to wait on some of these test activities until there is a complete system, but finding issues when a system is "complete" can cause waste by having to rework hardware, software, or Ops. There is no one right mix of early and delayed testing. Skilled critical thinking and experience are needed to get an optimal mix, but even then, only hindsight will be 20/20 as to what should have been planned.

Critical system patterns/features to test include

1) Usability

   Mandatory (Part 3, Chapter 11)
   ADA (Americans with Disabilities Act)/Disabilities
   Failure and recovery modes that users will experience

2) IoT security

   Mandatory (see Part 3, Chapter 13)

3) Connectivity

   Full connection available
   Max number of users or connections
   Partial/limited connection rates (dropouts, slow networks, etc.)
   No connection (offline mode and restore)
   Work with different interfacing vendors (e.g., different routers, edge devices, etc.)
   Transfer rates (performance)

4) Performance testing

Mandatory performance V&V/testing (see Part 3, Chapter 11) on performance, stability, response, loads, etc.

5) Compatibility and interoperability testing

Check the architecture, hardware, software, third-party vendors (OS, protocol, browser), comms, etc. Integration and interface testing/V&V

6) Pilot use – field testing

Mandatory (start in the lab, but the real world is full of surprises)
Who owns this testing (third party, IV&V, regulator, etc.)?
Consider having a prototype system put in the field to support chaos testing
Field tests get you ready for the release

7) Regulatory testing

What regulations, standards, and legal issues exist that must have V&V?
Will there be an independent assessment (IV&V, regulatory body, third party) that must pass (this is a risk)?
Has the independent assessor been involved in the project (device or system worked, so the efforts are successful)?

8) Upgrade testing (hardware and software)

Upgrade push or pull checks
What if an upgrade does not happen? What then?
What if an upgrade happens, but it crashes something or everything?
Regression and new testing
Combination of multiple protocols, devices, operating systems, firmware, hardware, networking layers, etc.
The detailed test plans for the software and system are still not what the individual tester thinks daily. Testers must learn to do "personal" planning in agile and even traditional daily activities. This is when testers start to build the planning and environment usage skills they need to improve.

Planning Individual Tests (What All Testers Should Do Daily)

I mention now the journey all testers should make into the world of test planning, which starts with planning the daily tasks and individual tests. Every tester should do this and build daily planning skills. The room for changes to plans for a particular test may seem limited. Still, many of the items listed in this book apply (e.g., how much time, the goals, how is testing to be done (scripted, exploratory, techniques), automation, etc.).

NEWBIE TESTER
When I started software testing, I was given a detailed assignment plan, individual tests I had to engineer, and a top-level schedule (e.g., be done in a week). I started learning about planning using individual tests. This learning expanded to detailed tests and, finally, master test plans. Newer skills and understanding of levels of testing detail were needed. I bought books, learned about standards, and reused information from other testers in my journey from the stand-alone tester to test lead to the test manager. This journey took years.

In the individual test planning, consider the following outline as a start:

1. Check email at the beginning of the day for any detailed or master plan changes.
   Or, have a daily agile test team stand-up meeting.
2. Estimate what test design was needed for today's tests (boundaries, constraints, etc.).
3. Conduct the test(s) research, looking for needed tools, resources, support, etc.
4. Produce a test design (select methods, attacks, tours, techniques, etc.).
5. Create detailed test implementation (based on test goals, requirements, designs, input data, expected output, manual/automated procedure steps, scripts, etc.).
6. Conduct first pass exploration or ad hoc testing.
7. Change the design and tests, if needed to meet goals.
8. Conduct tests for software acceptance or quality testing.
9. Gather outputs of tests.
10. Analyze test results for important information (passes, fails, bugs, lessons learned).
11. Report on testing to your manager.
12. Repeat.

Some of you may release items into the development library or even the real world using individual test planning. Several of these steps may be done in parallel or in a different order. Indeed, the parallel nature of how work gets done is a problem. *Communication on test status is essential in any approach.* Testers need an approach to communicate test status using the written word or verbally. Also, some may rebel at the mention of test "scripts and documentation." However, some test efforts may require written documentation and scripts for humans or computers, which may suffice as documentation.

NEWBIE TESTER CONTINUED
The message here is that, as a tester, I had to start learning planning and communication skills. I did it with a modified form of a daily personal software process. Your situation and approach should look different, but start improving your planning and communication abilities initially small and then build up to detailed plans. If you get good at the most straightforward levels of planning, then you can move on to more challenging levels of planning. I have helped teams where all planning seemed to be done daily and individually. They had levels of waste between team members and wanted more formality with master plans. Teams must balance between informal Agile and formal traditional.

The testers must participate in their daily plans for optimal success in development testing. Development testing success leads to operations activities which typically include software updates and maintenance. Again, test planning and testers themselves must be involved in these test activities.

Test Planning from Operations to the End Product Life

As the traditional functional and nonfunctional testing during product development nears completion, the project must think about the test activities that continue in the later project activities, such as operations, maintenance, and IoT device production.

Test Operations (Ops) Impacts on Test Planning

In IoT, operations may be significant for IoT systems. There will be user operations, data center operations, communication networks, and device/sensor/controller Ops – just to name a few.

Developers and testers must think "Ops" to have an integrated lifecycle solution in the test planning and execution. In some sense, it is a chicken and egg problem. In IoT, do I think development, test, security, or Ops first? Answer: Yes, all of these perspectives.

The Ops efforts need to be cyclically addressed with development and test, including

- How is the system going to be used?
- What are the Ops risks and features that must be tested on an ongoing basis?
- What data is used, owned by whom, and kept where – or protected how?
- What are the operation's reliability and other qualities?
- How is the maintenance of hardware and software going to be done?
- What about security and privacy?
- Does testing stop after the first delivery, or do I continue testing as a team and maybe consider using the user as a tester (e.g., chaos engineering, Part 4, Chapter 19)?

How Does Ops and Test Planning Change Over Time?

Again, there is no single best answer to the question, and, further, answers can change over time as an IoT system evolves. Here are a few operational items to consider:

- Do the users or stakeholders change over time?
- Do the requirements and needs change over time?
- Do the factors (Part 1, Chapter 3) change over time?
- Does the hardware change over time?
- Does the software change over time?
- Do the operations change over time?
- Does the development or maintenance team change over time?
- Does management change over time?
- Do security risks (and other risks) change over time?
- Do support processes need to change over time?
- Do regulations change over time?

Each of these may impact Dev, testing, and Ops over time. A new risk may be introduced. Success in these areas comes at a price, and that price may be more DevOps work to do in maintenance, including testing. Companies and projects want to be a success.

TEST ENVIRONMENT PLANNING

As a test manager, I got a new project assignment with new products, Dev teams, managers, and stakeholders. The job came with a test environment that had been in use for years and moved from one site thousands of miles to a new site. It was up and running. However, being a proactive test manager, I asked, "where is the test environment documentation?" The senior tester looked at me and said, "we don't have any." I was unhappy. So, I said, "how did you guys move the lab thousands of miles in parts?" The tester said, "We took pictures of everything and then used these to put it back together." I was still unhappy. I went to a senior Dev person I knew and said, "was there ever test environment documentation?" He said, "Oh, sure." He took me to a room that was about 50 feet by 20 feet, filled with moving boxes. He said, "in there." I was even more unhappy. So, I looked at him and said, "do you think you might know which box and what would it take for you to find it?" He smiled and said, "yes, and I like lunch at this restaurant." I accepted his deal. He found the test maintenance information, and it cost me a lunch. Later, when management wanted an audit to see the lab information, the team passed with flying colors.

System Maintenance, Security, and Retirement

 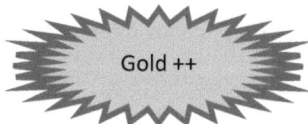

The significant three issues in Ops are likely to be maintenance, ongoing security threats, and product retirement. Testing should interface and support all of these.

IoT security and privacy are mostly agreed upon as top concerns for IoT. Users are becoming aware of this. Laws are being passed around the world. Your project cannot assume anything about security or privacy and that today's rules will work for tomorrow's issues. It may be best to develop and test with an open mind on security. Part 3 of this book addresses security testing advanced techniques in more depth, but it is a moving target. Whatever you decide on, IoT security and privacy are likely to change over the next ten years as the IoT system evolves. These changes bring up the topic of maintenance.

Hardware maintenance may be a little tricky. Teams will see IoT products returned for fixes or updates. How is this paid for? How do you deal with recalls? Is the IoT device a "throwaway" or something more expensive that can be updated?

Next is software maintenance. The IT and mobile device maintenance model has moved to constant software push updates. However, not all IoT devices may leverage this model and be associated with this way of doing business, plus there are security concerns to think about in any maintenance phase. Further, some IoT device software may be complicated to update unless designed for an update (think about how firmware or field-programmable gate array (FPGA) updates have to happen). The automotive industry has to have cars brought into the shop to update software. Will your IoT device have this model?

Finally, for each of these, what are the test plans?

Ops Maintenance in Test Planning

Like the current non-IoT industrial world, maintenance will span from an automated push model to the pull of the "you decide when to make an update" approach. The choice will be a development and maybe a marketing decision based on many factors. Some care is advised since this is a system risk area. Just know this, the software industry is littered with software updates on systems that failed.

Hardware maintenance options include

– Scrap the device and replace it, assuming the IoT is a line replaceable with some ease.
– Fix the device.
– Have hot spares (a redundancy) for an IoT device to step in automatically for the failed device(s), to be fixed or replaced (harder to do with an entire IoT system).

Software maintenance options include

– Software updates are pushed to the IoT device without user involvement (in the EU, there may be legal issues with this approach).
– Software updates are pushed following a user agreement.
– A user pulls all software updates following a user agreement.

Maintenance comes with various standards, statistics and measures (see IEEE 982.1 [5]), and other regulatory considerations. The testing must address these optional scenarios before maintenance updates can happen.

Planning Retirement and Disposal of an IoT System

What do the teams do with the IoT device when the project ends, and how does that impact testing? The answer, of course, depends on many factors, such as size, number of deployed systems, who owns the systems, etc. The primary inputs, actions, and results are shown in Figure 8-1, but there can be much more. The ownership is likely divided between the producing organization and the owner.

We live in a world where many products, including electronics and IoT, cannot (and should not) be just thrown out with the trash. There are batteries, chemicals, and construction materials worth money (if recycled), or they constitute a hazard when disposed of improperly. How many people have an old TV with a cathode ray tube (CRT), electronics, older model cellular phones and their accoutrements, or other equipment sitting around the house or property because the trash collectors will not take them? In the case of the owner of an IoT system, the end of life becomes their responsibility.

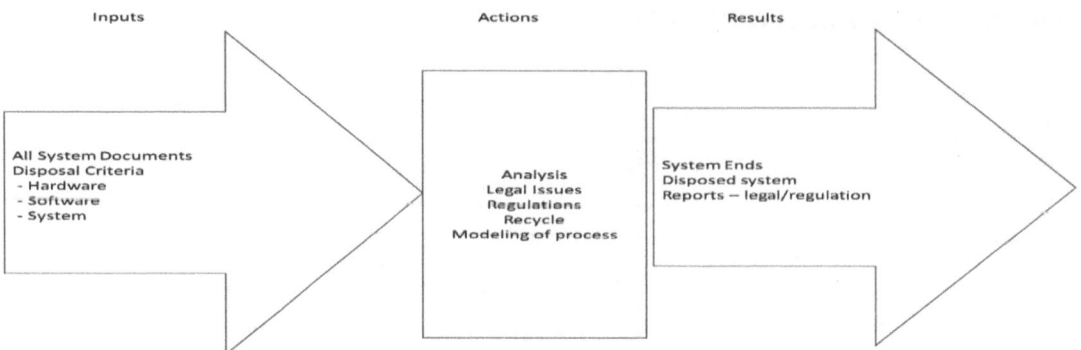

Figure 8-1. End of device life

Reference: Jon Hagar Built figure for IoT Classes 2016

When development has responsibility for IoT products, equipment, labs, etc., retirement and disposal must be considered and even tested. The legal and financial aspects of responsibility for the hardware and software must be considered. Further, there is the problem of the test environments and data that testers generated. Do you keep those items? Are there security or hazard concerns? Are there legal reasons to keep information? In some cases, the answer for test labs and data is that they must be kept and dealt with properly and legally and for a legally specified amount of time.

> **MANAGING OLD TEST LABS**
> I once ran test labs. An old system was retired, and the lab test equipment came available for a few parts to use in other test labs that I also ran. The owning customer wanted millions of dollars for this equipment. I refused to pay. The customer then abandoned the equipment in place. I then took the equipment I wanted/needed and charged them to dispose of the rest, essentially "junk." I won for my budget with both options.

However, this is just on the hardware side; what about software?

Well, it is just bits (0s or 1s). Right? Wrong. In some cases, data and IoT information need to be kept safe, secure, and private. The disposal needs to include software security and testing. I have removed and trashed memory and storage media to keep data safe. The planning and follow-through for development and testing of IoT must include software and hardware disposal.

Modern life is no longer straightforward or as simple as throwing things in the trash. Project teams and testers must consider "retirement."

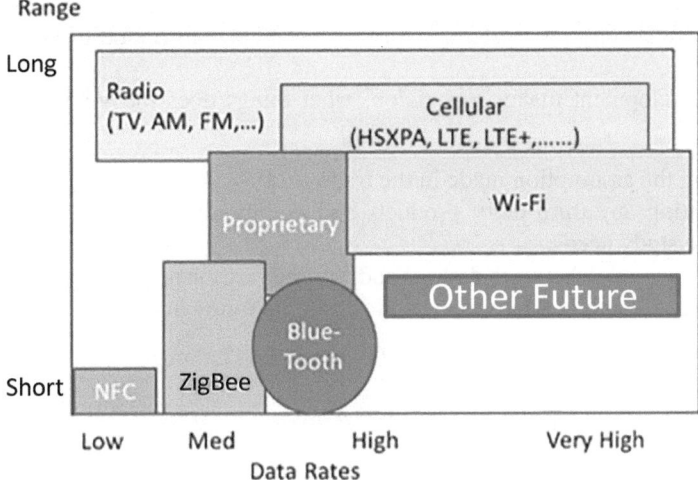

Figure 8-2. Example communication options vs. range (distance) and rates

Reference: Jon Hagar Built figure for IoT Classes 2016

Testing Integration Factors

One feature of IoT that separates it from traditional embedded systems is communication to the outside world, the amounts of data, and the related integration to other devices and systems that come with such communication. Figure 8-2 captures most of the current communication concepts while showing the range (low at the bottom to high at the top) with data rates on the bottom axis.

There is no "perfect" communication choice between communication options. For example, long range means more power will be needed, which can be a problem for battery-powered IoT devices. Going with Bluetooth may give you decent data rates, but not for long range (>75 feet). And some IoT devices may have multiple communication configurations. It is a trade-off that testers need to be aware of when it comes time to test these options. Each option and configuration will need plans for testing as well as integration. Figure 8-2 does not list all options, and new ones come available frequently. For example, narrowband, 5G (soon 6G), other board band options, LoRa, many proprietary systems, etc., will all fall on different parts of this graphic and change over time. LoRa and LoRaWAN have many consequences and implications if chosen or required. These include chirping, breaking signals apart, etc. These are used for IoT and small data rates, so they must be considered and tested. The point of the graph is that developers and testers need to understand where their system falls between range and data rates.

Further, my experience in systems with communication channels is that integration between system communications is often tricky. Many teams assume communication is just "plug and play" because vendors advertise communication integration "will be easy." However, in my experience, advertised capabilities often fall short, interface issues are common, and integration of the unplanned new testing work becomes burdensome during integration testing.

The development team will need to do decision analysis and trade studies to decide on the right communication lines and options. Testers, in turn, will need to confirm these choices with verification and validation.

Finally, once development makes a decision, what things does the V&V/test staff do? Testers' actions include

- Check by testing the assumption made in the trade study.
- Evaluate by testing any third-party products and communication lines (hardware or software) against the trade study needs.
- Assess product integration between devices and communication networks (see Part 3). Additionally, for more reading, I recommend the next section and the following link:

    `https://en.wikipedia.org/wiki/Communications_protocol` – accessed winter 2022

Planning Test IoT Operations and Maintenance (O&M) with Data Analytics and the Edge

Another aspect that will have differences for many testers in IoT is the operations side of devices. Since many devices interact with the real world, their use and data generated will be of interest to various stakeholders. Stakeholders here include users, the environment, marketing staff, help desks, troubleshooters, governments, companies, developers, and other testers.

Further, because devices have communication linkages, they process and create data that can be made available to these different stakeholders, often instantly. However, the data rates and the amounts being considered are mind-blowing. I hear about transfer rates in the Gigabyte level and storage in the Tera- and Petabyte levels (or even more).

No human can process such rates, and storage levels are not readily available for those sizes. This limitation will force operation teams, testers, and other stakeholders into sophisticated approaches to address data use and analytics. The IoT system industry will likely see

- AI systems (e.g., IBM's Watson) that support self-organization before humans "consume" the summaries
- Analytics tools using AI and neural networks
- Statistical analysis and visualization tools (e.g., spreadsheets)
- Advanced security and privacy features to protect the data internally and externally

Successful test teams will need to use these approaches to conduct ongoing test data analysis. Many of them seem reluctant to learn new tools and data analysis concepts in discussing this with testers. Test teams' use of Ops data analytics should include

- Assessment and validation of test data used in initial testing
- Updating test plans and designs with new test data and cases
- Performing test data taxonomy analysis (Part 4, Chapter 22)

Ideas such as test taxonomies and improving the test process using measurements (metrics) are not new. However, not many testers do data analytics or use taxonomies. So, I am wondering if the IoT data analytics wave will leave behind these reluctant testers.

Some essential high-level O&M activities, including tests, are shown in Figure 8-3.

Most IoT projects will spend years in the operations and maintenance phase. Inputs include updated plans, focused practice, Ops-driven requirements and data, as well as logistics considerations. The Ops team drives these inputs but needs support from development, V&V/test staff, and others. There are no simple system, software, hardware upgrades, and fixes. Small changes can crash systems and result in calls to the help desk, considered the first line of defense for the user, yet comes at additional cost for companies. In the engineering of Ops, there are regression test issues, ongoing data analysis, field support, security concerns, and integration of whatever is being maintained. It is not uncommon to have any part of the system become out of date and need to be replaced.

Further, the more complex an IoT system is, the more maintenance increases. The outputs of O&M include bug reports, upgrades, new documentation, more and more data, Ops security concerns, and, finally, more plan or contract changes. Since IoT vendors are looking to make money, O&M can be profitable, if proper planning takes place before jumping into it.

For more information on O&M, see the following links:

`https://en.wikipedia.org/wiki/Maintenance,_repair_and_opera-tions` – accessed winter 2022

`https://en.wikipedia.org/wiki/Operational_maintenance` – accessed winter 2022

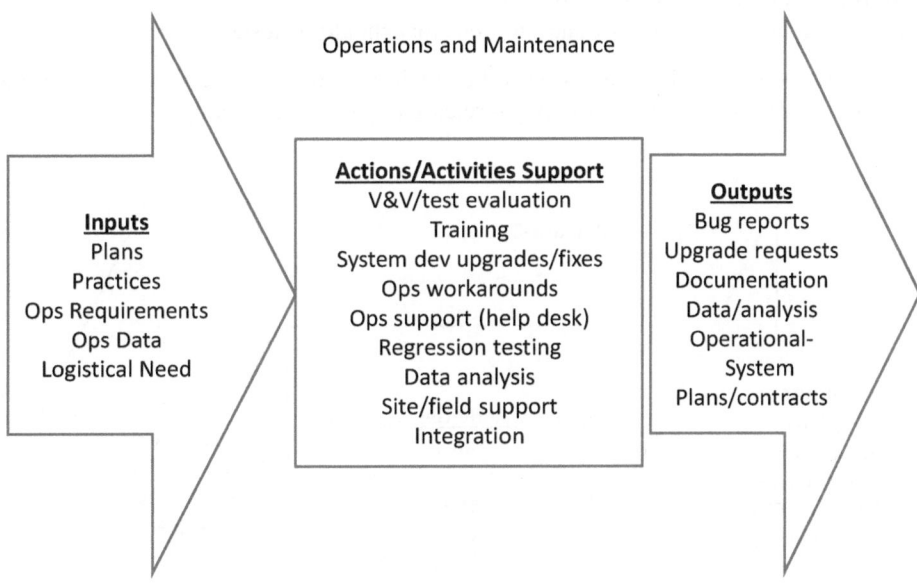

Figure 8-3. O&M inputs, tasks, and outputs for testing/V&V

Reference: Jon Hagar Built figure for IoT Classes 2016

Test Planning for Release Deployment

For years now, how software is updated has evolved. Once, users had to choose to upgrade by opting to do a "pull" onto their computer system. Now many software systems upgrade themselves frequently. Most smartphones and other IoT devices are scheduled and upgraded with the user's minimal consent or following a license agreement. Some users block or opt out of upgrades.

Not having a consistent version of software running on devices raises many questions and potential issues. However, there will always be users who circumvent whatever plans or processes companies put into place. This situation is why many companies try *not* to give consumers a choice.

There are legal and ethical issues concerned with release and delivery options. Some cases that may be special for IoT testing include

- The legal impact caused by release issues in the cyber-physical world
- Problems caused by communication connections between different software configurations (should these be tested?)
- Integration test issues in stakeholders' new or deployed systems
- Coordination and communication between stakeholders, including users
- Negative press from failed release deliveries
- Configuration management of any configurations that should be tested

Teams and testers need to think about these factors in planning their testing. Figure 8-4 shows a basic set of inputs, actions, and outputs to begin release deployment thinking.

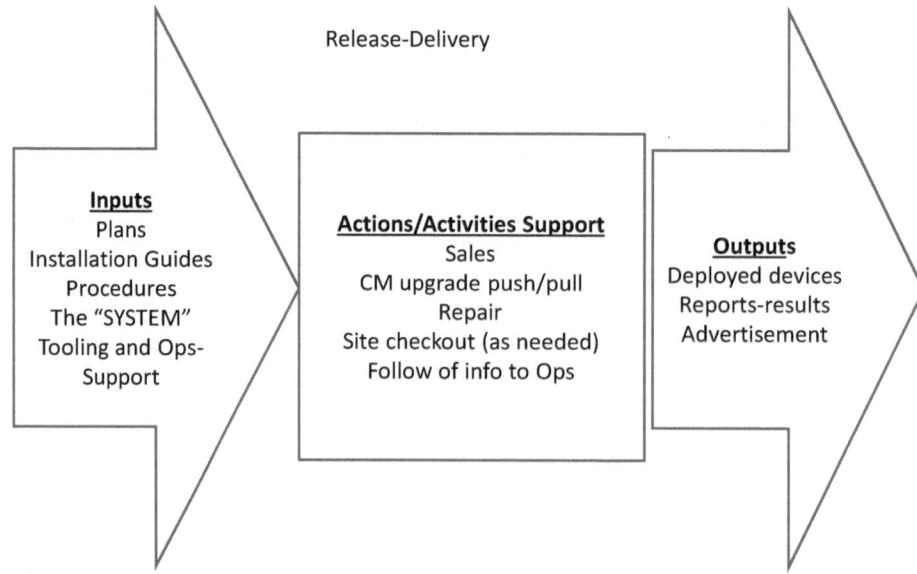

Figure 8-4. Release and deployment

Reference: Jon Hagar Built figure for IoT Classes 2016

As a sub-effort of O&M, software release and deployment involve plans, installation guides, and Ops security procedures that support the system, tooling, and any support efforts. These actions and activities include sales reviewing the Ops data analytics and CM/SCM with change control boards to ensure only authorized repairs (changes) and improvements are made. All changes must be tested, often at system use sites, and approved by a customer receiving the release and the changes. The outputs are (hopefully) improved devices, documentation, and happier users. There is a slight chance of one word or one line of script or code in software crashing systems. So remember, don't believe the developer who says "the change should not impact anything."

Factory Production: A Very Brief Introduction to Testing IoT Hardware

Unlike software that does not have a factory assembly and production line, IoT hardware must be made before the software can be installed to run. Software test teams may be asked or become involved in the production line. Hardware testing and quality control of mass-produced devices are different and out of the scope of this book. However, I hope Figure 8-5 will help introduce software testers to the new world of hardware production for completeness.

A reader can see Figure 8-5 on factory production lines for a fast overview. There are much more, in fact, whole books and engineering concerns on this topic. However, the figure should get you thinking and maybe looking at learning more.

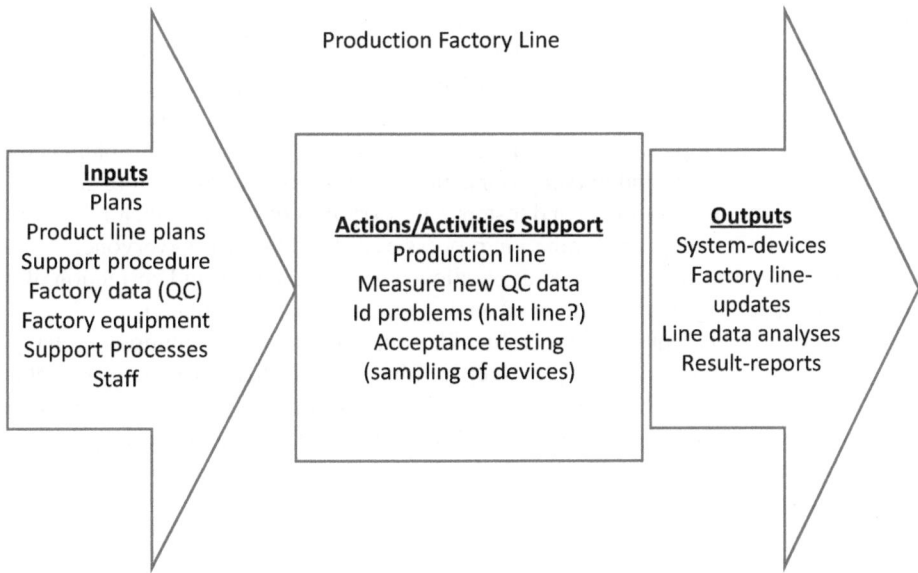

Figure 8-5. Production factory line

Note: QC is quality control and is very common in mass production lines but different from quality assurance (QA).

Reference: Jon Hagar Built figure for IoT Classes 2016

Test and Quality

IoT products may have longer useful lives than most development teams envision. Therefore, testing IoT devices to provide information on the qualities of a device will need to address the near and longer-term usage of those devices. For this short section, I would like for testers to think about the following during test planning:

- Testing provides information to the whole team about the qualities of a device.
- The perceived quality of a device may live forever in the user's mind.
- Testing may be seen as a necessary "tax" on device development that is to be minimized, but this can be a shortsighted view if the quality of the device is not good enough for sales to succeed.
- Comprehensive testing is more complicated than development, so testers deserve equal, "full engineering credit."
- Devices will often be used in new ways in environments that development and testing did not address. So those ways should be addressed during Ops and maintenance, which in turn may generate new test plans and risks associated with a new use case.
- Stakeholders want a device to succeed and have recurring sales and updates, but this will mean regression and new function testing (testing never ends).
- *Testers are not the sole owner of product quality*, and testing does not improve quality directly. *Testing provides information about the state of IoT quality for stakeholders.*

Evolving As You Go

Product evolution includes agile, process improvement, continuous quality, and many other slogans. My experience is that most projects, processes, products, and people must be evolved if they are to be successful in the long term. This is likely for a small startup and even a big company. This case is also valid if you are part of an agile group or a traditional hardware company. Those that do not evolve and change are likely to be overcome by market Darwinism.

The key is to strive for evolution by trying new ideas and taking some risks but do the change in smaller baby steps. *DO NOT change everything all at once.* I have seen agile teams trying to change everything at once, but then a number of these agile teams failed in some aspect. When you change ten things, and the system stops working, finding out which change caused the problem can be time-consuming, wasteful of budgets and person-hours, etc. Don't introduce new factors all at once. This is a common mistake among developers and testers alike. Try introducing one factor (one data point) at a time to make the fault and error elimination process easier and faster for everyone.

I like slow evolution over time. This evolution allows the identification and correction of mistakes as the changes are made. Any change will need testing (regression testing plans).

Yes, most change is good, but be very careful. This approach to slow evolutionary change works for projects, processes, products, and people. Keep in mind that the change (a quality) must be something people will pay for. As an example, too often, changes to the operating system in the high-tech world result in changes to the UI, which confuses users, violates the ubiquitous "easy" UI ideal, and sometimes corrupts or destroys valuable user data.

Minimum Activities Needed to Release an IoT Device into the Wild First Pass

 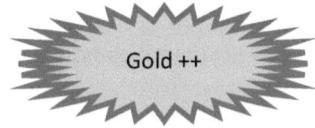

I get this question in classes: "what is the minimum I need to do in testing?" The answer is: "it depends" on

- Context (customer, product, use case, regulations, risks, etc.)
- What you can afford to lose (money, schedule, business or clients, company, life, etc.)

The full answer on testing/V&V takes some engineering and management work to find. If a project does nothing in the test, the risk of product surprise is higher. Failure to launch happens if a project tests and tests to the point of not releasing, which is terrible too. Finding the balance takes critical team thinking and solid engineering.

The checklist found in Appendix C might be considered a minimum for activities to release IoT products to the customer. It is provided as a starting point checklist. The list can be debated and changed.

Risk and Opportunity Management on an IoT Test Project

Therefore, there are many risks and opportunities in IoT for testers to consider. Examples include

- You lose your money.
- You lose someone else's money.
- Your product does *not* work (functional or nonfunctional qualities).
- You do not meet the prescribed or agreed to schedule.
- You kill someone.
- Your company goes out of business, and you lose your job.

There are the project types of risks outlined earlier and in other places in this book. Therefore, I recommend risk-based IoT test planning. Whether you do formal risk management activities or something less formal, a fundamental process is given in Figure 8-6.

The process of Figure 8-6 would be for the project but feeds into the test effort. For a detailed risk-based test flow, readers should refer to [1] and ISO 29119. Such a risk process should be ongoing over the life of a project. Risk analysis does not have to be heavyweight. I have used the preceding flow using sticky notes posted on a whiteboard. This approach guided one of my team's testing efforts and was very agile.

Finally, after ROM, testing, and generally speaking, the engineering staff do not like to do documentation, but it is essential for communication with stakeholders.

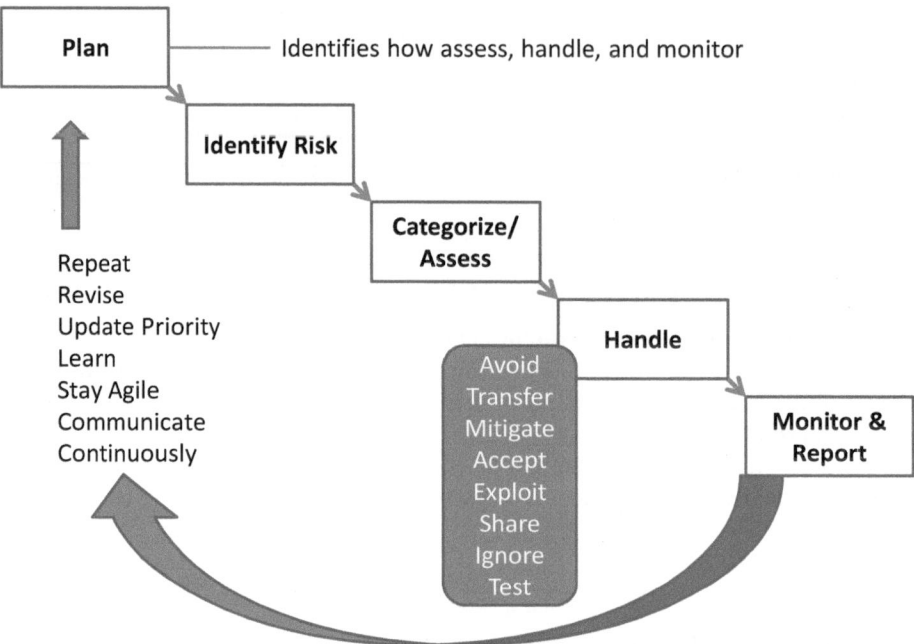

Figure 8-6. Risk and opportunity management (ROM) process

Reference: Jon Hagar Built figure for IoT Classes 2016

Last but Not Least, Test Documentation

The traditional standards-based side of systems and software engineering is often viewed as being "document-centric." Earlier, I advocated that test planning documents do not need to be heavyweight (no significant "thud factor" involved). Agile has whiteboards, storyboards, mind maps, and planning games. For many IoT devices, these can be sufficient documents. However, if you had a choice between IoT critical medical devices, all you knew about them was that, for one, there had been planning on a whiteboard while another had test plan documents reviewed and analyzed by experts, other testers, and regulators. Which would you feel safer using?

There are IoT devices and systems in which stakeholders will want more documentation. When I enter a new project for work or support, I typically ask to see things like test plans. If these documents are years old, look like boilerplate with generic words, and do not seem to be in use, I feel like the project lacks levels of control but, more importantly, discipline. I dig deeper to understand what kind of testing this team is doing. Next, I will look for project presentations on testing or quiz testers about what they understood their jobs to be. If the answer is different from what the formal plans said and not consistent with the team's understanding, I suspect the testing is out of control or not done effectively.

There is a fine line between creating and having too much useless documentation and creating and having useful documentation. I find that test documents can help solidify my thinking about the planning process. Agile teams produce "just good enough" documentation to aid in communication but also serve as a historical reference.

Another excellent guide for determining the proper levels and types of test documentation is considering the stakeholder(s). If the project is small and only has internal users, existing whiteboards, and some smartphone pictures, those may be good enough documentation. Suppose your stakeholders are regulators, where you will be subjected to audits, and there will be a long IoT device usage history. In that case, more heavyweight documentation that is followed and updated may be advisable.

I have recommended the ISO 29119 part 3 [4] software test documentation as one reference starting point. It is significant, generic, and reasonably complete with information. However, this standard is very heavily weighted regarding the number of document types and the outline of material content. I recommend that everyone tailor such a standard. In tailoring, a team arrives at the right "weight" level (page count or succinct outline) with stakeholders' buy-in. Additionally, keep in mind that as an IoT system matures, changes in test documentation may become advisable.

Tailoring documentation is not easy; it takes much critical thinking and effort. It also takes coordination and communication with stakeholders. This leads many teams and stakeholders to say, "we will do every document in the standard," which is wasteful, creates documents that are not used, wastes project person-hours, and, worse, leads to teams that decide to do *no* documentation instead of tailoring to valuable documents.

A TEST PLAN WHEN NOT REQUIRED
To solidify my test planning on one new project, I wrote a test plan following the template I had. At first, I shared it with no one but started my team's test activities (e.g., creating a test lab, writing high-level test designs, automation scripts, etc.). The team started testing, and after an extended period, organization management and customer stakeholders came and asked, "where is your test plan?" I happily printed them a copy. They reviewed it, reviewed my tester's work, and said, "gee, it looks like your test team has things under control." They left happy, my testers were happy, and management gave me a promotion.

There is no "best" level and type of test documentation for the testing process and techniques. Further, the levels and types of test documentation are likely to change over time, either increasing or decreasing. Getting the right balance of test documentation takes commitment by the entire test, support, and development teams, as well as commitment by project management because so much is at stake. Again, being agile and thinking things through are needed.

The daily test planning needs to be reflected in the higher-level test plans and documentation. This iteration is limited during early IoT activities such as a proposal but needs to become more detailed during actual test activities. The iterations need to be accounted for in the initial test planning and then used to update the test documentation and presentations as information flow from the test team to stakeholders. In an agile project, this can be very frequent (e.g., each sprint), and more traditional projects can be at major milestones.

Summary

This chapter introduced the test lab/environment planning. The details of creating and using a test lab are defined further in Parts 3 and 4 of this book, but planning is the first step. Creating a test lab/environment is a project inside of a project, and, therefore, careful and critical planning must be exercised long before the lab is usable. Each type of IoT development effort for Part 1, Chapter 4, will have different plans, environments, and architecture. There is no one type of lab or set of tools that will fit all contexts.

The next chapter completes our tour of planning for system IoT considerations, which is essential. Even though this is a book on software testing, IoT devices will be part of an overall system.

References

1. *Software Test Attacks to Break Mobile and Embedded Devices* by Jon D. Hagar, CRC Press, 2013
2. ISO 9000:2015 Quality Management Systems (series) – Fundamentals and Vocabulary
3. "IEEE Standard for System, Software, and Hardware Verification and Validation," in *IEEE Std 1012-2016 (Revision of IEEE Std 1012-2012/Incorporates IEEE Std 1012-2016/Cor1-2017)*, vol., no., pp. 1–260, 29 Sept. 2017, DOI: 10.1109/IEEESTD.2017.8055462
4. "IEEE/ISO/IEC 29119, ISO/IEC/IEEE International Standard – Software and systems engineering – Software testing" – Parts 1 to 5 series
5. IEEE 982.1-2005 IEEE Standard Dictionary of Measures of the Software Aspects of Dependability, in revision as of 2022

Chapter 9
System Engineering Concepts in IoT Test Planning

In Chapter 8, I looked at planning the software test environment and facilities. However, besides test planning, testers should understand and support many more IoT project planning activities, including software and system development. In this chapter, I provide a quick overview of how to do IoT system planning. This chapter addresses top-level system engineering activities, engineering interface/integration, various support processes, architecture considerations, trade studies, and impacts from other development approaches.

Reviewing Basic Software Engineering (SE) Concepts

As I have already mentioned, one thing that makes IoT different from other tech areas is that you must consider hardware and systems engineering (SE) activities [1, 2, 3]. Readers should have a book on SE and probably a book addressing hardware in their IoT product domain, as well as some prior training to fully understand the breadth of SE in the IoT domain. This section will outline some basic SE concepts to get you started or provide a fast overview for those that don't want to buy an entire SE textbook or reference standard. See the "References" section at the end of this chapter for pointers to books on these subjects. This chapter applies to every IoT project, as shown in Figure 9-1.

Not every activity needs to be done formally (documented), and some agile SE is possible. As a tester, if these are not familiar, you should do more reading and research. Table 9-1 lists many SE activities that teams should consider.

V&V/testing are defined in IEEE 1012 and ISO 29119 [4, 5]. Table 9-1 is summarized and paraphrased from ISO 15288; the basic concepts are shown in Figure 9-2. Again, the standards should be tailored, and they are only a starting reference point, but they do contain a base for a project with no history with which to start. V&V/test is integrated into SE at lower levels, and other project groups are shown in Figure 9-2. No single group or person is likely to do everything on a project except on the simplest of IoT devices.

Figure 9-2 shows that teams start a project with a proposal support effort – a proposal team bids on the job, cost, schedule, engineering, and support tasks. Proposals may be short, but I have seen proposals that were 1000s of pages. Management directly supports the proposal by providing the outline and scope of the IoT project. The scope includes hardware, electrical, software, system, and support engineering groups. As the project proceeds, engineering efforts create the IoT product that the "production" makes. Production may be done within the project or subcontracted out. If work is subcontracted out, the subcontractor's effort becomes vital since a project should not "trust" that other

J. D. Hagar, *IoT System Testing*, https://doi.org/10.1007/978-1-4842-8276-2_9

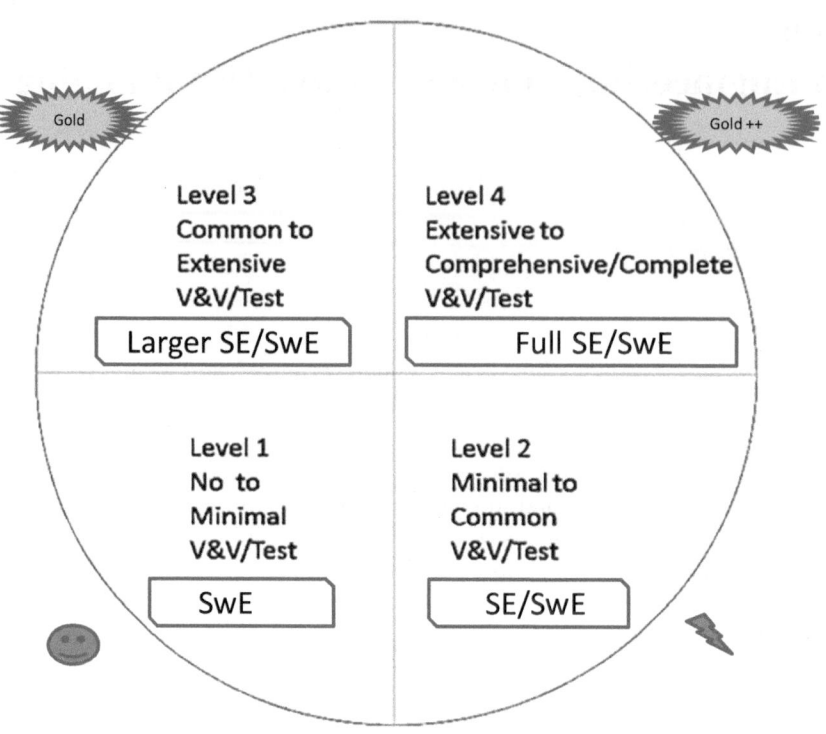

Figure 9-1. System IoT with engineering systems (SE), software (SwE), and V&V/test

Table 9-1. System Engineering Activities

System Engineering Activities (No Particular Order)		
State the problem	Design the system	Produce documentation
Understand customer needs	Do sensitivity analysis	Lead teams
Discover requirements/stories	Assess and manage risk	Assess performance
Validate requirements and device	Do reliability analysis	Prescribe and plan V&V/testing
Investigate alternatives	Integrate system components	Conduct reviews
Do the quantitative measures	Design and manage interfaces	Verify requirements and the system
Define models	Execute configuration management	Perform total system test
Perform functional decomposition	Integrate project management	Reevaluate and improve quality

Note: In my experience, the underlined items are critically important.

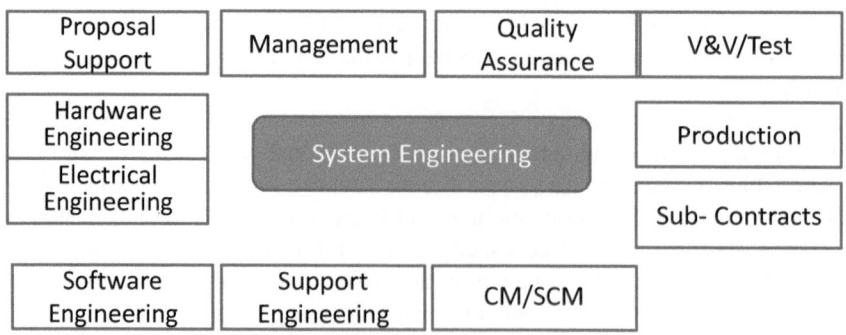

Figure 9-2. System engineering support for major project groups

Reference: Jon Hagar Built figure for IoT Classes 2016

contractors will perform as expected. Finally, the support areas of quality assurance, V&V/test, CM/SCM, and ongoing engineering continue over the whole IoT product lifecycle. Next, I look at the engineering, SE and SwE, support areas that are critical to V&V/testing.

Critical Test Enabling Engineering Support Processes

Besides the basic engineering efforts of SE, hardware, software, and test engineering, test teams should include a few essential enabling efforts to be successful and to ensure the viability of company products. These test support processes include configuration management (CM) and software configuration management (SCM), software quality assurance (SQA) and quality assurance (QA)/quality control (QC is traditionally used for hardware production lines), measurements, SE architecture, and design efforts [1, 2, 3]. The management side of a project will include the management teams, contracts people, planning-scheduling staff, and quite possibility logistics staff to address things including orders, production-manufacturing, shipping and receiving, help desks, distribution, security, and long-term support engineering. Some of these support elements will transition at some point into operations support. Some projects may have different or less support in some areas, depending on the nature of the IoT device.

SQA/QA

Many teams consider testers as part of SQA and QA. This is largely a convenience to "compact" organizational structures. There is nothing wrong with this structure, but in most international standards and classic project configurations, testing is separate from SQA/QA.

Testing and V&V actively assess IoT products as described throughout this book.

Quality is the value that some stakeholder is willing to pay. QA covers the activities, practices, and processes of monitoring the engineering and methods used on projects to help ensure quality work is performed. The monitoring includes checking activities and conformance to standards or models when required by IoT stakeholders. QA is done on the IoT system and hardware. SQA focuses on the software.

SCM and Testing

CM and SCM control and manage what goes into all baseline, deliverable products, whether that be hardware or software, working in conjunction with contracts and management (as required). Then, the design and requirements personnel perform work, as shown in Figure 9-3.

The CM/SCM personnel have processes, plans, baseline records of configurations for hardware and software, an issue tracking database that also tracks change requests, plus any driving contract factors. They may also assist with configurations for lab equipment (both hardware and software).

Now, the industry builds systems from off-the-shelf (OTS) components (both hardware and software). These IoT products must be configured and the "contents" of any configuration known. Inside of their processes, the CM/SCM person (or team) establishes and updates the baseline configurations, which offers assurances to any stakeholder, such as management, customer, government agency, auditors, and the test team, that what is stated is delivered. Its contents can be substantiated through a verifiable set of processes and records. This is done with tools, and documentation is

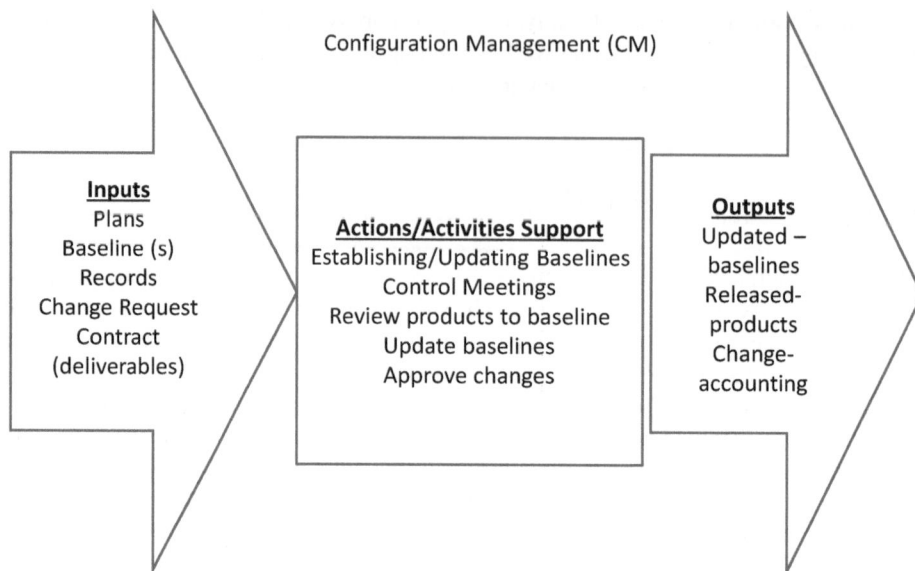

Figure 9-3. CM and SCM actions

Reference: Jon Hagar Built figure for IoT Classes 2016

driven by configuration control meetings, such as Change Control Boards (CCBs). CCBs review all inputs and changes to products (both hardware and software). Updates and their effectivity on products affected by configuration changes must be understood at the board level as to what the configuration will be before updates can be approved by any CCB (see Orion Malware in the following). The outputs of CM/SCM are known, recorded, controlled, and configurations baselined (i.e., changes approved by all concerned). However, CM/SCM are facilitators but have no voting power on these boards.

Aside from software build tools for SCM, CM/SCM tools are some of the first chosen on projects since these tools must provide reports to everyone as well as track issues with software or hardware and their effectivity level, any change requests and who has reported or requested the change and why, the testing status of changes or issues, if any inspections are done, as well as where changes or issues stand in the lifecycle, and, finally, what exactly is in *any configuration*. The CM/SCM tools also provide the critical thinking required to establish and examine lifecycle processes on projects since those processes are what define the logic of the issue tracking database and, thus, the reporting for stakeholders or others. This kind of support is well worth investing in skilled CM/SCM personnel and good CM/SCM tools [6, 7].

It is easy today to try and shorten these kinds of formal processes, particularly in Agile, or whenever the staff has an attitude as in the way we do things in school – "just get the project done as quickly as possible," and formal processes are avoided. Effective project managers and companies take CM/SCM contributions very seriously. For agency, safety critical, or government projects, this attitude on projects can devolve into very large trouble, quickly and for the longer term.

LEARNING THE IMPORTANCE OF CM/SCM EFFORTS
I have taught many college classes where teams of students did a group project for the first time. The teams have done SE, software, and testing. I warned every team that configuration management would be an issue because they worked in a group over a more extended period. Every team had CM plans. Every student team wished they would have created better CM plans and processes. The list of CM actions earlier is a minimal starting point involving complex processes that keep everyone out of trouble.

Here is what I mean by "running scared." I like what a company vice president told me years ago. She said, "I want everyone in engineering to run scared because *they* are out to get us." By "they," she meant other companies, bugs, hackers, stakeholders, auditors, or anyone you might think of and some you may not have thought of. You are not being paranoid that *they are out to get you.*

ORION MALWARE AND CM (REFERENCE TO THE HACK: HTTPS://EN. WIKIPEDIA.ORG/WIKI/SOLARWINDS)

The Orion software malware problem involved suspect software inserted into the baseline. Many of us think that if tight CM/SCM and code coverage testing had been done on the Orion project, the suspect software should have been detected before it was deployed to many customers. Now, does every IoT project have this testing/CM security risk level? No, but your local answer to this risk question is part of your project planning and security risk analysis. You should run scared in this area.

Software (Test) Measurements

Measurements support reporting to and for testers and other engineering domains as well as management. Most of us have a love-hate relationship with measurements. I use them, management and stakeholders want to know them, and at the same time, somebody in the development or auditing system may abuse them. Measurements of things like the number of errors a particular programmer has created (which can be extracted from the CM/SCM tool) can be used to punish that person for having too many bugs in their code compared to other programmers. *This is not good* and goes against the reason to use measurements first. After all, one must be able to measure where they are today compared to where they should be, and using measurements provides this perspective.

The thing to remember about measurements is that they allow not just a forward and backward look at development or testing efforts. Still, they also allow those interested in spotting significant trends found in the data. A negative ignored trend can cost a project a lot of money, eating into company profits or company viability. Therefore, I believe it is better to use and like using measurements.

Figure 9-4 shows measurement efforts, which almost every IoT project should have.

Story: Measurement Misuse

The Dev team had one programmer that seemed to create many high critical issues. Management was not happy with her. However, when I analyzed the part of the system she was programming, it was the most challenging and critical part. So, of course, no surprise, it would have more errors. She was a great programmer who got stuck with the most challenging part of the code.

Measurement and Status Supporting Systems Engineering and Testing

Measurement is the quantified and qualitative assessment of products and/or processes attributes. Attributes can be physical or logical. Typically, once a measure is achieved, it can be compared to expectations (plans) or a standard. In Figure 9-4, I showed the inputs to the measurement, which includes plans and standards and the measured data that may come from the engineering/testing or a database (repository). These inputs are analyzed, identified, and recorded. Other support activities

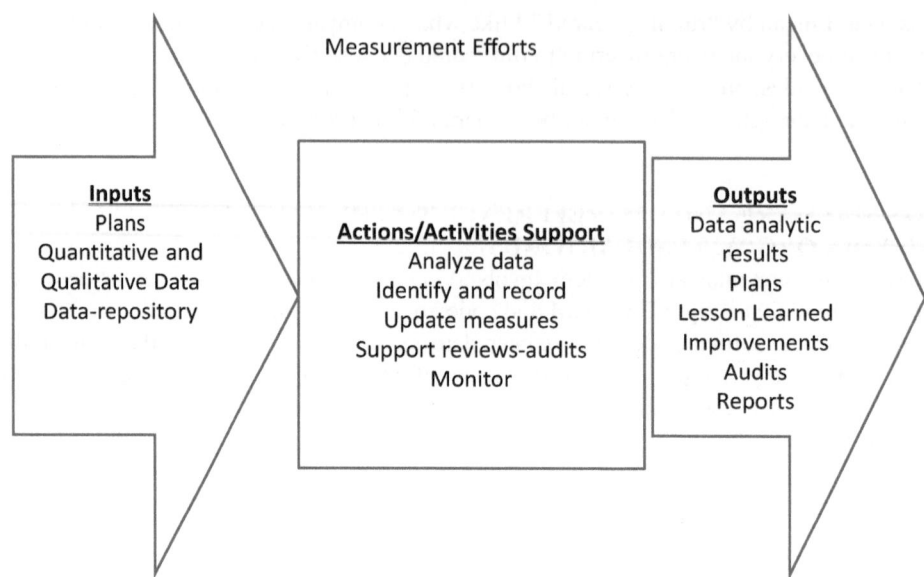

Figure 9-4. Measurement efforts

Reference: Jon Hagar Built figure for IoT Classes 2016

include updating measurement efforts as the lifecycle continues while reviewing and monitoring are done. The ongoing lifecycle efforts optimally will use edge, cloud, and AI (Part 4) analysis.

The output of this measurement is used by stakeholders, including data analytics/AI results, plan updates as the lifecycle matures, and conducting lessons learned or process/product improvements. Finally, measurements are used in audits and reports.

Here are some example measurements – good, bad, or ugly, yet directly applicable to testers:

Good – Tells teams where they are and where they are going and can communicate this information to stakeholders.

Bad – Teams get a bad reputation when a measurement looks "negative." Teams will make the measurement look good to avoid punishment when they do not understand what the measurement is telling them.

Ugly – Teams will have too many measurements to be helpful, or teams will not maintain or use the data.

Testers can have and use the following example measurements:

- A count of issues found in total reported through the CM/SCM tools.
- Measurements of each technique used.
- The aging of an issue (if it is not getting addressed timely especially given its assigned level).
- Number of tests to run.
- Number of requirements to test.
- How long does it take to create and run a test?

As a tester, I have used each of these. I have fought to avoid being abused by the measurements. As a test manager, I looked at this aspect as part of my job to keep and maintain the measurements but stood up to any abusers of the data. It is possible to use measurements successfully, but it takes work on both counts, which is outside the scope of this book.

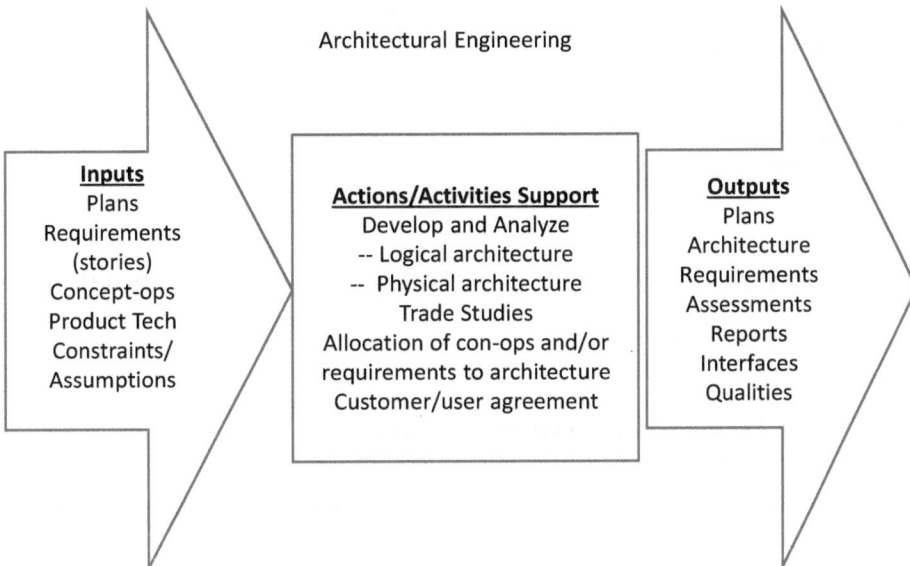

Figure 9-5. The top-level architecture

Reference: Jon Hagar Built figure for IoT Classes 2016

System Architecture and Design – Test Top-Level Support

The architecture of a tiny IoT device may not seem to be a critical SE task. However, when I start talking about systems of IoT devices and systems of systems (e.g., as in a smart city), the architecture is a critical piece that must be well defined during SE. The architecture applies to software, hardware, test, and operations. A high-level set of architecture engineering activities is shown in Figure 9-5.

The inputs to architecture engineering include IoT project plans and requirements, be they in the form of models, stories, concepts of operations (ConOps), product technology information specifications, constraints, or assumptions. These inputs are processed in activities and support actions, including developing and analyzing logical or physical structures (architectures). Actions to support architecture engineering are trade studies (see the next section) and requirements refinement to support architectures while supporting customer and user agreements.

To understand architecture, consider the company Apple Inc. Apple has been very successful because it provides architectural solutions for integrated hardware, software, and systems. Their products integrate and "play" well together and are easily used by almost worldwide users. They also have a consistency of usage that their consumers follow with loving devotion. This integrated and complete system architecture can serve as a model for other teams doing engineering for them to be successful.

Critical in the system, system of systems, smart cities, etc., are three areas in Figure 9-6. These are the essential system elements of hardware, software, and human operations that use the system.

The architecture details are fully defined in ISO 42010 [8]. Once the architecture has been considered, the development team can continue with planning the system and software activities. An excellent place to start planning is the concept of trade studies, where you decide what to make or buy.

Hardware   Software   Ops (human)

Figure 9-6. Top-level SE architecture elements

System Planning Trade Study – Decision Analysis

A part of SE and all engineering areas, including tests that will be very important in IoT, is called "make or buy" decision analysis. This can impact procuring third-party elements such as hardware, software, and support systems, such as testing. The make or buy decision should follow a process that is called a "trade study." Here, engineers are "trading" the positives and negatives of a product or set of products against each other to make a purchase (buy) decision. These studies can be formal (lots of written documentation) or informal (simple email with rationale or justification). The more money and risk involved, the more formal a decision should be.

Most IoT projects will "glue" elements together to create an IoT device system. Teams will leverage existing elements, add in some customization, and integrate a new IoT device. After a buy decision is made, the products should be tested, or V&V performed.

However, care must be exercised in making these kinds of decisions. A team may *not* want to buy the least expensive or most feature-rich element. I recommend that teams practice some robust decision analysis in the form of a trade study to evaluate candidates for their use(s).

Here are four reasons to perform trade studies:

1. Evaluate viable solutions against must-haves and wants (desired but not absolute).
2. Challenge requirements or resolve conflicting requirements.
3. Identify optimal engineering, including cost, quality, schedule, and design elements that satisfy requirements.
4. Obtain customer/stakeholder buy-in.

Figure 9-7 has an example trade study process that can be used as a starting point or guide. Many organizations will have a different and more formal trade study process that engineers should follow. Again, we are dealing with a topic and processes with books and classes to address the ideas. If the company is large enough to have a contract or legal team, these teams may already have company-approved processes for trade studies.

Figure 9-7 processes are general and further refined in references. There is much more to a trade study (see https://en.wikipedia.org/wiki/Trade_study – accessed winter 2022). Testers should be involved during this process because after step "V&V/test evaluate," a decision is or should be made based on testing information.

An Example Trade Study Process

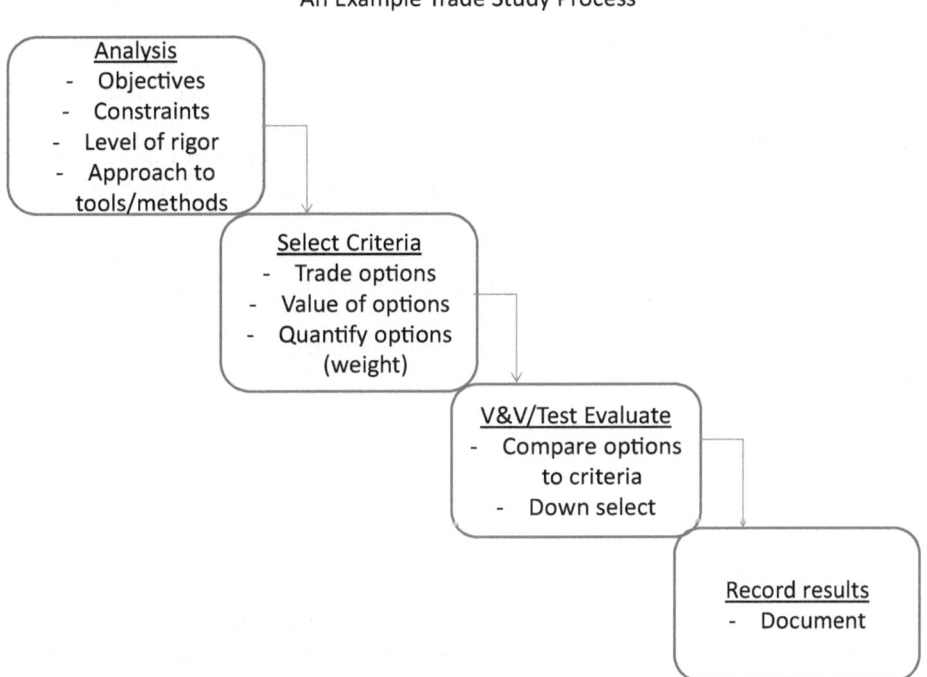

Figure 9-7. An example trade study process (recommended)

Reference: Jon Hagar Built figure for IoT Classes 2016

Warning

Many teams make the mistake of "trusting" what vendors tell them. Don't be gullible. Testers know how to "not trust" but verify. Testers are also helpful in doing the earlier analysis steps and then documenting the decision. Teams want to record a decision because it is common to "revisit" trade studies later in IoT project efforts due to changes, regressions, or failures. Further, when the final items are received from a third party later in the project lifecycle, they must be tested and integrated. Just because the best product won a trade study does not mean it will work for the project.

As trade studies are completed, the IoT project planning continues into system and software design. For many IoT devices, safety will be an essential design consideration.

Designing with Safety for IoT

Many IoT devices will have safety impacts. Safety is the land of regulations and lawyers. The IoT development and test team owns the safety of the device or system or system of systems. The team should *be cautious* (remember running scared?), and knowledge of standards is essential.

What are the industrial safety standards? Here is an important list of IEEE/IEC/ISO numbered documents that can be useful for testers but may be subject to change as time goes on:

- ISO/IEC/IEEE 29119 software test.
- IEEE 1012 Verification and Validation.

- IEC 62061 is Safety of Machinery – Functional Safety of Safety-Related Electrical, Electronic and Programmable Electronic Control Systems.
- IEC 62304 Using VectorCAST to Satisfy FDA Software Testing Requirements.
- IEC 61511 Functional Safety – Safety Instrumented Systems for the Process Industry Sector.
- IEC 61508 Understanding Verification and Validation of Software Under IEC 61508.
- EN 50128 Using VectorCAST for Software Verification and Validation of Railway Applications.
- IEC 61513 in Nuclear Power Plants – Instrumentation and Control Important to Safety – General Requirements for Systems.
- IEC 60335 is Household and Similar Electrical Appliances – Safety – Part 1: General Requirements.
- IEC 61508 is more of an industrial standard (e.g., PLCs, drives, and motors).
- ISO 26262 applies to automotive.
- DO-178C and DO-178D standards are for the avionics industry.
- IEC 60730 is for appliances.
- EN 50128 is for the railways.
- IEC 61511 is for the process industry.
- IEC 62061 is for machinery.
- IEC 62304 is for the medical device industry.
- IEC 60335 is for product manufacturing.
- UL listings standard requirements.

Ouch! There are standards that people might need to consider in testing and designing for IoT safety. After writing any book, this list will change, but it gets testers started with the SE safety thinking. During the design of the safety in the IoT system, the team selects the proper hardware and software.

Hardware Design Considerations

Hardware design is a big subject, with whole books, classes, and standards dealing with the subject. Testers should start with the following reference: `https://en.wikipedia.org/wiki/Hardware_architecture` – accessed winter 2022.

Immediate factors I consider for IoT include

- Is this a throwaway device?
- Is this a significant hardware project effort?
- Is security a high risk?
- Is safety a high risk?
- Are there significant financial risks?
- Is this hardware experimental (never been done before)?
- Is this historical hardware that is getting an "IoT" facelift?

There are not a lot of test activities here, but there are V&V activities to consider (see IEEE 1012).

Software Design Considerations

Like hardware, there are many books and references on software design and production, starting with
`https://en.wikipedia.org/wiki/Software_design` – accessed winter 2022
`https://en.wikipedia.org/wiki/Software_design_pattern` – accessed winter 2022

There are too many excellent references to list all of them here, but the subjects that a tester should have access to include

- Basic introduction and definitions
- Design engineering roles
- Project planning and estimation
- Support environment for engineering
- Process improvement
- Requirement identification and analysis
- Production and coding implementation
- Developer testing/V&V including reviews (inspection, walk-throughs, pair programming, etc.)
- Measurements
- Risk management

Testers need to be specialized in testing/V&V and be a generalist in the other aspects of the project. Testers do not need to be developers or designers but should have a keen understanding of the activities, practices, processes, and products other engineers create. This understanding aids in testing and team-to-team communication. Testers that do not understand engineering and software coding are at a disadvantage when dealing with other teams. These testers are not testing engineers but test technicians (see Part 1, Part 4, and Appendix on skills).

As the system, hardware, and software design take shape, the development team needs to start considering the integration of these elements. Often, the integration, in part, falls to the testers, and other times, a specialized integration team is needed for larger systems.

IoT System Integration

Communications and integration in IoT will be another high challenge area. There are three primary, ideal processes in integration approaches, as shown in Figure 9-8. A basic integration set of inputs, actions, and output is shown in Figure 9-8.

Bottom-up is typically used in hardware but can be used for software. In bottom-up, elements are assembled and integrated from the smallest part, building upward to assemblies, subsystems, and finally a completed system. The process is iterative, integrated, and test-integrated as the effort is made. The larger an IoT system is, the more complicated and necessary integration becomes. Early planning and ongoing integration activities can consume team resources.

In top-down, which is more common for software, a top component is created, then lower elements are assembled, integrated, and tested working downward before an assembly is reached. Once an assembly (component, object, class, etc.) of software is reached, it can then be integrated with the system or hardware.

Figure 9-9 shows common integration ideals (see `https://en.wikipedia.org/wiki/System_integration` – accessed winter 2022).

There are three "classic" integration process flows applicable to IoT: bottom-up, top-down, and continuous integration (CI) hybrid. In bottom-up, the integration occurs from the smallest part and builds upward to an assembly, then a subsystem, and finally a system. This approach is prevalent in hardware.

In top-down integration, an executive or component assembly is defined, and lower-level objects are created. This approach was pervasive in the early days of software. With modern programming, OTS software, particularly microservices, a hybrid approach is preferred.

For many IoT systems, the most common integration is the hybrid approach with continuous integration (CI) (see Agile in Part 3, Chapter 12, and [9]), where there is a mix of bottom-up and top-down with the constant ongoing CI and continuous testing (CT). This effort can be done with

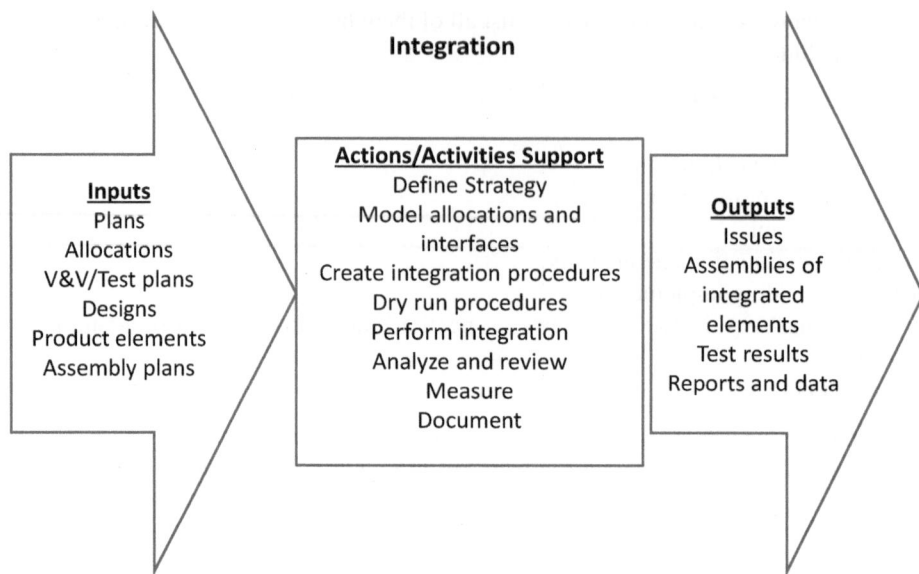

Figure 9-8. Integration of IoT

Reference: Jon Hagar Built figure for IoT Classes 2016

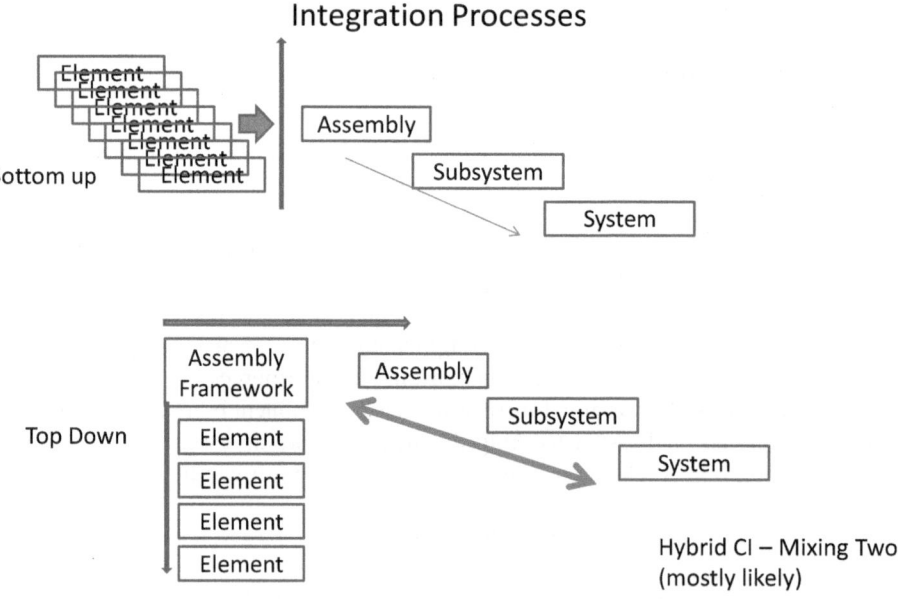

Figure 9-9. Integration process options

Reference: Jon Hagar Built figure for IoT Classes 2016

hardware and software but takes coordination and frequent replanning. A detailed integration plan is of benefit and important for larger efforts. Also, in systems, engineering, and software of complex systems, models (see SysML, UTP, and UML [10, 11, 12]) have proven to be of great value. Groups that had integration problems before using models found modeling tools solved many CI and CT problems.

In IoT integration, the factors and items to stay focused on include

- Who are communication providers?
- What are network configurations and interfaces?
- What are applicable communication protocols and their standards?
- How big is the integration (how many components, elements, assemblies, subsystems, etc.)?
- Who owns the integration (is there an owner or IV&V), and who is at risk – most or least?
- Is internal integration addressed (below the subsystem or system)?
- Are external connections and integrations fully identified?

Bottom-up and top-down integration efforts are classic, having been done for many years. IoT systems are expanding in the age of agile and DevOps system development models. This is where CI/CT and continuous deployment (CD) come into everyday use. It is worth considering other system impacts for these new models.

Agile and DevOps Development Impacts

So much of the world, including hardware, is evolving toward faster and agile lifecycles. A close concept to Agile is DevOps (see Part 3, Chapter 12, for more details).

Whatever the lifecycle is, when starting, the IoT team should ask, "What is the minimum for a viable IoT product at this lifecycle stage?" IoT unique factors that influence the answer to this question include

- Time to market
- Costs including money and schedule
- Ubiquitous UI
- Functionality
- Security
- Safety
- Communication
- (Other) Nonfunctional qualities

Agile seeks to balance these factors and has a goal of being ready to deliver a product at any point in time (maybe not to a consumer). DevOps philosophy includes the user to aid in testing and engineering. Both approaches include ideas of short steps in the development cycle, including continuous development, testing, integration, and delivery.

The context of the product will help determine lifecycle choices and evolution. IoT teams and management may be looking for the "one size magic way," but there is no such thing. If test and engineering jobs were easy, robots would be doing them. The jobs and lifecycle choices are challenging, fun, and engaging, and we get good pay. Be happy.

Summary

This chapter provided a very high-level "digest" covering the critical points of IoT system engineering for a software tester. Many projects and testers may find that they must look at other references or have expertise in areas not covered by this chapter. However, this chapter leads to Part 3 of the book, dealing in depth with IoT software test attacks, techniques, and security.

References

The following books and standards were used to prepare this chapter and are good general references:
1. *Software Engineering* by Pressman, 5th Edition, 2001
2. ISO Standard 15288 Systems and software engineering – System life cycle processes
3. ISO Standard 12207 Systems and software engineering – Software life cycle processes
4. "IEEE Standard for System, Software, and Hardware Verification and Validation," in *IEEE Std 1012-2016 (Revision of IEEE Std 1012-2012/Incorporates IEEE Std 1012-2016/Cor1-2017)*, vol., no., pp. 1–260, 29 Sept. 2017, DOI: 10.1109/IEEESTD.2017.8055462
5. "IEEE/ISO/IEC 29119, ISO/IEC/IEEE International Standard – Software and systems engineering – Software testing" – Parts 1 to 5 series
6. https://en.wikipedia.org/wiki/Configuration_management – accessed spring 2022
7. www.softwaretestinghelp.com/top-5-software-configuration-management-tools/ – accessed spring 2022
8. ISO/IEC/IEEE 42010 Systems and software engineering – Architecture description – www.iso-architecture.org/ieee-1471/index.html – accessed spring 2022
9. *Agile Testing: A Practical Guide for Testers and Agile Teams* by Lisa Crispin and Janet Gregory, Addison-Wesley Professional, 2009
10. SysML – SysML V2: The Next-Generation Systems Modeling Language – www.omgsysml.org/SysML-2.htm. Accessed spring 2022
11. UML – Unified Modeling Language – www.omg.org/spec/UML/2.5.1/About-UML/. Accessed spring 2022
12. UTP – UML Test Profile – www.omg.org/spec/UTP/1.2/About-UTP/. Accessed spring 2022

Figure Reference

1. https://image.freepik.com/free-vector/isometric-computer-hardware-parts-set-with-monitor-system-unit-electronic-components-details-isolated_1284-38504.jpg?w=740; https://image.freepik.com/free-vector/web-page-visualization-protocol-procedure-dynamic-software-workflow-full-stack-development-markup-administrate-system-driver-shared-memory-vector-isolated-concept-metaphor-illustration_12470220.htm#query=software@position=6&from_view=search; https://image.freepik.com/free-vector/delivery-logistics-shipment-set-four-isometric-images-with-colourful-icons-pictograms-human-characters-cars-illustration_1284-29112.jpg?w=740

Part 3
IoT Test Designs and Security Assessments

I start this part of the book by examining the brains and brawn of IoT testing.

This part's design techniques, patterns, attacks, and tours are all heuristics. While the team should know these heuristics, only a few test designs are applied to any IoT device system at any point in the lifecycle. Very few testers and experts can be expected to know by heart and be able to apply all of these techniques. But, if you can look the information up and apply it, the testing problem may be solved. There is no universal right idea, book, or standard in IoT system testing. I recommend this part of the book be used as a reference and referred back to throughout the lifecycle. I begin with some historical references, but they are only a start that should be used with other concepts in this part of the book.

Chapter 10
IoT Test Design: Frameworks, Techniques, Attacks, Patterns, and Tours

This chapter illustrates where the managers think testers earn their keep, yet I understand that testing is everyone's job and runs for the whole lifecycle. Every IoT team needs staff who know and can apply techniques and patterns to help produce a quality product, regardless of their job title.

Figure 10-1a and Figure 10-1b summarizes historical concepts of test design. These are classic names and technique processes defined in many books and industry standards on testing [1, 2, 3, 4, 5, 6, 7, 8, 9]. These references are used throughout Part 3, and specific references to them will be provided as needed. I will make specific recommendations to help you based on these references and my experiences. Also, there is no best test practice but only good advice in the context of your IoT project. As a starting point, one should use what you and your team feel most comfortable with and then learn to expand on other ones.

Many functional techniques in Figure 10-1a and Figure 10-1b [1, 2, 3, 4, 9], also known as black box testing, might be applied to IoT. The test team would apply these techniques supplemented with the "experienced-based" techniques of error guessing, attacks, tours, patterns, and exploratory practices, which many testers may be highly skilled at using. The "structural test design techniques," or white box testing, would be applied by IoT developers or testers supporting the code level assessment during test-driven development (TDD). Later in this part of the book, I will identify high-priority black box IoT candidates.

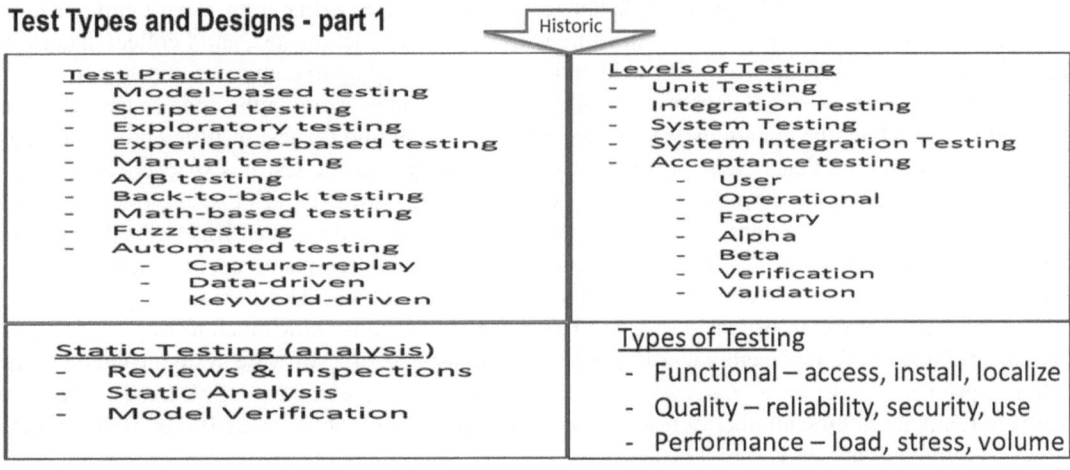

Figure 10-1a. Overview of Test Types and Designs (reference ISO 29119)

Reference: Jon Hagar Class

J. D. Hagar, *IoT System Testing*, https://doi.org/10.1007/978-1-4842-8276-2_10

Test Types and Designs – part 2 Techniques

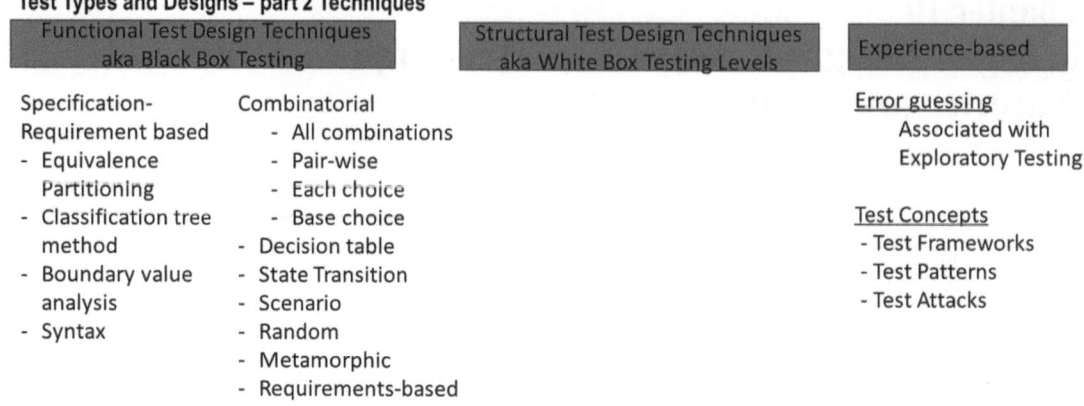

Functional Test Design Techniques aka Black Box Testing		Structural Test Design Techniques aka White Box Testing Levels	Experience-based

Specification-Requirement based
- Equivalence Partitioning
- Classification tree method
- Boundary value analysis
- Syntax
	Combinatorial
	  - All combinations
	  - Pair-wise
	  - Each choice
	  - Base choice
	- Decision table
	- State Transition
	- Scenario
	- Random
	- Metamorphic
	- Requirements-based
			Error guessing
			  Associated with
			  Exploratory Testing
			
			Test Concepts
			- Test Frameworks
			- Test Patterns
			- Test Attacks

Figure 10-1b. Overview of Test Types and Designs (reference ISO 29119)

For IoT specification-requirement-based testers, I recommend equivalence partitioning, combinatorial, state transition, use case, and requirements-based. For structural-based test teams, I would recommend coverage of statements, branches, and decisions. The structural testing leads to the "levels of testing," where, in the planning, the team needs to address low-level unit testing, integration testing (software and system), system testing, and, finally, the types of acceptance testing.

Finally, the static testing analysis using reviews/inspections, code static analysis tools, and/or model verification is an optimal and early activity for finding system issues before detailed testing begins. My practice was that any product created on an IoT project should be subjected to static analysis, at least by inspecting it with a thought to start my understanding of development efforts.

Testing is an exploration to address risk and provide information to stakeholders. Some testing is more rigorous and conducted with thinking akin to scientific methods. I have a theory or idea in the scientific method, define an experiment to provide data about the theory or idea, and examine the results to confirm, disprove, or modify what I have postulated. Product testing is very similar; I have various theories or ideas about a product with hardware and software. For example, the idea may be the device meets requirements, has some quality, such as time performance, or maybe has something negative, such as a bug, that I want to find. The tester sets about experimenting, what I call a test, to provide information about the product. There are activities to conduct the test, and I collect data from the test. From this, testers determine if the idea or theory I was trying to confirm is met or not.

A critical piece of information many testers are trying to find is errors in the code (bugs). If I find the product has a bug, the team may want to fix the error. I may find the test insufficient to find bugs or provide information that I want, so more is needed. I may find the test has succeeded (no bugs found), and thus I have some confidence the product is working. However, just because one test works under one set of conditions does not mean another test will behave the same. This is the problem of "the intractability of testing" [1, 6]. Testers of hardware and software must understand this principle. I will not treat this concept in depth here, so please read the references if you are unfamiliar with it.

Because of the intractability of testing, the next problem I face is "how much testing is enough?" "When am I done?" Testers cannot test 100 percent of all aspects of a software system (related to the halting problem [https://en.wikipedia.org/wiki/Halting_problem]), because there are really an infinite number of tests. If one test is not enough, it becomes 100 or 1000 or more tests. Therefore, all testing is sampling.

The test industry and standard IEEE 1012 for hardware, software, and systems have established a compromised approach to dealing with integrity levels and factors [Part 1]. These levels and factors are a compromise, meaning they are not perfect but accepted within many industry circles and companies. Following industry standard practices, such as ISO 29119 and IEEE 1012, offers testers and organizations some product acceptance in practice from a legal and ethical sense.

Different types of products pose different risks (Part 1, Chapter 4). I do not insist on following any particular standard or book on testing. However, following industry standards does offer some acceptance by many stakeholders.

STORE GETS HACKED VIA AC SYSTEM
IoT devices in the industry category, which can be used by many different factories to control heating and air conditioning, may seem low risk from a human danger standpoint. Still, buyers expect systems to work. Even worse, in the case of a Target store's data breach [www.usatoday.com/story/money/2017/05/23/target-pay-185m-2013-data-breach-affected-consumers/102063932/ – accessed spring 2022], the hacker gained entry into the Target "point of sale system (POS)" via hacking indirectly first into the HVAC system and then the POS. The cost may be millions of dollars, demonstrating that integrity/factors and risk can be easily misjudged.

WHEN IS A WATCH MORE THAN A WATCH?
For another example, it does not get much better in consumer IoT devices. Consider a wearable watch. It might be built to display time and measure heart rate for exercise. This seems low risk and integrity when first deployed to the users. However, another vendor can create an app that uses this information to track the watch user and adjust insurance rates based on the user's sleep patterns and exercise (or lack of it). The user may *not* want this information shared. Now a new possible risk is created that the watch creator had not foreseen. And it could get even worse for the consumer. The watch may be used to control security by passwords and biometrics, such as the user's speech pattern. This seems more secure until, because of poor testing, a hacker gains access to first the watch and then the personal information of this user.

So, with these stories, I illustrate that developers and testers must think long and hard about risk and integrity/factor levels. Further, the risks and factors may change over time. Thus, the first testing done to get the product in the field may not be sufficient because the use of the IoT product changes, evolves, and grows. The choice to do minimal (no testing) on version 0.1 of the IoT hardware and software may be acceptable for a newbie company or project. Still, testing will need to change as a company's product becomes more successful.

Test and Heuristics

Heuristics are engineering ideas that help solve problems but are not laws or even hard and fast "rules." Heuristics are common in engineering, and development testers need to understand that their use does *not* guarantee the software will work or that all errors will be found. Heuristics are based on the engineering history of the subject at hand, in this case, testing. The heuristic may vary depending on software domain, team history, methodologies, and company/industry regulations. As such, the use of an industry standard may be required on one IoT project by the company, but another project may use a different heuristic based on a different context. Testers start detailed test design by doing the following actions that feed into **testing concept heuristics**:

1. Identify the variables (e.g., data, sensors, users, inputs, outputs, etc.) as testing is about variables (it all depends).
2. Identify the inputs and outputs to the software, including public and private information, as well as users (e.g., human and nonhuman).
3. Know where the boundaries are in the IoT's requirements, design, and code.
4. Understand how the data flows in and out of the IoT software and device communications.
5. Know the IoT device's hardware (e.g., sensors, actuators, comm lines, power, processors, storage, screens, etc.).
6. Understand the interruptions to processing that the IoT system can experience (e.g., power, interrupt hardware levels, reboot, time marks, multi-users, etc.).
7. Identify resource limitations placed on the IoT system (e.g., time, CPU, memory, comm speed levels, etc.).
8. Outline dependencies and constraints for all of the preceding actions.
9. Conduct device state analysis (e.g., inputs, beginning, timing sequences, event transitions, outputs, end states, etc.).
10. Make maps, tables, or diagrams of the preceding actions (e.g., mind maps, state tables, flow charts, whiteboards, etc.).
11. Continue with test planning, strategy, and particularly risk analysis.
12. Continue with detailed test design and planning over the lifecycle.

If you think this is a lot, you begin to understand why test design is very complex. The preceding information will flow into test techniques, patterns, attacks, exploration, and/or tours. In many early test plans and designs, you do not need to know every piece of information or all the preceding details. Still, if you cannot answer some of these starting points, you have a heuristic that you may not know enough information about to have good testing results. You are blindly limiting your testing or making things "overly simplified." The answers to the heuristic concepts for IoT testing will evolve. As the IoT system matures, so will the test planning.

Test Patterns

A test pattern is a generalized way of test thinking. They are higher level and abstract than test techniques (see Chapter 11). Patterns can be reused from one IoT project to another IoT project and, hence, a good subject for this book.

Pattern problems can be solved in planning, processes, and meta-design. Examples are given in the following sections.

Example 1: Planning Pattern for IoT [8, 9]

Once you have some framework questions answered and understand heuristics, it is an excellent time to think about patterns. Since testing is exploration, this allows us to observe the software under test. For example, if you fire up your IoT device for the first time and the software crashes right after login, do not confuse this for lack of progress. You know your team has something fundamental to fix before proceeding to the next steps, which is a good observation. Likewise, if you power up and your first "hello world test" works, take the tester contrarian view, and do some unexpected action on login that a bad actor or hacker might do. This "hacker" pattern may give you "interesting" observations.

As a tester, simple patterns start with the earliest testing phases in the project. I am assuming some test planning and risk assessments have been done here. Also, the development team should be involved with these patterns whether you are using agile, DevOps, or even traditional lifecycles. If the development team is not supporting test exploration, a conversation within the team is likely needed. There are, of course, other time phase test patterns, but the following is one I have used on different projects.

Phase pattern 1 – Make a test with the basic, simple requirements on the first version of the software (early prototypes) and hardware (mock hardware, simulations, and emulations). Does it even "fly"?

Phase pattern 2 – Make tests of the critical risk features of the IoT system work (more hardware and software in the loop but still some prototype). This pattern may continue for a longer time. Can you retire any risks now?

Phase pattern 3 – Make tests using new techniques of the requirements and features of the IoT device (I assume needed hardware and software present). The time frame for this will depend on the nature of the schedule and device. Can you retire any risks, and can you break the software?

Phase pattern 4 – Start creating "stress" tests for the IoT device (mean = unusual cases, bad days, stresses, dependability assessment). This may not have much schedule time available, but I do like to try to break things. Have you retired higher priority risks, and has your list of risks changed?

Phase pattern 5 – Become a security hacker (attacks, hacks, tours, etc.). The amount of time you spend here depends on the security risks the team has identified.

Phase pattern 6 – Test critical quality characteristics (performance, reliability, safety, etc.). The amount of time you spend here depends on the risks the team has identified.

Phase pattern 7 – Test customer/stakeholder needs before delivery. If you have correctly made the other phase patterns, this pattern can be made in hours or days.

Phase pattern maintenance – Repeat testing for a new version of hardware or software as needed.

Example 2: Mind Maps – Test Patterning Tool for Process Selection for IoT

Since testing is about exploration and exposing risk to a project, capturing your observations and understanding is meaningful. A helpful pattern and observation capturing tool is to create a mind map. If you already use mind maps [https://en.wikipedia.org/wiki/Mind_map – accessed spring 2022], you may only need to scan this section.

When doing testing, you should watch for the following patterns of observation in IoT to create mind maps:

- Displayed device information (maybe part of the IoT device app).
- Functions and links the device offers.
- Overall color scheme, readability, and/or theme of any displays.
- Warnings and limitations the device produces.
- Placement of objects, displays, buttons, sensors, actuators on the device.
- Icon-object style and nature (if an app is used).
  Note: Some devices have guidelines for these display items (e.g., medical field, ADA, or accessibility constraints).
- Are there indicators to give users feedback when software is using on-device processing?
  Note: Feedback provided on timing, following user input, connections to other devices, etc.
- Do menus appear (visible), make sense to different users, and perform required actions? (If an app is being used.)
  Note: Try selecting all menu items even if they should not be selectable.
- Does the device and app's logic "flow" as well as make sense (logical, chronological)?

How do you get enough understanding to start a mind map pattern?

There are various open source mind mapping tools that one can use to assist you. First, you start with the list or something like it earlier and do a fast exploratory test [3, 6, 7] or test stories [4, 5]. Once you have done a basic exploratory test set, you can create a first pass mind map without formal written test scripts.

Next, you explore the mind map patterns. You can consider taking the next possible step as an example:

- Start (playing).
- Take notes and identify "coverage" of each mind map item as you go.
- After gathering as much information as possible and reasonable, stop and refine the mind map. Then, consider the following tests (more playtime, check documents, go hack, etc.).
  Conduct a first exploratory session (look at the device/app or user stories).

Note
Consider the earlier factors and risks (assuming risk-based planning has happened).

- After this first session and organization, you will have a better idea of what you are testing.
- You may now know additional information needed to continue testing, such as a user account, data, passwords, or other information to "feed" into the testing.

You now have a second- or third-generation mind map pattern to work with. You have documentation, which is not fancy but can be expanded, published, and used in test planning or test design. Documentation is good because development, testers, and stakeholders can participate in reviews and comment on the testing/inspection. And yes, tester information and products should also be reviewed and checked by the team during mind mapping and inspection. Options for expanding this pattern are shown in the following list:

- Add risk-based testing items (risk and mitigation).
- Expand the captured information:

  - Try options at each mapping point as the IoT device is traversed.
  - Stress inputs (in bounds, bounds, and out of bounds).
  - Introduce test modeling information.
  - Add combinatorial or other advanced test designs (see Chapter 11).

- Continue capturing information from sources, such as requirements, user stories, operational concepts, etc.

Mind maps start the test thinking process, including defining risks necessary for test planning, but I have not addressed test attacks, tours, frameworks, or techniques. These ideas are next, but I have you started on the critical test thinking process.

CASE STUDY: USING A MIND MAP PATTERN AS A PLAN DOCUMENT
You can build a mind map pattern supporting testing, as shown in Figure 10-2. This mind map is for a heat system controller that does not exist in hardware or software but has conceptual requirements (user stories). For example, I want to do "device on/off" controls in testing. A second mapping exists for "set comm connection" to select between a Wi-Fi, a smartphone with an app, and "other." These would be considered tests. However, here it gets interesting. Are the connections between the "Smartphone" and the "Set user actions" and "Security" an area needing more detailed "hack" testing? This would be something to explore in a future mind map pattern.

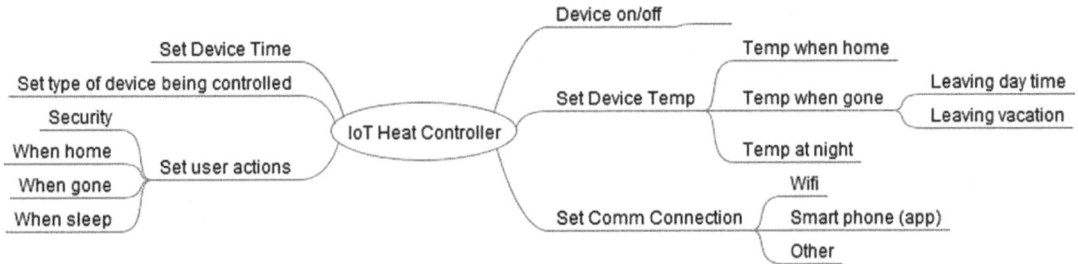

Figure 10-2. IoT heat controller device mind map

Reference: Jon Hagar

> In this example, I start my planning by identifying a series of "nominal usage" tours to address the left side of my heat controller system. Then, I defined a specific environment tour on the right side to set target temperature variations of "Set Device Temp." At this point, I am trying to see what happens with extremes in hot and cold. I would go to the required ranges doing boundary testing and then tour outside of the required ranges in stress "rain day" tours. These "stress" testing tours may find unexpected information on the hardware-software system. When doing "rain day tours," you will get development saying, "no, you are testing outside of requirements, and nobody will ever do that." But if you read the more detailed security testing chapter (Chapter 13), you will find that hackers love to do "what nobody will ever do." This small mind map can now be a simple test plan for a project wanting a document with minimal words.

Example 3: Attacks for IoT – High-Level Test Design Pattern

Test attacks have been popularized in the agile world and written about by authors like James Whittaker [10, 11]. *A test attack is a pattern of testing based on a common mode of failure and information gathering, seen repeatedly in areas of software.* Some may see an attack as negative when it is really *positive.* An attack goes after the "bugs" and other useful information in the software. When the attack does not indicate an issue in the software, the tester has increased confidence and information about the software qualities.

An attack is a pattern because it may include or use classic test techniques, test concepts, and other test ideals. An attack pattern is more than a test procedure because it is modified for the IoT context and likely includes various techniques, approaches, or concepts to do the testing. Attacks are focused and guided with an exploratory approach to testing.

Those who work with patterns see experienced testers learning mental attack patterns when working over the years in a specific domain or product. These are the testers that seem like they use magic to be able to "break" the software and find information. It is not magic, but they have gained mental test attack patterns through the years. By learning predefined attacks, you can jump-start your mental patterns.

Attacking your software is, in part, the process of attempting to demonstrate a system (hardware, firmware, software, and operations) does *not* meet requirements, functional and nonfunctional objectives. This is different from just showing that a requirement is met [11] but can show that the software under test meets a need. For IoT attacks, the software under test must include "the system" (i.e., hardware, software, operations, users). Addressing software risk from test planning and finding common modes of failure, especially where the user's IoT system is visible, are significant benefits of software test attacks. When I do not find issues but retire risk, I have useful IoT project information that I can share.

Support for software test attacks includes using tools, conducting attacks at different device levels, and applying techniques [11]. Whittaker offers a good starting point for software attacks in general that can be applied to IoT systems:

- User interface attacks
- Data and computation
- Filesystem interface
- Software/OS interface

Whittaker's *How to Break Software* series [10, 12] lists over 30 test attacks, some of which are shown in Table 10-1, which can be applicable to IoT security testing.

My book [11] lists specific ideas in Table 10-2 that can be used in IoT testing to extend test attacks further.

Table 10-1. IoT Security Applicable Attacks from Whittaker and Thompson

Attack #	Name/Description
1	Block access to OS internals
2	Manipulate the IoT app's registry values
3	Force the IoT app to use corrupted files
4	Manipulate and replace files that the app uses
5	Force the app to run with low memory, storage, or slow network access
6	Force overflow of input buffers
7	Examine all common switches and options
10	Use a tool to expose unprotected inputs
11	Connect with "bad data" IoT device ports
12	Input "fake" sources of data
15	Force the system to reset values
17	Create "bogus" files with the same name as protected files
18	Force all error messages (conditions)
19	Scan for temporary files that are left in the system – scan their contents

Table 10-2. Attack Patterns from Software Test Attacks to Break Mobile and Embedded Devices

Attack 4: Finding Hardware-System Unhandled Uses in Software	Attack 14: Breaking Digital Software Communications
Attack 5: Hardware-Software Interface Bugs	Attack 15: Finding Bugs in the Data
Attack 6: Long Duration Control Attack Runs	Attack 17: Using Simulation and Stimulation to Drive Software Attack
Attack 7: Breaking Software Logic and/or Control Laws	Attack 18: Bugs in Timing Interrupts and Priority Inversion
Attack 8: Forcing the Unusual Bug Cases	Attack 19: Finding Time Related Bugs
Attack 9: Breaking Software with Hardware and System Operations	Attack 23: Finding Missing or Wrong Alarms
9.1 Sub-Attack: Breaking Battery Power	Attack 24: Finding Bugs in Help Files
Attack 10: Finding Bugs in Hardware-Software Communications	Attack 25: Finding Bugs in Apps
Attack 11: Breaking Software Error Recovery	Attack 26: Testing Mobile and Embedded Games
Attack 12: Interface and Integration Testing	Attack 31: Attacking Viruses on the Run in Factories or PLCs
12.1 Sub-Attack: Configuration Integration Evaluation	Attack 32: Using Combinatorial Tests
Attack 13: Finding Problems in Software System Fault Tolerance	Attack 33: Attacking Functional Bugs

Example 4: Test Meta-design Pattern – Tours for IoT

A close cousin to attacks is tours. Tours are higher abstract than attacks. They are stories that I have used on many test projects, though by different names. The basic idea is a high-level story and archetype pattern to test a specific aspect or set of attributes for the product. The attack is a specific pattern to "break" the software based on common risks or concerns. A tour provides information on software qualities beyond just finding bugs. James Bach, James Whittaker, and others refer to tours as a quick, higher-level approach (or heuristic) to test planning while conducting experience-exploratory testing. Tours provide general information and observations and do not necessarily target specific concerns, risks, or bugs. Tours should be combined with frameworks, concepts, approaches, and test techniques.

Table 10-3 introduces some tours that can be used for IoT testing.

Table 10-3. Tours for IoT Related to the Level of Project

Tour Name	Example	Level
Rainy day tour – what happens when many things go wrong	A "Friday the 13th" type of test where everything that can go wrong in a system during the test goes wrong in a single test	All
Efficiency shortcut tour – reach user actions by a series of inputs	A novice user takes actions as a set of "longer" steps, while a normal user takes a different set of actions, and an expert user takes even different actions to the same end	All
User feature set tour	Test each feature your IoT device has for users including human and nonhuman actions, possibly following a mind map	All
Documentation tour	IoT documentation is used to determine what tests to run following online, paper, maintenance team, or other documents	All
High-low input/output performance usage tour	What happens when an IoT device is visited or used during low and high user levels of traffic (in an IoT system) or network times (in a communication system)	2, 3, 4
Security	See Chapter 13	2, 3, 4
Configuration tour	Test each configuration version of hardware and software that is in use or planned to be used	3, 4
Sensor-actuator tour	Test the sensor-actuator subsystems with a variety of use cases	All
Network comm tour	Test the IoT device on different network configurations, e.g., Bluetooth, cell, different carriers, devices not connected at all times, different Wi-Fi systems, etc.	2, 3, 4
Fog/cloud tour	Test the IoT device on different fog and cloud systems with busy, dropouts, slow, fast, etc. connections	4
Edge tour	Test the IoT device on different edge devices, busy, dropouts, slow, and fast	4
Local data storage tour	IoT device memory never full, full, and corrupted	4
Environment tour – on software, hardware, firmware, and systems	Test under different conditions of temperature, humid, weather, salt water, noise, vibration, etc.	All limited
Ops tour	Test under different Ops team usage scenarios	2, 3, 4
Key function tour – control, nav, guidance, recovery, OS	Test key internal functions of the IoT system, e.g., calibration, control, navigation, guidance, recovery	3, 4
Performance tour	See Chapter 11	3, 4
No-code/low-code/OTS tour	Test, tour, and attack code that is not developed by the IoT team from vendors, GIT, open source, microservices, your grandmother, etc.	4
Multiple system IoT devices in use	Test the IoT device in the lab with other IoT devices as a system	4
Full system IoT	Test the full connect system (if possible)	4

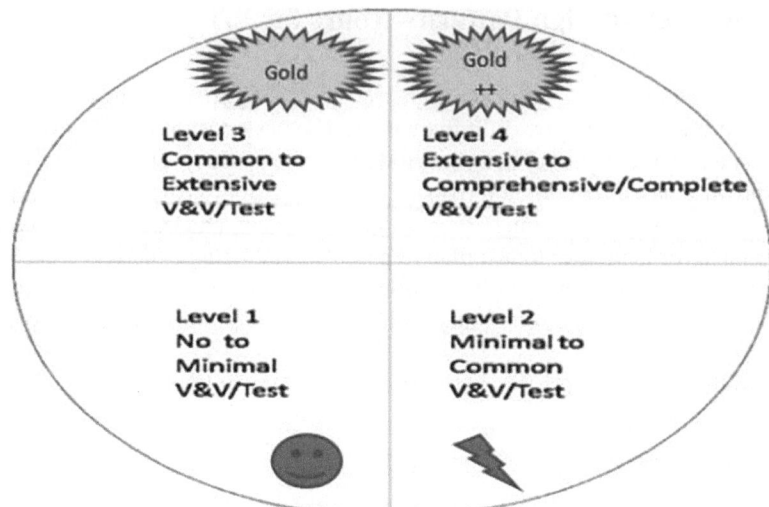

Figure 10-3. Example division of the component integrity determination

The levels column indicates what type of project integrity/factor level might be best applied in tour planning. To use this table, a test planner should look for a tour name and description to decide if the tour might apply to their context. Of course, testers may identify their local project and domain tours to include in this table. There are no limits on the number of tour names and classifications, as this table is just a start of common ones found in various parts of the testing industry.

Applying the Specific Tours to Project Factors

The tours of Tables 10-1 and 10-2 can be applied to most IoT test frameworks, depending on the nature of the system. The allocation starts during test planning and continues repeatedly over the life-cycle of the IoT system. Table 10-3 lists levels in the far right-hand column that correspond to the levels of Figure 10-3.

Example 5: Frameworks – The Top Level of Test Design Planning

Finally, the heuristics and patterns (attacks and tours) lead to frameworks for testing. In frameworks, I apply the judgment of test observations to select test techniques and approaches. In my experience, beginning testers use one or two test concepts. Experienced testers may have 10 or 20 test concepts. However, many testers limit their thinking to concepts that they are familiar with and have used. Finally, expert testers know there are many more test concepts and frameworks for planning, design, testing, and analysis. They will expand the test planning by learning more and repeating test cycles as the project matures. All testers need to remember they should consider many concepts over the test techniques, patterns, attacks, and tours, even if they only start to use one or two in any given phase of the IoT testing lifecycle.

A testing framework is a set of structures and rules used for planning and high-level designing tests. A framework comprises a combination of practices, approaches, and possibly tools to help testers be more efficient and gives a starting point for IoT test technique, attack, and tour selection. IT frameworks are often associated with development and test team automation, but I take a bigger test viewpoint. Different types of frameworks to consider in the test planning include

- Scripting, manual testing [9]
- IoT attack, tour, and technique-driven testing (Chapter 11)
- Object-based testing [13]
- Data-driven testing (for the database testing, when the project has one [7])
- Keyword, automation, and test lab–driven tests (Chapter 16)
- Test-driven development [14]
- Behavior-driven development and testing (BDD for agile testers [4, 5])
- Functional, structural, and quality-driven testing [9]
- System-software testing [8]
- Hybrid mixed testing (mixing the preceding frameworks)

Most test IoT projects will select one or two frameworks and start the planning-testing cycle. Some extensive IoT system of systems will use more frameworks since the number of parts and contractors will be greater. The frameworks change as the project unfolds, and the test provides information. Frameworks are not static. Test plans and strategies are driven by changes to risk as iterations happen.

A generic framework structure for test planning is given as follows:

1. Test planning and strategy for risks, structure, processes, and SCM.
2. Understand the IoT system using stories, requirements, contracts, etc.
3. Determine the test environments, test architecture, and gather detailed information.
4. Support development testing and/or
   A. Conduct attack/exploratory tests first and produce a mind map (or another notation).
5. Expand testing by adding tours, techniques, levels, approaches, and patterns.
6. Apply BDD and/or standardized testing processes (ref: 29119).
7. As needed, add V&V.
8. Repeat until stopping criteria (cost, schedule, plans, risks) are met.

Summary

This chapter presented the test planning to meta-design processes in reverse order to what many might expect because this is the way I learned testing. The meta-design process can be applied to any framework. These ideas can and should be mixed and matched in many IoT projects.

First, in my career, I started work using test techniques (structural code coverage) but had test leads and managers who had solved the upfront activities of this chapter and test planning before I started work for them. I was told, "Here are the test plans, frameworks, and techniques to apply. Get to work." And now, I have set the stage for the next chapter of classic test techniques.

References

1. *The Art of Software Testing* by Myers (the grandfather of them all)
2. *A Practitioner's Guide to Software Test Design* by Lee Copeland
3. *The Domain Testing Workbook* by Cem Kaner
4. *Agile Testing: A Practical Guide for Testers and Agile Teams* by Lisa Crispin and Janet Gregory, Addison-Wesley Professional, 2009
5. *More Agile Testing* by Crispin and Gregory, Addison-Wesley Professional, 1015

6. *Testing Computer Software* by Kaner, Falk, and Nguyen, Van Nostrand Reinhold, 1993
7. *Systematic Software Testing* by Craig and Jaskiel
8. "IEEE Standard for System, Software, and Hardware Verification and Validation," in *IEEE Std 1012-2016 (Revision of IEEE Std 1012-2012/Incorporates IEEE Std 1012-2016/Cor1-2017)*, vol., no., pp. 1–260, 29 Sept. 2017, DOI: 10.1109/IEEESTD.2017.8055462
9. "IEEE/ISO/IEC 29119, ISO/IEC/IEEE International Standard – Software and systems engineering – Software testing" – Parts 1 to 5 series
10. *How to Break Software: A Practical Guide to Testing* by James Whittaker
11. *Software Test Attacks to Break Mobile and Embedded Devices* by Jon D. Hagar, CRC Press, 2013
12. *How to Break Software Security* by James A. Whittaker and Hugh Thompson, Addison-Wesley Professional, 2004
13. *Testing Object-Oriented Systems: Models, Patterns, and Tools* by Robert Binder, Addison-Wesley Professional, 1999
14. *Test Driven Development*: By Example, Part of the A Kent Beck Signature Book Series and The Addison-Wesley Signature Series, 2002

Chapter 11
Classic IoT V&V/Test Concepts, Techniques, and Practices

For this book, testing provides information and makes observations, particularly product issues, about an IoT system for use by stakeholders. Chapter 10 introduced frameworks, patterns, tours, and attacks, while this chapter addresses the classic, complex technique concepts, which can each be used within a framework. We will start with functional, structural, and experience-based testing, considering the directly applicable ones to IoT. This chapter concludes with industrial related techniques, math-based and model-based testing. This chapter supports IoT test designs derived from classic test books and ISO/IEC/IEEE 29119 [1, 2, 3, 4, 5, 6].

Moving from Simple Testing to Providing Information on Quality

I find in many cases that testers apply exploration testing [2, 4] to learn about the software before they become interested in adding the classic test techniques. Once they get some "test time," they start to want more information, metrics, and the insight that specific techniques provide. Thus, they should become interested in classic test techniques.

At the same time, information about the product's qualities is what nontest stakeholders are most interested in. Quality testing is addressed in this chapter and the next (Part 3, Chapter 12). Many people have observed that quality is the value that someone is willing to pay for. Products have many qualities that we investigate during testing. Testers do not make a quality product by themselves, but we assess the product's quality characteristics. Testers provide information on the qualities of the software. The first qualities that most stakeholders think of for any product are functional and structural in nature. However, other qualities are crucial, including performance, safety, security, usability, operability, reliability, and others.

There is one negative quality that stakeholders want to avoid which is that of serious errors or bugs. Many authors on testing advocate that the only good test is one that finds an error/bug, which is far too simplistic. Many tests that do not find a bug still provide helpful information about a device, such as qualities. Testers would like to find errors when they exist, so they should plan, design, and modify tests to optimize error detection. I address this "bug hunting" aspect of V&V/testing throughout this book. Many of the test techniques, patterns, attacks, tours, etc., are based on their ability to find errors, and I will note this, but keep in mind, there are other kinds of quality information V&V/testing should provide.

© Jon Duncan Hagar 2022
J. D. Hagar, *IoT System Testing*, https://doi.org/10.1007/978-1-4842-8276-2_11

Testers are also stakeholders. Testers provide information to other stakeholders of the project. The success or failure of the project affects the test team as much as other members of the project team. Developers are stakeholders and thus should be interested in the information that testing provides. Management is a stakeholder who wants to know if the product is "good enough" to ship to make money. Customers are also stakeholders who want the product to work and meet a perceived need. Regulators, as stakeholders, want the product to be safe and meet legal restrictions or constraints. Other IoT stakeholders include other computing machines since IoT often has machine-to-machine communications.

Projects should test products during initial creation. This testing is often included as part of development efforts and quality assurance (QA). Products are also tested as they are mass-produced, and this is often called quality control (QC). We do not address factory hardware production QC in this book. It is a big subject necessary for mass production runs of IoT devices. There are standards on the subject of QC [7]. This book considers classic software QA and testing updated for IoT devices.

In exploratory testing, testers investigate the qualities of an IoT device using test design concepts. Testers use names like exploratory testing and test techniques and call things that we do verification and validation (V&V) when providing information on IoT qualities. We expect V&V to be done by the whole development team, covering subjects such as analysis, demonstration, and modeling with test supporting V&V.

In the past, many of us separated hardware and software V&V/testing, but with IoT, that division blurs. Given the preceding viewpoints, testers in IoT should understand and apply classic software testing. Further, I believe many software testers will also need to take on system testing, including assessing the unique aspects of the hardware. In the following sections of this chapter, I summarize the classic software techniques and ideas IoT teams should consider in designing tests.

Techniques, Practices, Levels, and Types of Testing to Apply to IoT

Figure 11-1 summarizes the classic test designs of functional, structural, and experience-based approaches. There are many subvariations for most techniques, but this section only details the first ones I consider when applying test design to IoT. These techniques are defined in detail in references such as ISO 29119-4 [1, 2, 6]. They are classified as test techniques because they have precise inputs, processes, and outputs. These ideas are just a start, and you are encouraged to explore other tests that may apply to your IoT devices and systems.

The following subsections of this chapter provide more information on these IoT-specific test techniques.

Functional Test Design Techniques

Functional test design techniques are based on the requirements and functions of the software. The most useful starting points for IoT are the following test techniques defined in [1, 2, 5, 6] and presented in no particular order here.

Functional Test Design Techniques aka Black Box Testing	Structural Test Design Techniques aka White Box Testing

Specification-Requirement based
- Equivalence Partitioning
- Classification tree method
- Boundary value analysis
- Syntax
- Combinatorial
    - All combinations
    - Pair-wise
    - Each choice
    - Base choice
- Decision table
- Cause-effect graphing
- State Transition
- Scenario
- Random
- Metamorphic
- Requirements-based

Structure-based (code / design)
- Statement
- Branch
- Decision
- Branch condition
- Branch condition combination
- MC/DC
- Data flow
    - All-definitions
    - All-C-Uses
    - All-P-Uses
    - All-Uses
    - All-DU-Paths

Experience-based

- Error guessing
    Associated with
    Exploratory Testing

Figure 11-1. Test techniques based on ISO 29119

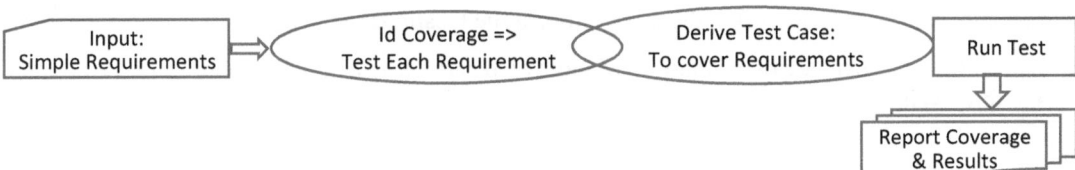

Figure 11-2. Requirements for testing

Requirements Based

Figure 11-2 highlights the processes for requirements-based testing. Requirements-based testing is the classic approach to test that the code meets the requirements (or acceptance stories if you are agile):

1. Input – For the requirements-based technique, I identify requirement statements. On many projects, this is done by looking for statements that contain the word "shall," since this is contractually binding and must be demonstrated in the testing. However, some projects will not follow this "shall" rule, in which case the team must work a little harder to know what is required of the software. Yet other projects will have a requirements database, making identifying and allocating requirements to testing much more manageable. Requirement "items" can also be called stories, use cases, needs, etc.

2. ID coverage – Each item is identified to be covered by one or more tests. Depending on the testing and item level, the test coverage may be done at a structural or functional system level.

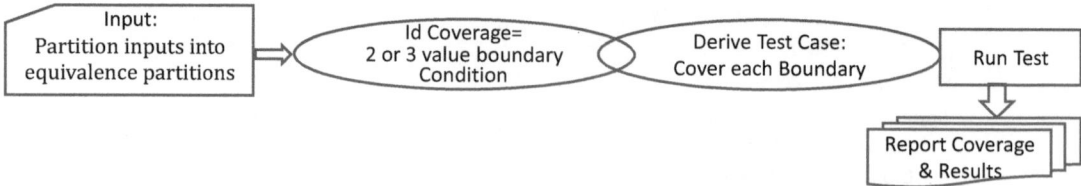

Figure 11-3. Boundary value analysis

3. Derived test case (inputs) – Next, the tester must identify an input or set of values that will drive the test case. Inputs imply the expected results (outputs). Test cases may be combined into test suites and later procedures. Automation can also be used.
4. Run test – Once we have the input data and test case, we can run tests with the software at the appropriate level of the system.
5. Report coverage and results – After the tests are executed, the tester obtains the results, looks for the test's information (success, errors, failures, etc.), and checks if the testing met the coverage expectation. If coverage is met and all information needed is present, the test can be recorded as complete. If not, the test can be modified and repeated.

Boundary Value Analysis (BVA)

Figure 11-3 highlights the process steps for BVA testing. BVA checks around the data value boundaries where there is logic in the code to change how the code works at these locations. Since the code is more complex at boundaries, often there are errors at the locations in the code:

1. Input – In BVA, there are two variations of input associated with a range of input values: two-value and three-value.

> **Note**
> There are two classic approaches of two- and three-value conditions for boundary testing.

2. ID coverage – Each boundary is covered by two or three test sets depending on the variation, so the tester must pick which variation is to be used.
3. Derived test case (inputs) – Next, the tester must identify an input or value that will drive the test coverage. In the two-value, the tester identifies boundary conditions. Then, the tester picks a value of the boundary condition and then a second value just outside the boundary range. In the three-value, the tester identifies boundary conditions. Then, the tester picks a value at the boundary condition, a second value outside the boundary range, and a third value just inside the boundary value. Test cases may be combined into test suites and later procedures.
4. Run test – Once the tester has the input data, the tester can run testing against the software.
5. Report coverage and results – After the tests are executed, the tester obtains the results, looks for the test's information (success, errors, failures, etc.), and checks if the testing met the coverage expectation. If coverage is met and all information needed is present, the test can be recorded as complete. If not, the test can be modified and repeated.

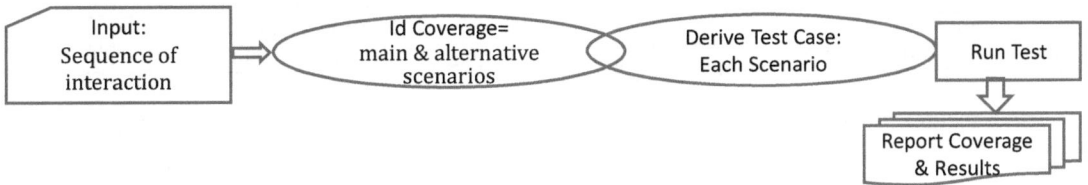

Figure 11-4. Scenario testing

Scenario (a.k.a. Use Case and Stories) Testing

Figure 11-4 highlights the process steps of an IoT scenario testing. Scenario testing tests the code against usages that are more complex than individual requirements and relate more to how a user will interact with the IoT software.

1. Input – The inputs to this testing are scenarios and usage, also known as use cases on some projects. They are often called stories in the agile world and used in acceptance test-driven development (ATDD). At the beginning of a project, testers like these inputs to be small (e.g., logon to the IoT device, check mail, logoff). However, they can take long usage scenarios with multiple cases, inputs, and outputs later. It is best to start small and build up [3, 5]. Also, the use case may employ a model with primary (normal) and off-normal usage (e.g., normal = logon is normal usage, while being hacked at login with a password cracking attack is off-normal). Test models can grow and become complex, so many projects will use a modeling tool within model-based testing (MBT).

> **Note**
> The sequences of possible interactions between the test item and users are of two types: main (nominal usages) and alternative (cases that are not typical, errors, threats, failures, etc.). IoT "users" can include humans, software, hardware, and other systems.
>
> IoT has more alternative action logic than nominal logic.

2. ID coverage – Identify each scenario, model element, and data with normal usage and off-normal to be covered.
3. Derive test case (inputs) – Inputs for each scenario are created. There likely will be many test data inputs to meet coverage in scenario testing, and the tester identifies the following:

   a. The main usage scenario of a typical set of actions.
   b. A set of alternate use cases that are different from the typical set of actions includes failure, attacks, stress, and extreme out-of-bound cases (see Chapter 13 of Part 3).

4. To drive the scenarios, the tester creates test cases to exercise each scenario case, identify test case data inputs, and determine the expected test results for each case. Testers will create more test cases until test plan coverage is met (e.g., testing each requirement).
5. Run test – Once the tester has the input data, they can run testing on the software.

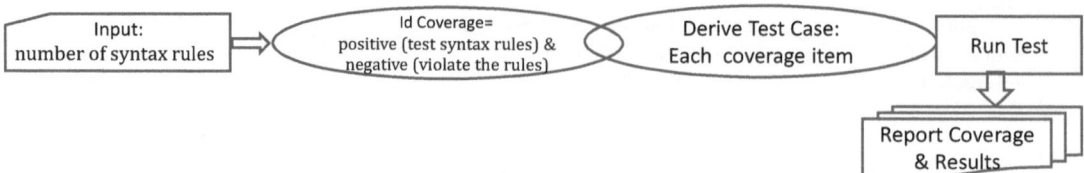

Figure 11-5. Syntax testing

6. Report coverage and results – After the test is executed, the tester obtains the results, looks for the test's information (success, errors, failures, etc.), and checks if the testing met the coverage expectation. If coverage is met and all information needed is present, the test can be recorded as complete. If not, the test can be modified and repeated. The coverage will likely take several passes of test execution, depending on the complexity of the scenarios or models.

Syntax of IoT Command Structure Testing

Figure 11-5 features the steps for testing the syntax of the command structures of an IoT device. Many IoT devices have logic associated with commands. Commands have syntax structures which should be tested:

1. Input – Many IoT devices will have a variety of input and output streams, which will have syntax structures including human input/output streams, external device inputs/outputs (inputs from other software systems), sensor inputs (data read from the sensor), actuator commands (send instructions and data to the actuator), and others. The tester must identify each input or output stream. Each stream will have a syntax command structure of "formatting rules" for the information which must be tested. This format information will be contained in development documentation or information from other teams. Some of these streams of commands in certain syntax areas will be human-readable, while others may be computer data (binary, HEX, ASCII, etc.).

> **Note**
> All sources of syntax documentation should be tested (e.g., user documents, online help, design documents) since each may contain the added or wrong syntax.

2. ID coverage – Each set of stream data and rules (commands input or output) will be covered by this testing.
3. Derived test case (inputs) – Test inputs for each syntax rule are created, and in the case of test, IoT software output syntax, the inputs to trigger these commands must be defined. There likely will be many test data inputs to meet coverage. The syntax definitions can be found in project documentation (e.g., user guide command string rules, diagrams, or a use case model). Inputs can be positive (e.g., following the rules of the syntax) or negative (e.g., violating the syntax rules).

> **Note**
> Negative cases must be limited since an infinite number of them are likely to exist. Also, options and iterations of the syntax rules must be limited for the same reason.

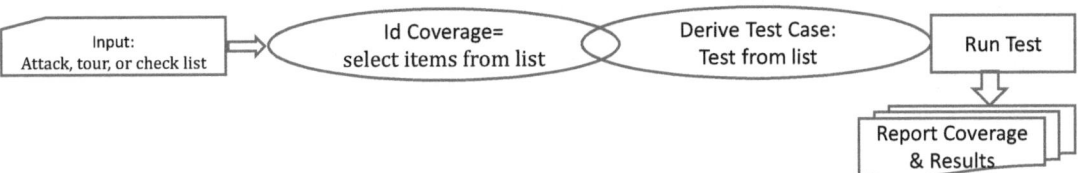

Figure 11-6. Experience-based exploratory testing with error guessing

4. Run test – Once we have the test input data, we can run tests with the software.
5. Report coverage and results – After the test is executed, the tester obtains the results, looks for the test's information (success, errors, failures, etc.), and checks if the testing met the coverage expectation. The coverage will likely take several passes of test execution, depending on the complexity of the syntax input/output data. If coverage is met and all information needed is present, the test can be recorded as complete.

Exploratory Testing

I believe that exploratory testing is something all testers do. In exploration, the tester is trying to understand how the IoT device works and maybe does not work under approaches like attack patterns. I highly recommend experience-based exploratory testing (Figure 11-6). So much so, I offer a process overview of this idea and then provide more information on the attack and tour testing (Part 3, Chapter 10).

Note
An exploratory list can be formal (such as in this book), project, product, or tester based. As the tester gains experience, the exploratory list will grow and mature.

Testers can use exploratory testing with attacks, tours, checklists, and patterns. I practice this technique when I first get onto an IoT project, get a new device that I am trying to "learn," or even at the end of the project when I have time to assess untested qualities. Guessing at errors is something experienced testers have in their brain. But, as I "explore" the device, I learn more by trying "unexpected" things, maturing my experience, and expanding my skills. .

Finishing testing early
I was on a test team that finished ahead of the planned schedule. The testers were ready to stop testing and go home early. Being the mean evil test manager, I said, "stop; remember those exploratory tests we did NOT have time for?" "Yes," the team said. "Well, we have time now to run them now." The team did more testing, and management was happy because testers were not sitting idle, and we continued to prove the system would work.

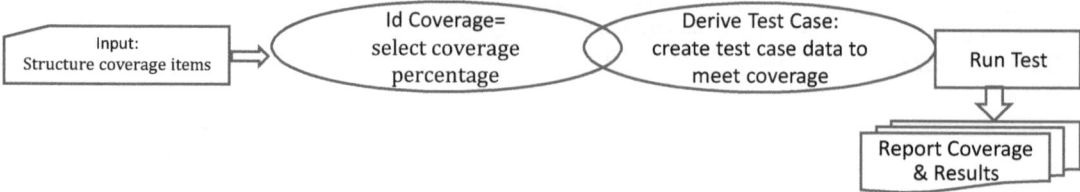

Figure 11-7. Structural (a.k.a. white box) testing

Structural Testing

For developers, functional testing is a secondary concept. Since they want their code to "work," they will focus on code testing (structural). The developer team more typically drives this test, but they can be supported by testers who understand the structural techniques and support tools (Figure 11-7). Since coverage measures are defined in step 2 of this testing technique, many project people want to believe that getting a high (100 percent) level of coverage means the testing is complete. Still, given the other techniques of this book or references, readers know how silly this belief can actually be.

Nonetheless, since it can be done in conjunction with the coding effort, structural testing is an excellent first filter for issues and information. In fact, in some IoT devices, it will likely be the case that there will be mandated levels of structural coverage. The industry sees these mandates with safety-critical systems, security, and other high-risk software:

1. Input – The inputs to this technique are the code. We can easily measure each code item in Figure 11-7 with static analysis tools. Once we have the code factors measured, we have the data for coverage of the factors.

> **Note**
> Testers can be developers, testers, and/or V&V staff. Coverages on: code statements, branches, decision logic, or others. Some IoT devices may have regulatory "legal" coverage requirements.

2. ID coverage – In this step, the tester defines the type of structural coverage measure that needs to be addressed. Measures are typically addressed as a percentage up to 100 percent (full and complete coverage). For example, "100 percent of statements" means that one or more tests exercise all logic code statements. The following is a list of standard structural coverage metrics [6]:

   - Statement
   - Branch
   - Decision
   - Branch condition
   - Branch condition combination
   - MC/DC
   - Data flow coverage

     - All definitions
     - All uses

> **Note**
> For an exact definition of these metrics and how to achieve them, see [6].

Table 11-1. Math-Based Testing Practices

General Technique Concept	Examples of Where the Math Technology Can Be Used	Specific Subtechnique Examples
Combinatorial testing	Medical, automotive, aerospace, information technology, avionics, controls, user interfaces	Pairwise, orthogonal arrays, three-way, and up to six-way pairing are now available
Design of experiments	Hardware, systems, and software testing where there are "unknown" factors to be evaluated	Taguchi [8]
Random testing	Chipmakers, manufacturing quality control of hardware production lines	Testing with randomly generated numbers includes fuzzing, use in model-based simulations, and randomly selecting hardware to be tested off of a production line
Statistical sampling	Most sciences, engineering experiments, hardware testing, and manufacturing	Numerous statistical methods are included with most statistical tools

Note 1: No tools are listed here. It should be recognized there are numerous tool vendors in these areas. Note 2: While math-based practice goes back 40 years, most software testers do not fully understand depth and usage.

3. Derived test case (inputs) – Inputs are created to test the software factors to meet the coverage. Table 11-1 identifies factors for coverage metrics.

> **Note**
> Some tools will report this coverage, but care must be taken since many of these tools do this by code instrumentation and automated execution reporting. A concern for IoT software is that adding code during instrumentation may have side effects such as slowing performance or impacting memory.
>
> For an exact definition of these factors, see [6].

4. Run test – Once we have the input data, testers can test the software. Tools are typically used to support the test execution and measurement (Part 4, Chapter 20).
5. Report coverage and results – After the test is executed, testers obtain the results, look for information the test provided (success, errors, failures, etc.), and check if the testing met the coverage expectation. If coverage is met and all information needed is present, the test can be recorded as complete. If not, the test can be modified and repeated.

After functional and structural testing, IoT teams need to consider additional testing practices. There is no one correct answer to which practices to use. Management may want the testing to be curtailed at this point, but IoT has a variety of product risk areas, which industrial test practices will address. The selection of these added techniques should be driven by test planning, risk, and information gathered during earlier testing.

Industrial Test Practices

Many industrial test practices are helpful for IoT. The following is a list of standard industry test practices:

- Model-based testing
- Scripted testing
- Exploratory testing
- Experience-based testing
- Manual testing
- A/B testing
- Back-to-back testing
- Math-based testing
- Fuzz testing
- Automated testing

    - Capture-replay
    - Data driven
    - Keyword driven

This section highlights practices that I use most often for IoT, though IoT teams can certainly use others of these practices. Summary overview information of these is

- Model based – Refer to the following section. This defines where models are used to define test designs and aid in test automation.
- Scripted – Used for formal testing and instruction to new testers. Scripts are typical human language narratives of steps to take in the testing activities. The scripts can be diagrams or computer code and not just human language. However, the scripts may be required in IoT human-readable form for traceability, repeatability, and (legal) documentation. The downside of scripts is the inability to respond to interesting thoughts during testing. Thoughts can include sensing errors, attractive risk reduction, and creativity on the fly. However, ad hoc, nonrepeatability testing may have the issue that when an error happens, the first thing a developer wants to see is "how did you create that issue?" To avoid this, using input loggers (capture tools) can avoid the "lack of repeatability" problem.
- Automated testing – Refer to Part 4, Chapter 20. Automated scripts are detailed step-by-step instructions for the test.
- Exploratory – Exploratory testing does not typically use scripts but allows the tester to take unexpected actions when a possibility arises. James Bach points out a continuum between free exploratory and tightly scripted following this exact set of test practices. Each practice has its benefit and cost. A well-rounded IoT test plan will likely include the whole continuum.
- Fuzz and security testing – Refer to Part 3, Chapter 13. Fuzz and security testing will be part of most (if not all) IoT test plans.

For other test practices not defined here, consider reviewing [1, 2, 6].

Math-Based Testing for IoT System and Software

An underutilized set of system and software testing techniques are based on mathematical concepts. I think software testers should know about and practice these techniques, as needed, on IoT projects. Table 11-1 provides a sampling of my favorite math-based test support techniques for IoT hardware, software, and systems.

To support the justification by increasing usage by projects, there is evidence from various environments where math-based testing techniques have been used successfully, including IoT software, hardware, electronics, manufacturing, and systems development. These practices help because math supports large numbers and computational processes of IoT. Software situations where math-based testing can be of assistance include large amounts of data, a variety of combinations, where sampling heuristics are needed, and places where the "variables" of testing are unknown. Kaner and others [1, 2, 3, 4, 5, 6] shed insight and education on these situations.

Model-Based Development and Testing/V&V – A Highly Automated and Integrated Test Practice

The system, software, test, and model engineering have grown in use and interest. Specific industry segments have active model-based engineering and testing efforts, such as in telecom, medical, IoT, automotive, and aerospace. Additionally, many European countries are actively participating in updates to modeling standards, and their support is ongoing (e.g., the Unified Modeling Language (UML), test profile (UTP)). While interest and use are growing, model-based testing is still immature in general IoT usage. Therefore, another test area to consider, for groups seeking to improve IoT project activities, is the use of model information to support testing.

Model-based testing can support

1. Generation of test cases from models into test automated execution engines directly using scripts or through the use of keywords
2. Improved understanding of the system and risks
3. Use of models to support simulations to drive test environments
4. Verification via a comparison between development and test models
5. Generation of test result oracles or judges
6. Support independent testing such as independent V&V (IV&V)
7. Model(s) analysis using formal methods

These features can improve the testing process and V&V lifecycle.

Many of these advantages have been used on a customized basis. Others are being developed with supporting tools and processes. The rigor in producing the model aids in the removal of errors. Then, the models can be used to generate test information, including data, execution scripts, and, in some cases, test oracles.

Figure 11-8 shows two modeling paths, one for the system-software tool developer usage. The other is the modeling by the tester. Each team generates separate models. An IoT system's independent test and development models avoid missing errors caused by using a single model. The developer model is transformed into code and used in the model analysis (a form of static testing). The tester model, created independently, can be compared to the developer's model and then transformed into test cases. The code and test cases are exercised in a test environment where static analysis and comparison can occur. These methods yield test results and two-way comparisons. The highly automated system checks requirements, verification, and validation results using the different models in an optimal test flow. Finally, this test and development information is stored in a project database.

The expense and efforts involved in model-based testing may not be justified in all IoT contexts. The concept and expense of having separate development and test models can be justified for high-risk test planning. However, test models have been shown to find errors and serve as an oracle. Model-based testing may offer project improvements for organizations dealing with critical software, complex software where risks are high, or software with long-term regression problems. It has been applied with specialized support tools in automotive, medical, and aerospace.

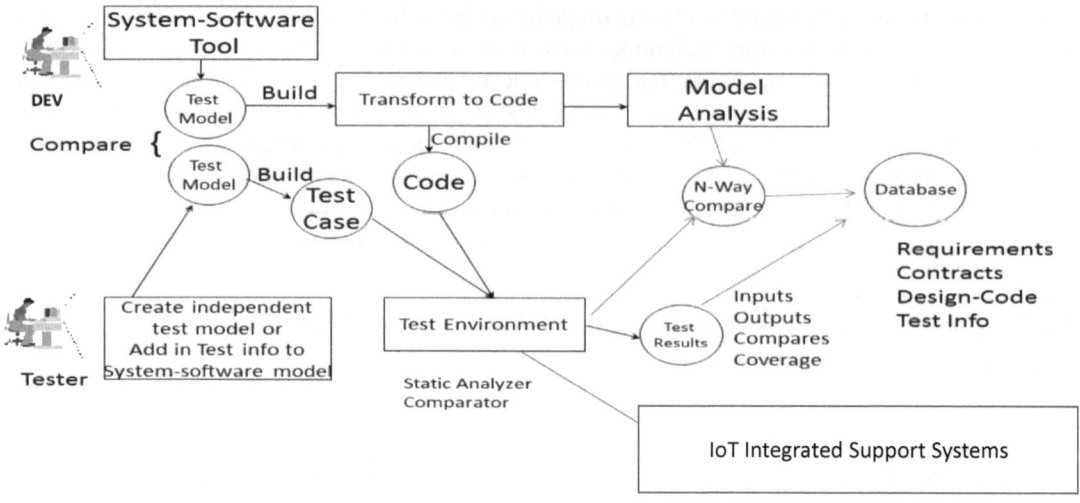

Figure 11-8. An example model-based testing (MBT) flow

Model-based IoT testing will likely grow as tester modeling skills improve and IoT projects use more model-based engineering. Model-based testing must be conducted in suitable environments (computer-aided development or labs with test automation tools), with skilled staff (people who know modeling languages and testing), and model-based tests. These support resources are becoming more widely available.

The levels of testing relate to the lifecycle of development phases. Depending on the development lifecycle used, these levels may be more or less applicable. Industrial test practices are chosen during test planning.

IoT Levels of Testing Related to Lifecycle Phases

Figure 11-9 shows the testing's common levels (lifecycle phases). The techniques and practices found earlier in this chapter can be applied to many of these levels, as determined by test planning.

Levels of Testing
- Unit Testing
- Integration Testing
- System Testing
- System Integration Testing
- Acceptance testing
    - User
    - Operational
    - Factory
    - Alpha
    - Beta
    - Verification
    - Validation

Static Testing (analysis)
- Reviews & inspections
- Static Analysis
- Model V&V

Figure 11-9. Common levels of testing for lifecycle phases

In the lifecycle levels of testing, I see test design and cases being started as soon as some code exists, even if that code is off-the-shelf (OTS). The levels I practice in my planning are

- Unit testing by the developer team
- Integration testing
- System testing
- System integration testing
- Some types of acceptance testing depending on the project nature and criticality factors (Part 1, Chapter 4) of the IoT

I think the idea of static testing/analysis is essential, particularly with OTS. Having a tool to do static code analysis should be viewed by the programmers and testers like the developer having a compiler. Teams use compilers without a second thought. Almost no one writes assembly or binary code by hand these days, although IoT is where these levels of coding will be. So, team members should run a static code tool also.

Many development programmers see this as a burden, because one must review and understand tool messages. Every tool produces messages that identify a problem to be fixed, but a percentage of messages are called "false positives." Development people tend to hate these messages but still have to review them. So, I suggest that the IoT team use a tester person to analyze and only work on actual issue messages with the individual programmer. I have done this with teams and found coding issues that would have been hard to find otherwise.

A CALL USED FOR DEBUGGING MAKES IT INTO PRODUCTION CODE
I used a static analysis tool to look at calls to subroutines and objects in the software. We presented this to the development staff as a call tree map. It showed one call going across objects. A programmer said, "that tool is wrong; there is no linkage between those objects." The tester said, "check your code." On checking the code, the programmer found she had left a link during debugging. This could have caused a significant problem in production, so the link was removed for code improvement.

The use of face-to-face peer code reviews and/or inspections is the first line of "error" finding defense. Some teams in the open source community use a form of peer code review. Teams I have worked with had face-to-face, line-by-line code inspections of critical software with the programmers, system, hardware, test, and stakeholder teams. Although time-consuming, a higher-quality code was produced.

Many engineers are nervous about showing their modeling work, but highly cooperative work teams (Agile) do not have a problem sharing. If the IoT team uses modeling at any level (software, hardware, system), model analysis with tools and/or inspection will find issues and inform the whole team of work design concepts. Teams and people that are "dysfunctional" do not want such group thinking. Such dysfunction is a management and tester warning sign to notice and resolve.

Finally, unit-level testing should be done by the IoT team. However, I have been on teams where schedule pressures led the team to cut unit testing. As the test leader, I volunteered my system-level testers to do unit- and integration-level testing for the overworked programmers. This made development happy by taking some of the load off them and letting the tester learn more about the software before testing it. Everyone on the team was happy with the outcomes.

Summary

This chapter addressed the classic detailed testing techniques that can be used for IoT, including functional testing and structural testing design techniques. I find in many cases that testers apply exploration testing to learn about the software before they become interested in adding the classic test techniques. Once they get some "test time," they start to want "more information," metrics, and insight into specific techniques and then become interested in these classic techniques. The next chapter addresses more IoT test approaches for agile and quality characteristic assessment.

References

1. *The Art of Software Testing* by Myers (the grandfather of them all)
2. *A Practitioner's Guide to Software Test Design* by Lee Copeland
3. *The Domain Testing Workbook* by Cem Kaner
4. *Lessons Learned in Software Testing* by Kaner, Bach, Pettichord
5. *Systematic Software Testing* by Craig and Jaskiel
6. "IEEE/ISO/IEC 29119, ISO/IEC/IEEE International Standard – Software and systems engineering – Software testing" – Parts 1 to 5 series
7. ISO/IEC 9126-1:2001 Software Engineering – Product Quality – Part 1: Quality Model
8. https://en.wikipedia.org/wiki/Taguchi_methods – accessed spring 2022

Chapter 12
Test Approaches and Quality Assessments for IoT Agile/DevOps

In the previous chapter, I reviewed and provided pointers for details on test concepts, techniques, practices, and levels applicable to IoT. These were offered with some tailoring to IoT. In this chapter, I explore necessary practices for agile/DevOps test planning.

Working Without a Formal IoT Test Plan (or Any Other Plans)

Many projects and companies expound on the benefits of Agile. Many agile ideas should be understood and practiced even when project management has more "traditional" leanings. As has been observed, Agile does not say "no" testing [1, 2], but many projects, even "traditional" ones, have minimal or no formal (separate) test plans. It is optimal if test plans have a strategy and are in written form. The reason for this is that writing ideas down helps testers solidify their thinking and mostly sets project thinking on the same page.

Do you need an outline's large page count number, such as what ISO 29119 part 3 defines? For me, the answer is "it depends." A well-documented test plan is essential for mission-critical IoT systems where there can be loss of money (direct or indirect), life, possible hazards, and others. If working on IoT devices that will be subject to regulatory investigation (e.g., the medical sector), having significant levels of documentation may make sense to address legal concerns. However, if you are working on an IoT device to monitor a pet's location and status, maybe all you need is one or two pages of plans and strategy such as a mind map or an outline on a whiteboard that can be snapped using a phone picture. Thus, the range of test plans can be from very small/informal to large/formal.

Test plans help your test thinking and make communication with management, the team, and subcontractors less complicated. In just one or two pages, test teams can generate (1) a "master test plan," which does not have to be a large document, and (2) create detailed product test plans for each IoT release. Test plans should cover end-to-end testing approaches, schedules, strategies, organizations, risks, and the division of test efforts between levels and subcontractors, environments, and test design concepts. Agile test plans can include

- Essential things that do not often change, such as stakeholders and products.
- Having a short "planning game" on a whiteboard or mind maps for a product plan on a sprint or short test cycle.

© Jon Duncan Hagar 2022
J. D. Hagar, *IoT System Testing*, https://doi.org/10.1007/978-1-4842-8276-2_12

- Having a test library (online) of historical plans.
- Having test plans for any outsourced service providers who should be doing testing. If the service providers don't have any test plans, this is a problem by itself that deserves careful consideration before moving forward.

> **Note**
> I always create a test plan, even if they are just for my use to clarify thinking. On one agile project, I had a master test plan that I created for myself, but the project didn't require it when I first got there. They said, "we just run tests." However, later in the testing cycle, when things got tough, upper management watched everything and asked questions, such as "where is the test plan?" Well, I had one to show.

So, what does a short agile test plan look like? How long does it need to be? In Figure 12-1, I show an agile test plan in a mind map format. It is just a start, but hopefully it starts you thinking about the testing that must be done.

Figure 12-1 as a one-page test plan is for a new startup company for a small IoT device (heater control). The plan addresses the product, test team, schedule, test design, stakeholders, strategy, STA, and risks on the map. Other teams might want to see additional items, but this gets the thinking started. So, a critical step in planning can be seen in the six risk-based statements of the plan. The risk-based planning has started but is probably nowhere near complete. The map illustrates product-to-support testing as identified, including stories (agile requirements), the software, and the fact that there are off-the-shelf (OTS) software items to be used. OTS software promoted the test planning to identify security testing as an activity (Part 3, Chapter 13), which they seem to believe negates this item being on the risk list. The high-level test designs defined so far in the planning include software development (Dev) team test-driven development (TDD); functional tests of the 500 stories concepts (SC); security testing; acceptance test-driven development (ATDD) with several stages based on the products in the list of microservice verifications; low-code V&V audit; vendor testing (vendor X); and Dev team sprint test. Off-the-shelf (OTS) testing, a low-code V&V audit, and verification checks of microservices are of interest in this product list. While many teams might ignore such checks, this team believes untested software is a risk that can be mitigated. While this is an early test plan, it is done on one page, supports Agile, and can be taken to the stakeholders for review and buy-in.

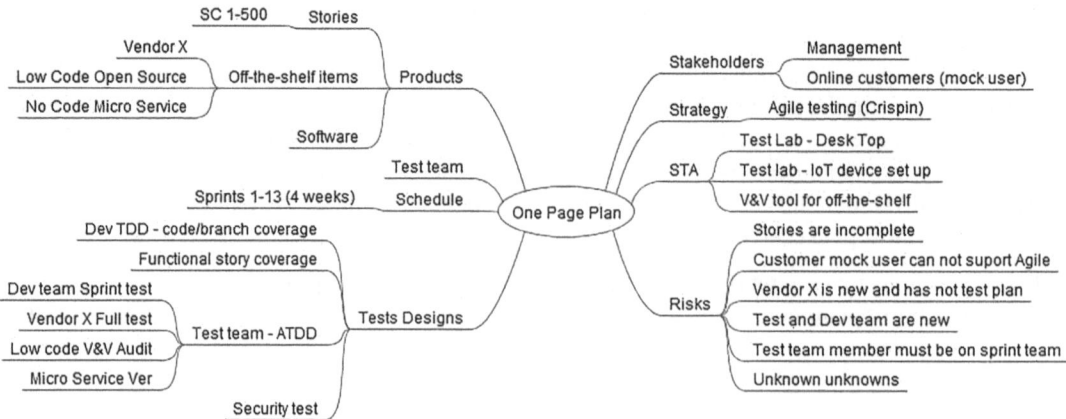

Figure 12-1. Example one-page mind map test plan for agile testers

Test Planning for the Agile Teams

A first step for many in the agile world is test-driven development (TDD). TDD has the goal of "clean code that works," according to Ron Jeffries. In test-driven development [1], Kent Beck advocates developers should be driven by creating tests first and then the code. He says, "create a little test, then a little code for that test and then repeat." This can go a long way to having much "cleaner" code that works as defined by the developer. Is the total project done then? No, but it is a good start for developers to engage with testing.

> **AN AGILE PROJECT IN NAME ONLY**
>
> I consulted with a development and test team doing Agile, or so they claimed. I came in and asked, "well, are you doing TDD?" Developers said, "No." Are you doing agile test planning? The team said, "No." "Well," I said, "what makes you Agile?" The team said, "We just don't document anything." They missed the point of Agile and were using the name for ad hoc software hacking. I recommended the developers read Kent Beck and the testers read "Lisa Crispin." When I came back months later, the code and testing started to take shape, and the team was starting to succeed by following some guidelines and based on experience found in the references provided in [2].

In Crispin and Gregory's *Agile Testing* book [2], they start with guidelines on how an organization must change with organizational challenges. The organization had issues with the "name only" story and did not truly understand agile methods. Their book defines technology and business-facing tests with supporting toolkits. Their testing aims to provide information and a critique of the product. They advocate test automation while recognizing the roadblock to full automation and strategies that enable automation success. This automation approach does not mean that every test had to be automated. Tests that were to be run only once or test a seldomly used function were manually tested. However, tests that were to be repeated in regression testing were automated.

Work to produce small amounts of working software in short time periods (weeks and not more than a month). Another key for [2] and most agile advocates is the idea of rapid small iterations in the form of sprints. Teams should work on what is most important to the user-customers first. Many teams never deliver software that works since the temptation is to work on what is easy (but not necessary) first, so the formal plan-schedule progress looks "good." However, this approach leaves the hard important (to the stakeholder) code to last when often teams run out of time and resources. This is why, historically, many software projects deliver no working software.

While working with the stakeholders in the sprint idea, the teams work on the essential behaviors or features first. This allows the agile team to continue on the next important behaviors/features. By taking small steps, working on the important code with frequent deliveries and ongoing stakeholder feedback, the agile team can get something that is working though maybe not be complete.

AGILE PROJECT DELIVERS CODE TO FIELD USAGE WHERE NOT EVERYTHING IS WORKING

On one project where I was the test manager, the development staff created code that only worked for one customer. There were many customers and usages. We tested the function and software for that one customer to ensure the system worked for them and only them. The testers made sure that other parts of the software and functions could not be accessed since we knew the hardware and software for those parts were not coded. This was risky, but it allowed the first users to be happy and management thought the first cycle was a success, so we got funding to complete the other parts of the hardware and software. Later, we delivered the other parts to different stakeholders. The first customer stakeholder knew we were doing this, but they just wanted their system to work. Everyone was happy.

Crispin and Gregory's *Agile Testing* book [2] lists numerous success factors, which I like:

1. Use the whole team approach (I add hardware, systems, software, support, management, stakeholders, and testers).
2. Adopt an agile test mindset (use only enough test planning, make changes rapidly, listen to the team and stakeholders, and make changes when needed).
3. Automate regression tests (which are repeated many times).
4. Provide and obtain frequent (daily/weekly) feedback (from development, testers, support, management stakeholders, etc.).
5. Build a foundation of core practices (Parts 2 and 3).
6. Have many agile development cycles, as shown in Figure 12-2.
7. Agile system testing is an important test activity after the cycles of Figure 12-2.

Software teams now have a long successful history using Agile. Figure 12-2 cycles for each iteration are fast (weeks). A plan is done, followed by software design, development coding (TDD), and testing acceptance test-driven development (ATD). These development efforts are followed by deployment (delivery to a stakeholder or even customer-users for trial use) and a team review (a.k.a. a team retrospective [3]). There is no one way to do agile testing, though [2] gives a lot of good starting points. Agile has now been used on small and large projects with successes.

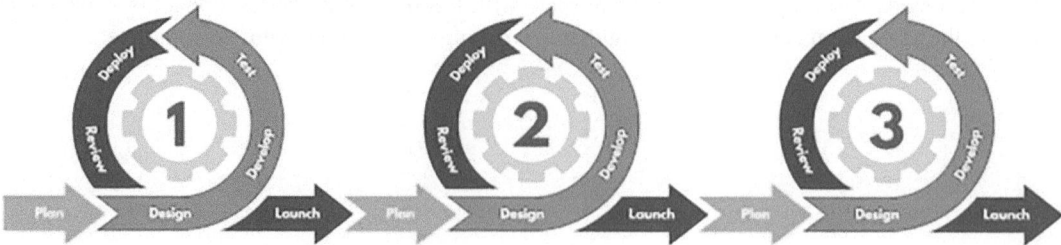

Figure 12-2. Agile development cycles

Agile Grows and Evolves (and Will Continue to Change)

So Agile has been around for a long time. Is it passe? Well, I was an early practitioner of Agile. It was never a single or set of practices for the project I worked on. We used what worked and changed what needed to work for IoT. Hardware could be agile. Behavior-driven development (BDD) and testing drove our teams and testing. We wanted the device to work. Did we have requirements? Sure, our stakeholders wanted them. Did we have test plans? Yes, management and regulators expected them. This was our context. However, we changed when needed. For example, we were able to change requirements right up until product delivery, which is very agile.

Crispin and Gregory's second book [4] related similar experiences on small and large projects. They list

1. Learning better testing (skilled tester)
2. Planning for the big picture (sound familiar)
3. Testing business value (how I see BDD)
4. Investigative testing (I practice exploratory testing. See Chapter 11)
5. Test automation (software test automation [5])
6. Understand your context

A popular idea for IoT these days is the idea of development-operations (DevOps). For me, DevOps is just a natural progress of Agile. Here, development, which includes testing and security, must consider how the system operates. The operations, with stakeholder's activities, must be considered during development. And during operations, the Ops team must remember more development is likely during maintenance. It is a unified infinite continuum, as shown in Figure 12-3. Ideas of this figure look pretty agile to me and match the infinite diagram I have created for testing (see Part 3, Chapter 13). In the figure, planning, continuous testing (CT) is performed with continuous integration (CI). The integrated product feeds to continuous delivery (CD) of the product first to stakeholders, users, and customers during IoT device deployment. The device's operations, deploy, and integration are shown in the loopback in continuous feedback for the next planning, doing, and testing loopback.

In the IoT world, Agile and DevOps are common. Agile IoT teams are more likely to succeed when working on hardware, software, and systems in small integrated cycles. If the IoT teams do not

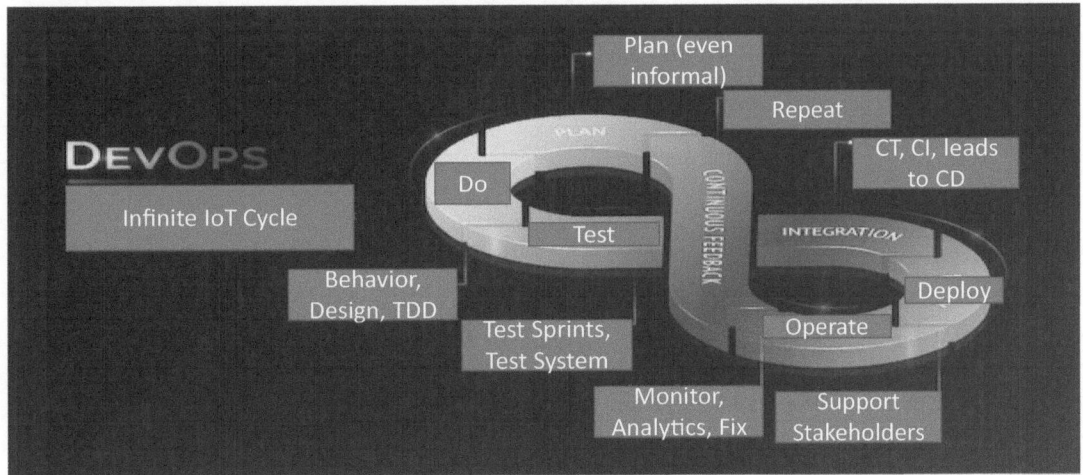

Figure 12-3. DevOps IoT cycle – CI, CT, CD, for continuous development

Reference: Jon Hagar Class

consider DevOps and testing analysis during operations, a significant source of data analytics (Part 4, Chapter 22) will be lost.

As I have noted, traditional ideas such as standards, documentation, and V&V/IV&V will place in IoT projects, even while teams still focus on Agile. In all cases, test context thinking will be necessary. Agile IoT teams must think agile software context, but also hardware and security.

Agile Hardware

Agile is typically thought of as used in software development, but it has also been used in hardware development, as noted in the following list. The ideals of software are applied to hardware. In the case of hardware, a developer may apply the following:

- Rapid prototyping – `https://en.wikipedia.org/wiki/Rapid_prototyping` – accessed spring 2022
- Set-based design – `www.sebokwiki.org/wiki/Set-Based_Design` – accessed spring 2022
- Rapid FPGA prototype creation – `https://en.wikipedia.org/wiki/FPGA_prototyping` – accessed spring 2022
- Fast hardware production – `www.manufacturinghub.io/prototyping/7-rapid-prototyping-methods-you-need-for-your-hardware-startup/` – accessed spring 2022

Again, this is not a book on hardware or system engineering, but software testers should be aware of these preceding concepts since they may use some of these hardware elements to support the software testing.

Agile Security Testing

As mentioned already, I believe software security testing is essential for IoT. In the book, Chapters 13 through 15 are dedicated to security testing. On an agile team, testers may be deployed to a development team and have their own test sprint team or a combination. In any case, I recommend the agile team be well versed in security testing. Doing TDD and ATD for IoT is necessary but not sufficient unless security is included.

Agile Quality V&V Assessment

There are many qualities that agile IoT test and/or V&V teams may want to consider assessing. I list these as V&V, but they can be done within any agile or traditional project. These assessment levels will depend on IoT factors (Part 1, Chapter 4). Some agile/DevOps projects wonder what quality characteristics to assess. Figure 12-4 presents a start based on ISO 9126 [6].

The boxes found on the right side of Figure 12-4 are my additions for IoT/DevOps projects. They include IoT device recovery testing (V&V), localization testing (will the device work for somebody in a different country from where it is made), and interoperability V&V (does the device work and interface with other IoT devices, the edge, fog, and cloud). Teams should include disaster recovery, in case something catastrophic happens to the IoT system. Interoperability of the IoT software system

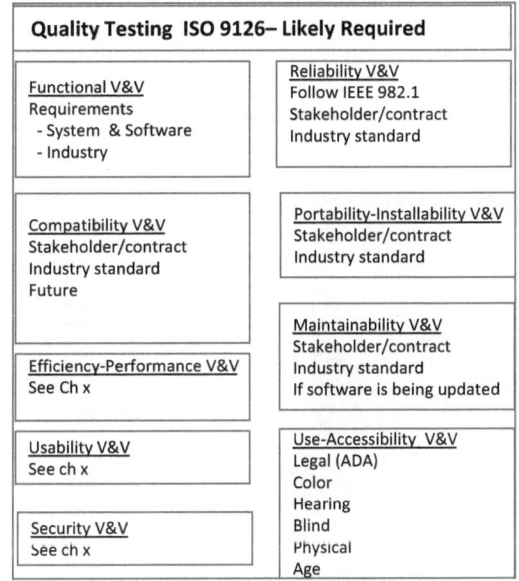

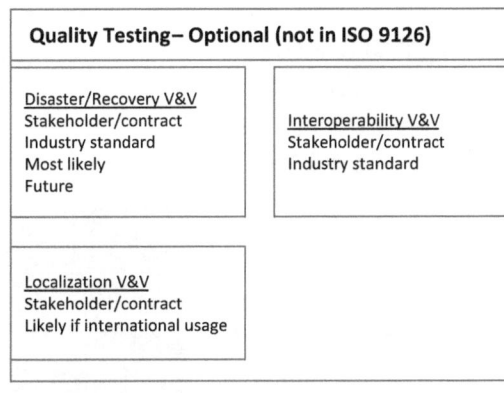

Figure 12-4. Quality traits to assess

Reference: Jon Hagar Class

when used with other systems upon which the IoT device is dependent should also be considered. Finally, a localization test should be planned if the device is used worldwide with other languages.

The items on the left side of Figure 12-4 should be considered for assessment by most IoT projects. These qualities include

Functional testing.

Check for compatibility within the IoT system.

Performance testing, which is part of efficiency.

Usability testing.

Security testing.

Reliability following ISO 982.1.

Check that the software installation is working or can be maintained with software updates, when needed, as part of assessing portability.

If the IoT device meets legal guidelines or laws (see Americans with Disabilities Act (ADA) in the United States) and contractual accessibility.

The degrees of assessing and testing these qualities will vary based on the nature and maturity of the IoT project. These items are refined in the ISO 9126 quality model standard.

While the left side should be considered in planning any project, the items on the right side are often optional, depending on the IoT device and project. These right-side items are not directly drawn from ISO 9126, but relate to many of the items in that standard. Besides the items in Figure 12-4, IoT projects may want to review standards such as the ISO 9000 series for more qualities to assess.

Summary

There is much more to Agile and DevOps, so teams should follow up with and then provide a reference when practicing the ideals and methods of this chapter. Figure 12-5 provides a graphical representation of these aspects and interfaces into the agile/DevOps cycle.

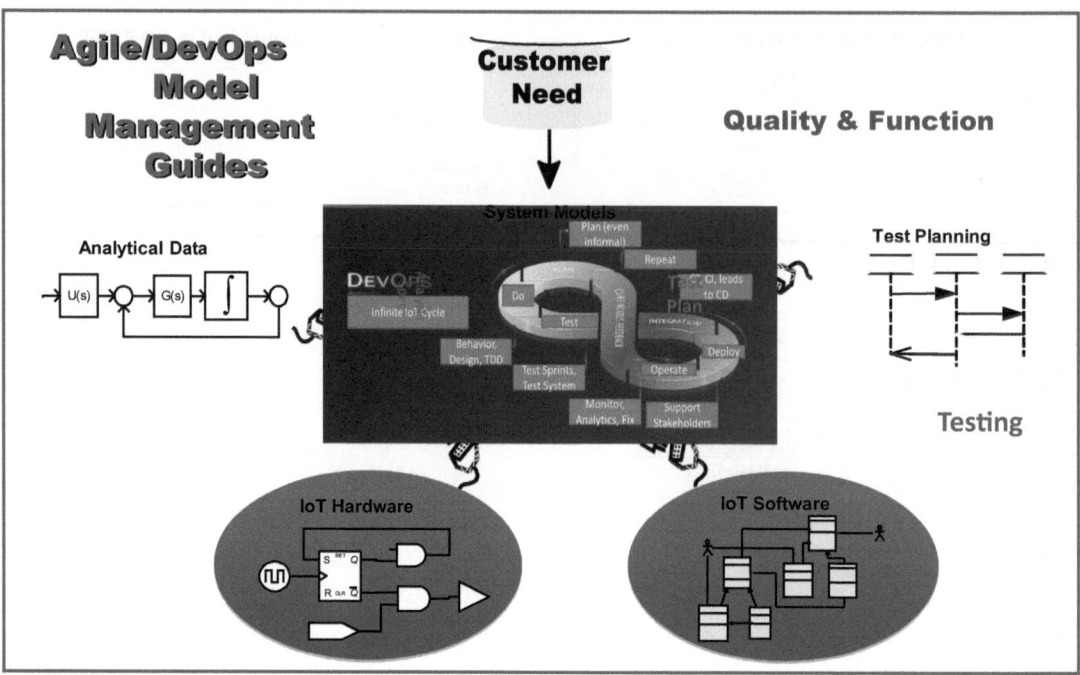

Figure 12-5. Agile/DevOps brings together all the pieces for IoT success

Reference: Jon Hagar Classes

In this chapter, I pointed out that even for agile projects, testers must remember the hardcopy version of a test plan document is not as important as the planning for testing success. Plans can be short (e.g., one page or a mind map). Teams should make sure that their test plans address functional and nonfunctional quality assessments with testing and/or V&V. In the next chapter, I will move on to IoT software security test attacks and designs, which, like the items in this chapter, can be overlooked by IoT teams in test planning and design.

References

1. *Test-Driven Development* by Beck
2. *Agile Testing: A Practical Guide for Testers and Agile Teams* by Lisa Crispin and Janet Gregory, Addison-Wesley Professional, 2009
3. *Agile Retrospectives* by Esther Derby, Diana Larsen, Ken Schwaber, July 2006, O'Reilly Media, Inc.
4. *More Agile Testing* by Crispin and Gregory, Addison-Wesley Professional, 2015
5. *Experiences of Test Automation: Case Studies of Software Test Automation* by Fewster and Graham, Addison-Wesley, 2012
6. ISO/IEC 9126-1:2001, Software Engineering – Product Quality – Part 1: Quality Model

Figure References

1. www.freepik.com/premium-vector/agile-development-methodology-software-developments-sprint-develop-process-management-scrum-sprints-illustration_8636934
2. www.freepik.com/free-vector/landing-page-continuous-devops_9292918

Chapter 13
IoT Software Security Test Attacks and Designs

Security quality will be paramount to many IoT device users. Since IoT devices will be everywhere, smart and connected, they have become the target for malicious users (i.e., hackers). There is no single set of answers or attacks to ensure adequate security quality. This chapter details the techniques on how a team should assess IoT device resilience vs. security threats. The security techniques are defined as test attacks and designs that expand on the information found in Chapters 11 and 12. The world of security testing is evolving rapidly, so the information presented in this chapter should only be viewed as a starting point. The amount of security testing needed on the project will be functional of the planning, risk factors, and legal requirement areas to consider including access protection, encryption features, hardware and software controls, and how security monitoring is done. This chapter starts with a warning and continues with security attack surfaces, planning, and vulnerabilities to get teams started. Next, the chapter considers the security kill chain, key activities, support tooling, and finally some historic security test books. Skilled engineering and critical thinking are strongly suggested and are desperately needed in the IoT area.

Before We Get Started – Security Data Points and Warnings

Black Hat vs. White Hat Hacker Warning
In this chapter, I advocate that the testers play hackers (malicious users), but you wear a white hat (a good guy) for your IoT system. This is because the stories happen every day where hackers who wear black hats try to hack into our systems. Good security defense starts with white hat offense attacks, where testers try to hack into the IoT system. Making attacks a security test activity is good planning when performed legally. However, I know of a story where a tester tried to hack a system without the proper permissions. **This is illegal. DO NOT DO IT.** The tester who did the unauthorized hacking lost their computer and dealt with the police.

1. Security testing of IoT systems is driven by the following kinds of statistics from a 2020 and 2021 report from Forrester Research and `www.globenewswire.com/en/news-release/2021/04/08/2206579/0/en/Global-IoT-Market-to-be-Worth-USD-1-463-19-Billion-by-2027-at-24-9-CAGR-Demand-for-Real-time-Insights-to-Spur-Growth-says-Fortune-Business-Insights.html`:

- 69 percent of enterprises have more IoT devices on their networks than larger computers.
- 84 percent of security professionals believe IoT devices are more vulnerable than traditional computers.
- 67 percent of enterprises have experienced an IoT security incident.
- 96 percent of companies continue to experience challenges protecting data from insider risks.
- 93 percent of tested networks are vulnerable to breach.
- 16 percent of enterprise security managers say they have *adequate visibility* of the IoT devices in their environments.
- 93 percent of enterprises are planning to increase their spending on security for IoT and unmanaged devices.

> Security of the IoT concerns everyone on the planet. Any IoT device (networked or not) can be hacked. We have all seen news stories about IoT devices, such as baby monitors and security cameras, being hacked. How much of an issue do these stories reflect? Well, it depends on what is being threatened: your child, the privacy of your home, your health, your finances, etc.

Definitions Used in This Chapter

- Attack surfaces – The points of failure in the cyber environment where a malicious user (the "hacker") can attempt to gain unauthorized access to the system [1].
- Assessments and inspections – Review of any software work products by trained individuals who look for risks and defects using a defined process in ISO 20246 [2].
- Breaking software security – Using software test attacks to hack into the software system; see books by Whittaker, Thompson, and Hagar [1, 3].
- Cybersecurity testing – Assessment focuses on assessing the security risks, vulnerabilities, and faults in a software computer system [1].
- Social engineering – The use of deception related to computer systems to manipulate users (human and nonhuman) into divulging security information, which, in turn, can be used in hacking.

Using Attack Surfaces in Security Test Planning

Having a large attack surface for an IoT system means many avenues from which malicious users can attack. Figure 13-1 lists starting points for these surfaces at a high level, which could be used as a beginning checklist. The surface points from which the malicious users attack could begin, starting at code levels, moving into issues such as "zero days" and malware code [1, 4]. Commence security testing here. The next level to attacking is the inability of the IoT software or the operations to detect a security threat or attack. After assessing the overall system's resilience and dependability characteristics [5], this level is followed by assessing hacker threats. Finally, the IoT system's universal security to all other systems and society's ability to evade hackers, attacks, and cyber threats are at the top, which is outside the direct scope of this book.

To address the attack surfaces, testers start with security test planning, including

- How data breaches happen – Identify the sensitive and at-risk information in the IoT system, including user, source code, sensor/actuator data, history, threats, etc. (see Part 3, Chapter 14).
- Identify the IoT attack surface – Start with Figure 13-1, because the bad guys will look to leverage every technology and then add local IoT attack surfaces during your security tests.

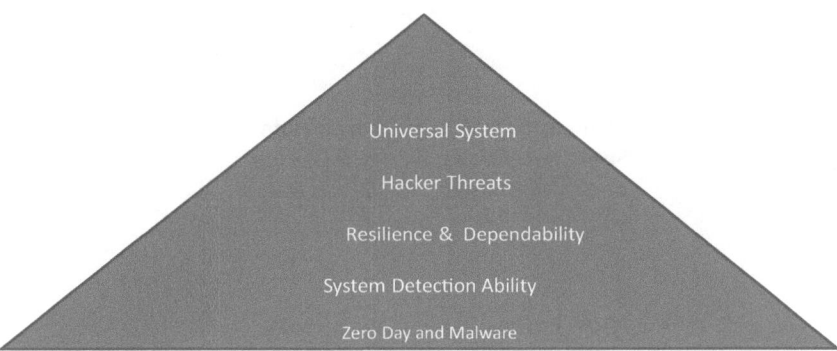

Figure 13-1. Examples of IoT attack surfaces

- Assess – What are the actual risks to the attack surfaces identified for the IoT system? Rate risks from low to high. Conduct assessment of risks by doing the penetration testing of Part 3, Chapter 14.
- Security test and mitigate (working with development) security risks – Conduct the activities that will be discussed in this chapter to address risks identified for security testing, for example, to ensure medium and high security risks are tested.

Cyber threats are ever-increasing, and cyber criminals find new ways to exploit systems almost daily. As more usage scenarios for IoT increase, so will the attack surfaces for bad guys and security testers, who will go after the weakest attack surface link. You and your IoT system do *not* want to be the weakest link. One only needs to watch almost any news feed or website to see the explosion of threats and what those threats entail. These news stories will give you items to add to the model, as it expands and the threats grow.

> **FTC warns companies to remediate Log4j security vulnerability**
> Log4j is an example of how wrong malicious code can go. This malicious code enables the hacker who finds a vulnerability to do bad things. It can perform arbitrary code execution (ACE), reverse-engineer the product, falsify warranty claims, enable locked features, compromise a user's private data, destroy information, and even use the product for criminal activity. The article at the link is provided as a warning tale of why testers and managers ***must*** take security testing and threat surface testing very seriously.
> Reference: `www.ftc.gov/news-events/blogs/techftc/2022/01/ftc-warns-companies-remediate-log4j-security-vulnerability` – accessed winter 2022

Cybersecurity and penetration (Pen) testing are complex and resource consuming. There are multiple internal and external organizational activities over the continuous IoT production lifecycle. The ethical hacker-tester needs access to tools, vital analytic/test skills, and perseverance. Not every person is suited to this type of security engineering and testing. Further, differing security regulations and legal considerations affect security concerns, depending on what kind of IoT device you are creating. It may be more cost-effective for many companies and teams to have proper security testing vs. getting sued.

Based on NIST 800, in Figure 13-2, I present the top level of cyclic planning activities in ethical IoT hacking.

Begin with security planning, including considering the security threat risks and proper test design. This leads naturally to risk-based testing with Dev support, high-level attack test design, and Ops

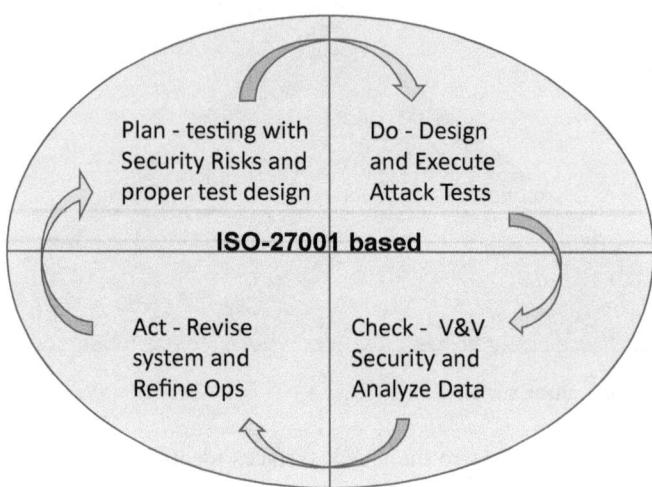

Figure 13-2. Ethical top-level plan-do-check-act cycle

support for testers. The next step is a detailed attack test design and execution, starting as soon as code is available and continuing throughout operations since threats continue to evolve over the entire IoT device lifecycle. Management may not want to continue security testing, but keep in mind that *the bad guys don't stop*. The security vulnerabilities evolve for both hardware and software, and projects must constantly assess the risks.

Data analytics and AI (see Part 4, Chapter 22) for many IoT systems will constantly look for security issues. V&V results and analysis check the incoming IoT data, whether a development team does the V&V, a separate project security test group, or an outside organization. New threats and issues will emerge, which requires the last step in the cycle: revision of the IoT system and refinement of Ops. This leads back to new plans for testing and risks. As Figure 13-2 illustrates, this will be an infinite loop.

In planning, the ethical hacker-tester needs to understand the IoT system environment and landscape by

- Researching the IoT device and organization during planning
- Attacking the IoT user processes and software systems
- Checking the social engineering threats of the development team and users

These are the first substeps you will need to perform successfully security planning work. The essential social engineering plans in security testing should include

- Research the organization to help find the technologies used in a test attack (e.g., the software, OS, no-code/low-code, hardware, networks, edge/fog/cloud, Ops procedures, etc.). This information is used in specific test designs.
- Check external resources, including Google, social media, CAICS[note 1], and other open information sources on the Internet. The hacker-tester gathers this information to plan test attacks. It may even be worth conducting a trip or a social media visit to meet members of the Ops organization and user base.
- Note 1: Cybersecurity and Information Systems Information Analysis Center (CSIAC) is an element of the US Department of Defense's (DoD's) Information Analysis Center (IAC) enterprise. It is a good place to start for cybersecurity information.

– Investigate those organizations and users who have processes and security checks in place. In gathering this information, the hacker-tester wants to identify these processes because overconfidence is a vulnerability that can be exploited.

> **Social engineering in a banking device** An external V&V hacker company was hired to do security testing on a bank's new IoT system. In the kick-off meeting, the bank executive claimed to the V&V team, "You can't get in. I did a phone interview with the developers who told me how good our system is." The V&V company found the executive's phone number, spoofed it, called the developers saying they were the executive who needed access, got the executive's password to the system reset, and left a "you've been hacked message" for the executive. She changed her tune. It took the V&V hacking test company less than a week to crack the bank's system.

In attack test planning, the hacker-tester gathers information from everywhere. They should map out what is known about the company development and operation teams, particularly any "observations" that seem strange, illogical, or out of place.

Questions that should be asked include

- On the software side, what kind of authentications are used?
- Are there defaults of passwords in use (use Shodan)?
- Is the data encrypted?
- Are incoming and outgoing commands to the hardware secure?
- Where are the weak, vulnerable points of the system's input and output?
- Who and what are the users of the IoT device?
- What happens if the IoT device is hacked, and how bad can the damage get?
- What are the boundaries of any hack?

Knowledge is power. In planning, hacker-testers do research to gain knowledge as a goal. I like to create mind maps (see Part 3, Chapter 10) or even use yellow (hack) stickies on the wall that I can play with, move around, or even add to my security risk analysis.

Planning Security Tests: IoT System Level Inspection and Assessment

Figure 13-3 presents a full flow for security testing, which I step through in this chapter. The process is never-ending and repeated over the IoT production lifecycle. In the middle of Figure 13-3, let's begin with a new threat risk detected before doing attack planning, preparing security attacks, conducting the attacks, and analyzing the test results. These activities and supporting sub-activities are examined in this chapter. Still, before tackling them, a security test team must conduct a series of assessments and inspections on the IoT system(s) (their activities are given on the left side of Figure 13-3). (And, will those efforts be in the current budget?) The outcome of the steps is to find an IoT vulnerability, in which case new software will likely need to be designed, coded, tested, and produced.

If the quality of a secure system is met, the current software can be released, yet threat monitoring continues. These steps start with inspecting the system and gathering intelligence about the system. The results of the assessment are to identify new and old threats. The threats indicate risks to explore, leading to actual IoT system vulnerabilities. This inspection and assessment should be a constant security test team activity. We learn as we grow.

Key Activity: Risk and Vulnerability Analysis

Intrinsic to most test concepts is the idea of risk-based testing driven by risk analysis. This remains true for security risk analysis for IoT testing. Figure 13-4 presents a tailored risk analysis model derived from NIST 800-30.

The model starts by identifying threat sources and characteristics. Sources might be hackers, naive users, unhappy internal employees, or outside organized attack teams. These sources create target characters such as IoT unsecure capabilities, commands, communication channels, and bad intent. These sources lead to initiating testing events such as an attack, spoofing, or communication hijack, covered later in the chapter. The test team must assess the likelihood that an exploit will be successful and its resulting impact. Exploits of low likelihood or minimal IoT system impact may be disregarded or considered for later testing. Such security risks may be subjected to later risk reexamination. Next,

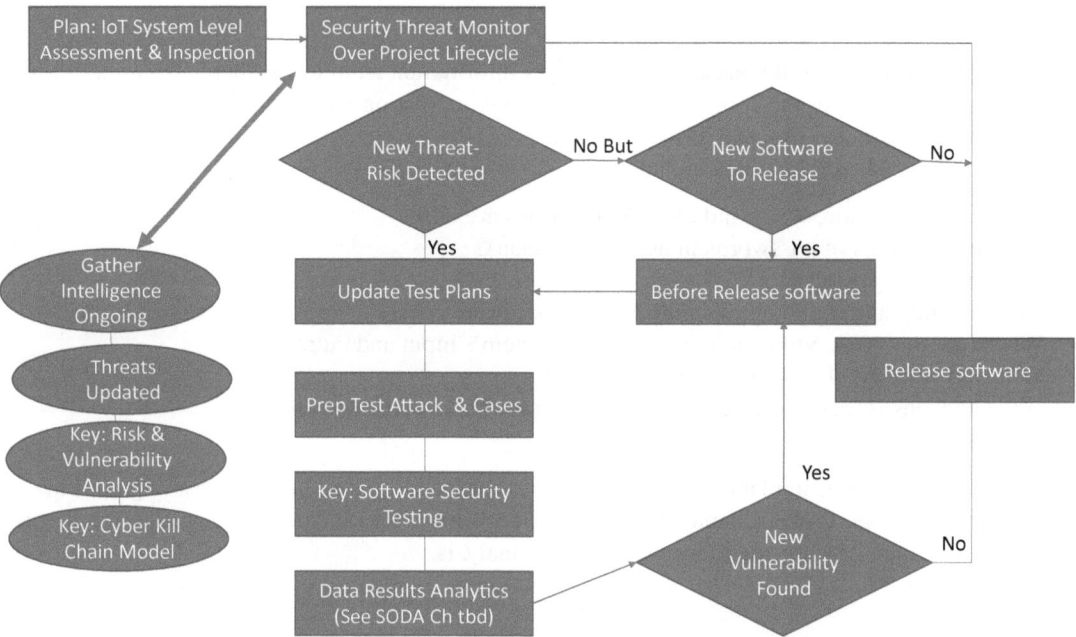

Figure 13-3. Full security testing lifecycle starts with assessment and inspection

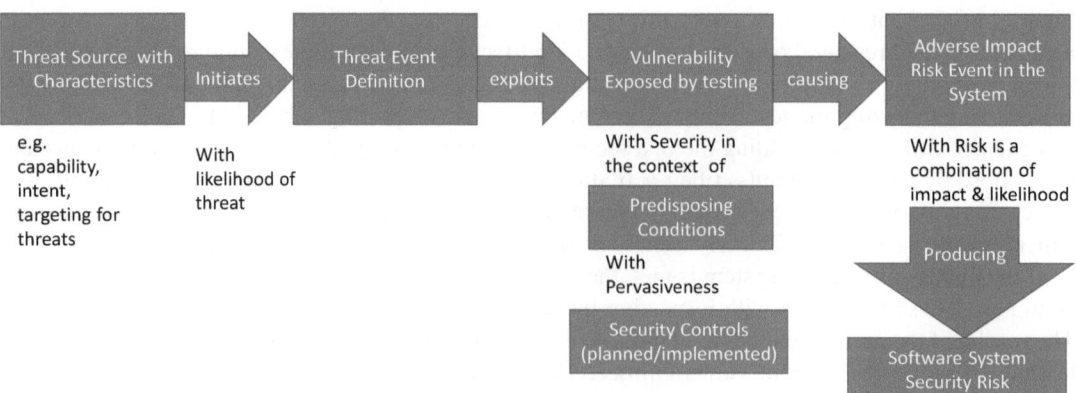

Figure 13-4. Security risk model derived from NIST 800-30

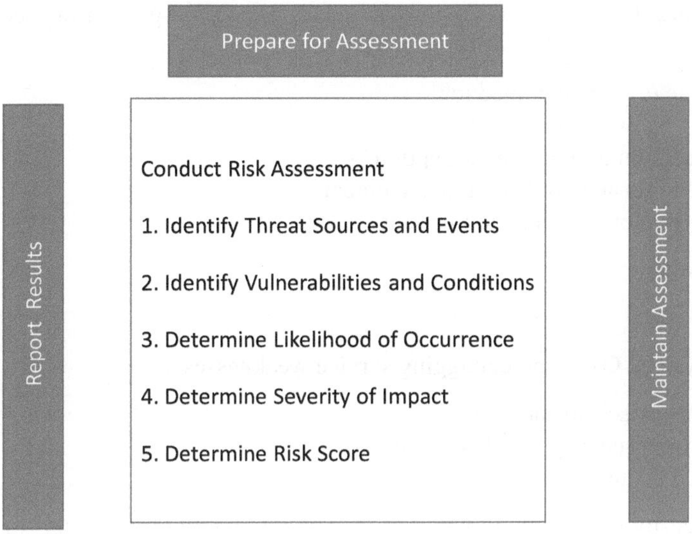

Figure 13-5. Risk assessment process steps

the IoT system vulnerabilities should be identified relative to the threat event severity. Higher threat and severity risks will likely need to be tested. Vulnerabilities are considered concerning for existing IoT conditions and functions while looking at the pervasiveness and effectiveness of the IoT system security controls (e.g., user controls, encryption, and qualities). Finally, the risk analysis model concludes with a combination of impact and likelihood producing risks at an organizational level.

The risk model of Figure 13-4 is driven by a risk assessment process, as shown in Figure 13-5.

Step 1, "Prepare for Assessment," involves gathering information to prepare for the assessment. In step 2, "Conduct Risk Assessment," the team conducts the assessment over meetings or interviews. This step identifies the threat source, threat events, actors, targets, etc. The next substep is to identify the risk model's vulnerabilities, predisposing conditions, and security controls. Once these things are defined, the likelihood of occurrence and magnitude of their impact are estimated. These are taken together to determine a total "risk score." The team should expect to iterate over these assessment items, determining the total "risk score." The risk analysis assessment should be communicated to the stakeholders in step 3, "Report Results," and iterations into the earlier steps are likely. Finally, in step 4, once an agreement is reached with stakeholders, the security risk assessment should be maintained and updated throughout the IoT lifecycle until product retirement. Testers should consider the contents of Figure 13-5 as ongoing efforts, that is, an infinity process. Many IoT devices should consider the security risk assessment process "forever."

IoT Security Threat Risks and How to Find Them

This book assumes risk assessment is a project-level activity and that management and engineering teams will do some or all of the activities to address the threat risk space. Many of the IoT project activities to mitigate security risk must be performed by developers. However, in some cases, managers have that responsibility. We all *assume* that management will fund security testing.

As we get into the details of test attacks, the book addresses some specific activities or items the engineering test team and/or independent V&V team should be considering. The following sections of the chapter begin your journey as a security hacker-tester good guy (white hat). These sections are only the beginning.

Here are the things to consider as a security hacker-tester good guy during security hacker-tester risk planning:

- Support language (e.g., SQL) injection
- Cross-site scripting, when applicable
- Insecure file/data/communication encryption
- Weak access controls and credential management
- Vulnerable nondeveloper source components

  - Hidden backdoors

  - Malware

  - Operating system, OTS, and debugging service weaknesses

- Breaking security attack surfaces [1]
- Embedded software security attack surfaces [3]
- IoT project–specific attacks

Once initial security risk planning has started, a tester should quickly start making security test designs and planning attacks using the models of the following sections.

Cybersecurity Team: Test/V&V Supports Developer-Operations (DevOps) Security

The security test team has parallel activities for the DevOps security team and the test/V&V team, each with some degree of independence between the teams, as shown in Figure 13-6.

In this security testing activity, the test team attempts to evade and defeat the measures of the DevOps team. The testers "play" the bad guys using black hat concepts, but with the testers' white hat on. The DevOps team designs IoT defenses to evade hackers, and the test team monitors the IoT product for risks using test planning and social engineering. These development activities lead to a product entering testing usage, even if the IoT system is not ready for release to the public.

On the tester side, monitoring and planning lead to designing test attacks with monitoring information such as probes, hacking tools, and code trace systems. As the attack testing goes on, the test group evaluates the controls and defenses for gaps, leading to new attacks. The end goal is to show that the system can be compromised by security faults that must be considered for fixes or remediation. Once hacker-testers have designed tests, they get to simulate bad guy attacks using all the vectors, risk threats, and vulnerabilities they have defined. At the same time, Ops continues to try to deny, defend, and disrupt the attacks, just as they will have to do once the IoT system goes live. This is a fun game of "cat and mouse," but who is the cat and who is the mouse?

The creativity of both development and test processes requires yet more thinking and models and close teamwork. Teams will often want to use a cyber kill chain model.

Key Activity: Cyber Kill Chain Model

The cyber kill chain is a conceptual cybersecurity model created by Lockheed Martin [6] that traces the stages of a security attack, identifies vulnerabilities, and helps security teams to stop the attacks over the chain.

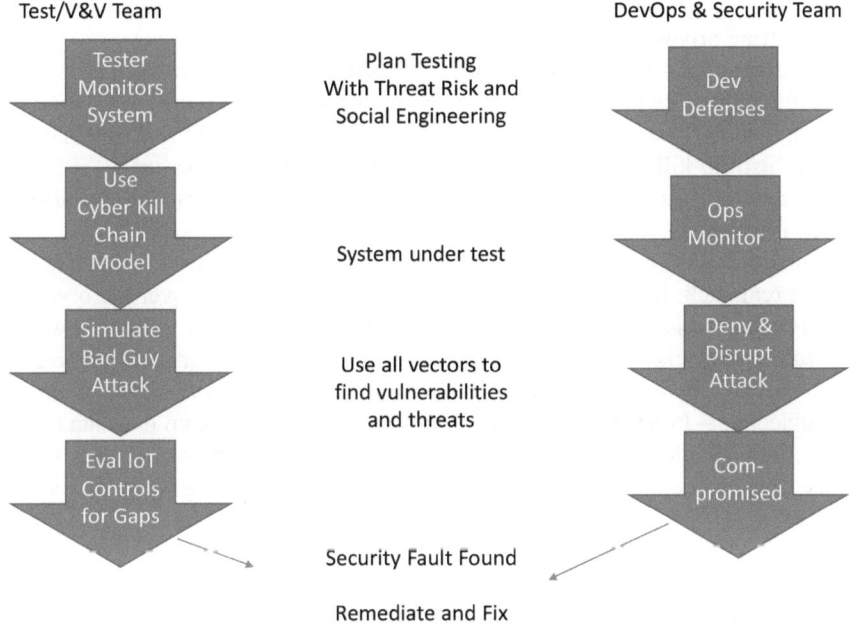

Figure 13-6. Test vs. DevOps and security processes

The term "kill chain" is adopted from the military, who use this term related to the structure of a battle. The military battle chain historically consists of five steps: identifying a target, dispatching the resource(s) to conduct an attack scenario, the decision to commit to the attack, orders to execute the attack, and, finally, destruction of the target.

The cyber kill chain consists of seven distinct steps:

1. Reconnaissance – The attacker collects data about the target, strategy, and tactics used in the attack. This includes harvesting data such as email addresses, contacts, user, and data analytics information. For example, automated data scanners are used to find points of vulnerability in the system. This includes scanning firewalls, intrusion prevention systems, login passwords, communication ports, etc., to get to the point of entry for the attack.

2. Weaponization – Attackers develop malware or attack tooling to leverage security vulnerabilities. Attackers program malware or tooling based on the needs and intentions of the attack. This process also involves attackers trying to reduce the chances of being detected by the security solutions of the IoT system(s).

3. Delivery – The attacker delivers the weaponized malware, attack data, or social engineering information via a phishing email or other media. The most common delivery vectors for weaponized payloads include websites, removable disks, USB devices, downloads, and emails. Or, the hacker-tester uses tooling to probe the vulnerabilities. For example, try attacking user login and passwords with a tool designed to crack passwords. This is a critical stage where the IoT security operations team can stop the attack before damage is done, so the hacker-tester must try to prove they can evade the defenses. If the hacker-tester evades successfully, the development and security team has more work to prevent real hackers, and the process continues.

4. Exploitation – An application or the operating system's vulnerabilities are often attacked next. In the case of the malicious code, it is delivered into the organization's IoT system. In the case of using an attack tool, success is when the hacker gains access to the system. In both attacks, the security perimeter is breached. And then, the hacker-tester gets the opportunity to exploit the orga-

nization's systems by installing tools, running scripts, and/or modifying security certificates, wiping data, formatting drives, installing trojans, or something worse.

5. Installation notification – The hacker-tester has penetrated the system, so the development staff should be notified. This can be done by filing an issue or problem report, or it may be helpful to leave a message on the system that the white hat team has hacked it. In lieu of this notification and following the historic kill chain model, a backdoor or remote access trojan is installed by the malware that provides access to the intruder. This is another essential step because the risk of a security hack is exposed.

6. Command and control – This step should be skipped for the hacker-tester kill using the kill chain. However, in a real hack, this is when the attacker would gain control over the organization's systems and network. For security testing, in this step, we do step 5 and follow up with the team to assure the vulnerability is fixed. Bad things can happen once hackers are in the IoT system or control it. *Remember the warning about illegal actions at the beginning of this chapter.*

7. Actions on objective – In a hacker-tester kill chain model case, the team has data that a threat risk exists and actions should be taken. The exact nature of the action(s) depends on the risk and consequence of the white hat hack. Based on these steps, the following layers of control action can be considered:

1. Detect – Identify attempts to penetrate the IoT system.
2. Deny – Stop the attack before it happens.
3. Disrupt – Detect and intervene in the communication done by the attacker before a hack happens.
4. Degrade – Limit the effectiveness of an attack (e.g., firewalls, monitor systems, etc.).
5. Deceive – Mislead the attacker using misinformation or misdirecting them to "safe" IoT data.
6. Contain – Direct the attacker to a restricted or "safe" parts of the IoT system.

Ref: `www.computer.org/publications/tech-news/trends/what-is-the-cyber-kill-chain-and-how-it-can-protect-against-attacks` – accessed winter 2022

Here's how simulating, in test, a cyber kill chain use can protect against cybersecurity attacks:

1. Simulate cybersecurity attacks via hacker-testers using simulated "test" attacks on all vulnerabilities.
2. Evaluate the controls to identify security gaps – Run test simulation attacks to identify risks and score (high, medium, low) the threat level.
3. Remediate and fix the cybersecurity bugs following a software maintenance-update lifecycle.

Key Activity: Using a Zero-Trust Security Architecture Model to Support IoT Testing

Many IoT development teams will use or should be using a zero-trust security architecture model. The process of implementing a zero-trust security architecture can be broken down into nine steps (based on the US Department of Defense [7]):

1. Identify users who need access to the network or IoT system.
2. Identify the IoT devices that need network access.
3. Identify the data and/or information inputs that need network access.
4. Identify all key processes, including testing.
5. Establish policies, including security risk analysis.
6. Identify solutions (Dev, test, and management – who needs to do what, when, where, how, and budget).

7. Deploy solution(s) and then test with test attacks (repeat).
8. Monitor controls and testing over the lifecycle.
9. Expand the zero-trust architecture, as needed.

The test/V&V team should verify this architecture model has been followed and implemented. Each step should be verified and, as needed, probed with attack tests. When such an architecture is not used, a hacker-tester can still use the preceding model to design security tests, possibly to show management and development that the model should be implemented, which is justified by risk threats.

Key Activity: Software Security Testing Cycle

Figure 13-7 presents my favorite basic approach and attacks for system-software security testing, with subsections that follow briefly describing them. References are provided for more details.

Social Engineering

Testers think and act like the "bad guys" in social engineering. They ask: "How can we trick the users and system?" An excellent model to follow is the cyber kill chain model. Testers think the way the hackers do, but before the bad guys can cause problems, they ponder their thinking. Again, *I remind you to be legal and ethical* (white hat). Also, readers will note that in the IoT security cycle model of Figure 13-7, there are blanks for other security test activities that may need to be customized to your local threats and risks. Finally, for more information, see

- `www.datto.com/blog/cybersecurity-101-intro-to-the-top-10-common-types-of-cybersecurity-attacks`
- `https://cisomag.eccouncil.org/10-iot-security-incidents-that-make-you-feel-less-secure/`

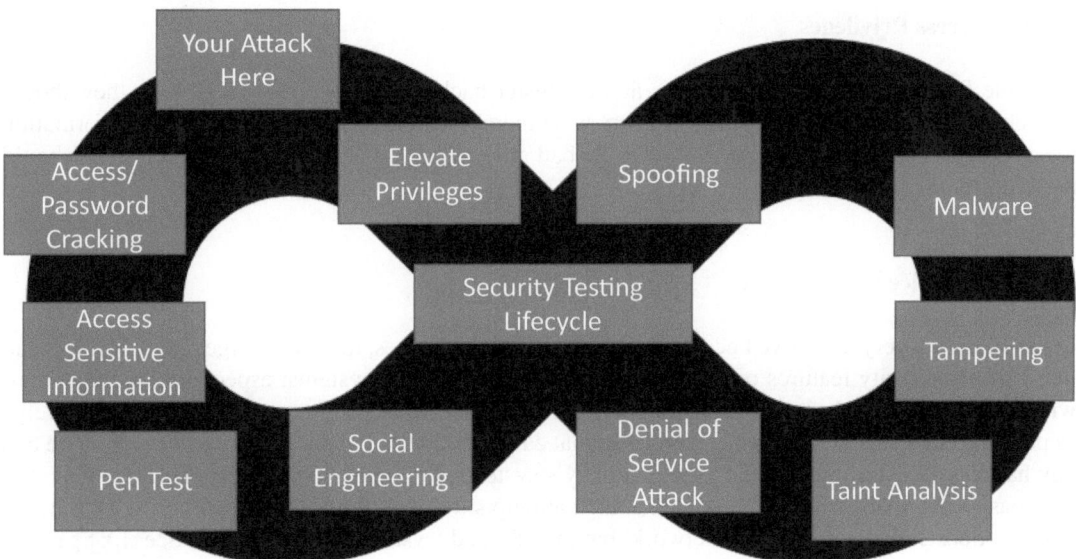

Figure 13-7. IoT security test cycle

Penetration (Pen) Testing

Pen testing is defined in detail in Part 3, Chapter 15. Pen test results include testing software, issues, reports on what was done, and prioritization or guidance on untested areas since testing is never 100 percent complete. In summary, the tester should think of this security Pen testing as a service on software and a collaboration between the teams doing development. When the Pen results come through, developers can seek clarification and advice from testers and V&V and the security team on what the issues are and how best to resolve them. I also recommend outside testing groups or V&V conduct Pen testing when risks justify it.

Access Sensitive Information

This activity is a follow-on and continuation of social engineering. This is part of gathering information to support security risk assessment and later testing. The hacker-tester tries to access information about the system for which they are not privileged to access. They may "play" the role of a user with limited access and then try to gain privileged access or even play an outside hacker trying to get in. Testers should try to access the IoT device, edge, fog, development areas, and even the cloud. Testers look for data and information used in later testing and attacks in this information-gathering phase.

Access Control and Password Cracking

A favorite process that is a very productive subset of Pen testing is password cracking or gaining system control access. To do this, there are tools such as pkcrack (see `https://en.kali.tools/all/?tool=1042`). Even worse, many IoT systems use just default passwords and tools such as those found on the website Shodan (see `www.shodan.io/`), exposing these systems to anyone looking to do a hack. Why? Well, the bad guys will do this. Here, I start simple (using defaults), move to cracking, and then social engineering.

Elevate Access Privileges

After the Pen or cracking testing works and the tester-hacker is within the IoT system, they should gain elevated access privileges. This will expose vulnerabilities and can provide yet more information to use in security attacks and testing. Again, the bad guys will try to do this, so the hacker-tester should try to do it too.

Denial of Service (DoS) Attack

IoT systems are very attractive because there are many IoT devices, making a large attack surface, and many of the security features of IoT devices are lacking. All IoT systems, especially the huge ones, will be attacked by bad guy hackers, and, therefore, you will want your system to respond as "safely" as possible (safe engineering is a design issue that can be addressed in an STA with DevOps). The bad guy hackers will start by attacking the company's systems, including IoT devices. Hackers will start with easy kills, like DoS on IoT. A DoS attack attempts to block the IoT site operations by sending many requests that exceed the network bandwidth and system responsibility. See `https://en.wikipedia.org/wiki/Denial-of-service_attack`, which devotes many pages to the DoS attacks and more information.

DoS simple example

Start sending data packets to figure out the open ports, IP addresses, operating system information, etc. Try to interrupt the target system(s) by sending large numbers of data packets to get a denial of service on them. Next, scan a network with a tool such as Nmap, as shown in Figure 13-8. Record information and findings. A tool like Nmap will help you figure out the network map and provide a lot of information about the systems. As you get more information, the speed of sending the packets can be increased until the system's performance level is exceeded. At this point, the team starts looking for a crack in the system's defenses.

Most network scan tools offer a number of command-line options, including scan exports that you can then import into the exploitation tools. You may need some performance test simulation tools to continue the saturation.

Tampering Attack

In tampering, testers try to "break" into the hardware and/or software to gain access. This is closely related to password cracking but takes other approaches such as inserting code, data, or hardware into an IoT configuration (CM/SCM tools and/or processes). The IoT tampering attack can modify parameters exchanged between the device and the edge, fog, or cloud. The goal is to manipulate IoT data such as passwords, output values, permissions, etc. The tester is trying to understand the following:

- Does the external CM/SCM system or internal version checks (e.g., did a realtime MD5 scan or checksum e.g., `https://en.wikipedia.org/wiki/MD5`) of the IoT device detect the tampering?
- Does the system itself notice and inform Ops or provide notification of the attack?
- Can bad data be output or used?

Reference: `https://csrc.nist.gov/glossary/term/tampering` – accessed spring 2022

```
Starting Nmap 7.80 ( https://nmap.org ) at year-mo-day hh:mm
Nmap scan report for site.domain (xx.xx.xx.xx)
Host is up (0.01s latency).
Not shown: 80 closed ports
PORT      STATE SERVICE
21/tcp    open  ftp
25/tcp    open  smtp
80/tcp    open  http
110/tcp   open  pop3
143/tcp   open  imap
443/tcp   open  https

Nmap done: 1 IP address (1 host up) scanned in 7.77 seconds
```

Figure 13-8. Nmap example

Malware Attacks

Finding inserted bad software (malware) in your production code before the code is released to a customer or out in the field is significant, as the SolarWinds hack demonstrated. *Your first line of defense should be your CM/SCM processes and tools* (Part 2, Chapter 9). However, many projects will have weak defenses in this first line. The following actions may help.

Security Attacking the Off-the-Shelf (OTS) Software for Malware

The use of tools and analyses is a good idea. When you do not trust the OTS software, conducting these scans and analysis may help, in addition to running security test attacks. In this analysis, the tester will need access to the source or binary code of the OTS. The OTS code can, of course, be visually checked during an inspection meeting by a team, but humans will still miss things.

There are three check columns of action in Figure 13-9. Each one starts with disassembly or scanning the software code to generate information. The tester identifies the system, software, and external calls using a static analysis tool in the first flow. Many tools will generate call tree maps and even visual graphic trees [10]. Here, I have identified a call outside of the class/object structure on one system. It was not a hack because it was a debugging aid, but it could have been a hack. *Any anomalous or untrusted call should be fixed.*

The middle path uses the scanned code to construct a logic flow graph and possibly do static taint analysis. Again, the tester is looking for anomalous logic, which should be corrected. The final path uses capable static analysis tools to identify issues. Programmers can do the review, but I find that many test teams use the testers to do the first pass on the generated information to remove what is called "false error" messages before programmers review them. Programmers hate doing this, so programmers are much happier if the test team provides a "clean" list of issues; see `https://en.wikipedia.org/wiki/List_of_tools_for_static_code_analysis`–accessed spring 2022.

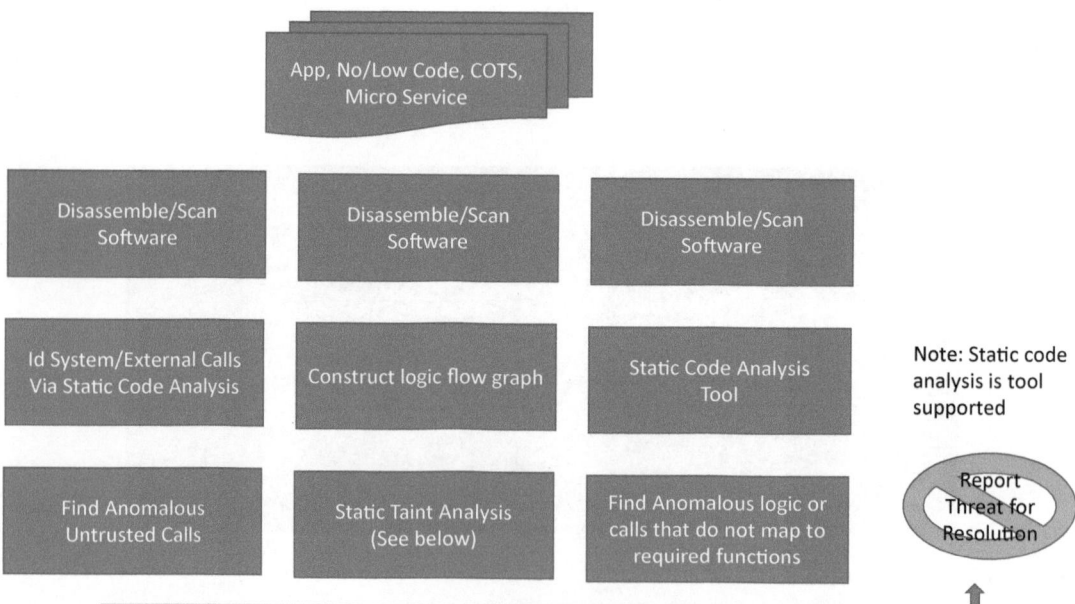

Figure 13-9. Securing OTS code analysis (looking for possible malware)

Using a Table for Detection of Malware in the IoT Off-the-Shelf (OTS) Software

Malware gets created for many reasons by different actors (good and bad). Table 13-1 is a starting model if you need to find or attack malware in your OTS software, which you should do if you have OTS items.

Security Taint Analysis

Security taint analysis is associated with malware, faults in the code, and, to some extent, a DoS. It is the process of staff checking the flow of user input in the IoT inputs to determine if unanticipated input can negatively affect program execution during each functionality step. The staff doing this analysis can include developers, security staff, or testers. The analysis staff will "play" different roles and attempt to detect a normal input processing flow disruption. Taint analysis can help check the IoT attack surface and define new avenues of attack. The flow of inputs is "tainted" with unexpected inputs, and then the output is monitored for an improper response, as shown in Figure 13-10.

The code, including all OTS parts, is loaded into the IoT system processor in the taint analysis. The staff doing the analysis assumes a "user role" (e.g., a bad guy hacker, a novice user, superuser, etc.). The staff knows the following kinds of information, which can be tainted: inputs, data values, sequence of inputs, message outputs, file outputs, response actions, and links to other elements of the IoT system such as the edge, fog, and/or cloud. During the analysis, the staff "user" tries to corrupt (taint) the inputs and review outputs for improper actions and outputs. The output could be reviewed in real time, but the output must often be postprocessed. Postprocessing can be enhanced if a dynamic runtime taint monitoring system is included in the taint configuration. Such a monitor can be one of the tools listed in the section "Tools for Software Security Testing." After postprocessing, taint analysis may indicate more actions and outright vulnerabilities.

Spoofing [11]

A spoofing attack uses the information to impersonate an authorized device or user. Once in the system, spoofing can steal data, insert malware, or bypass access control systems for later access. Most of us are familiar with spoofing phone numbers, where a phone number appears to be a local number

Table 13-1. OTS Detection Table

Malware Inserted Because	Description of the Malware Capabilities
A hacker does it because they can hack something for "fun"	Software malware created to show the hack of the IoT device is possible by somebody who thinks this is "fun"
Selling IoT data from a hack to make money	A user creates malware to get data and pass IoT data to interested third parties for money or some other gain
Capture IoT user login information to sell to hackers	Login access information from the IoT device may be helpful to control or access other systems and then "sell" to make money or some other gain (see above)
Manipulate the IoT input/output data to support hackers	Control IoT inputs or outputs for nefarious purposes. This manipulation can be done on the edge/fog/cloud also
Change IoT inputs to allow the hack	Control IoT sensor data for nefarious purposes
Conduct middleware attacks to gain access to the IoT system	Control the IoT system to gain access or impact other systems in the IoT network for nefarious purposes at the edge/fog/cloud (known as a "man-in-the-middle attack")
A hacker monitors the IoT device hacks to understand how to attack better	Establish a monitoring capability in the IoT device for nefarious purposes

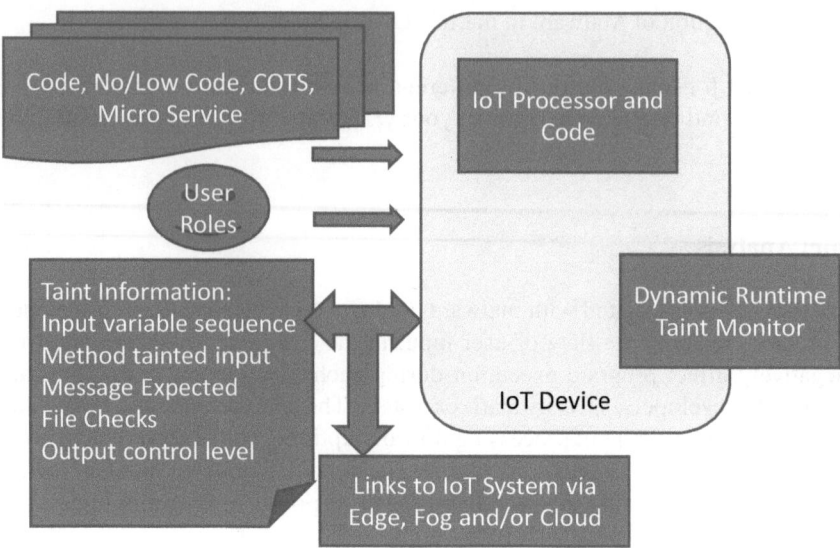

Figure 13-10. Security taint analysis flow

or even your own when, in fact, someone uses a random number generator to dial numbers from a call center for illegal reasons. For IoT systems, connections can be spoofed, including

Website or domain name for IoT edge, fog, or cloud services
IP addresses
Address Resolution Protocol
GPS spoofing of the location
User (man, system, hardware) in the middle
Identity of user or login

Here are some examples of spoofing sub-attacks.

GPS spoof – Hacker-testers apply this sub-attack when the IoT device uses GPS information. This attack uses technology or alters data so that an IoT device appears in a different location or even time zone than it actually resides in. GPS spoofing output fakes location coordinates or time zones. A spoofed GPS IoT device can increase cybersecurity risks in other areas or bypass other security features.

Identity fraud spoof – Hacker-testers try to fool the IoT device into gaining an identity that provides better access (e.g., regular user to privileged user). For example, hackers use the Wireshark tool to sniff and decode data being broadcast. Here, the hacker-tester should see if the identity can be "hijacked." Items to consider for identifying fraud spoofing include

– Is authorization or access to the system and its files temporary or permanent?
– If temporary access, the hacker-tester should check for remnant data files to glean information from.
– Use TestOps tools and/or the basic operating system to poke around in the filesystem.

Note
The file should be encrypted if the IoT device is secure, in which case you may need a file encryption cracker for that type of file/encryption (e.g., www.firewalltechnical.com/encryption-cracking-tools/).

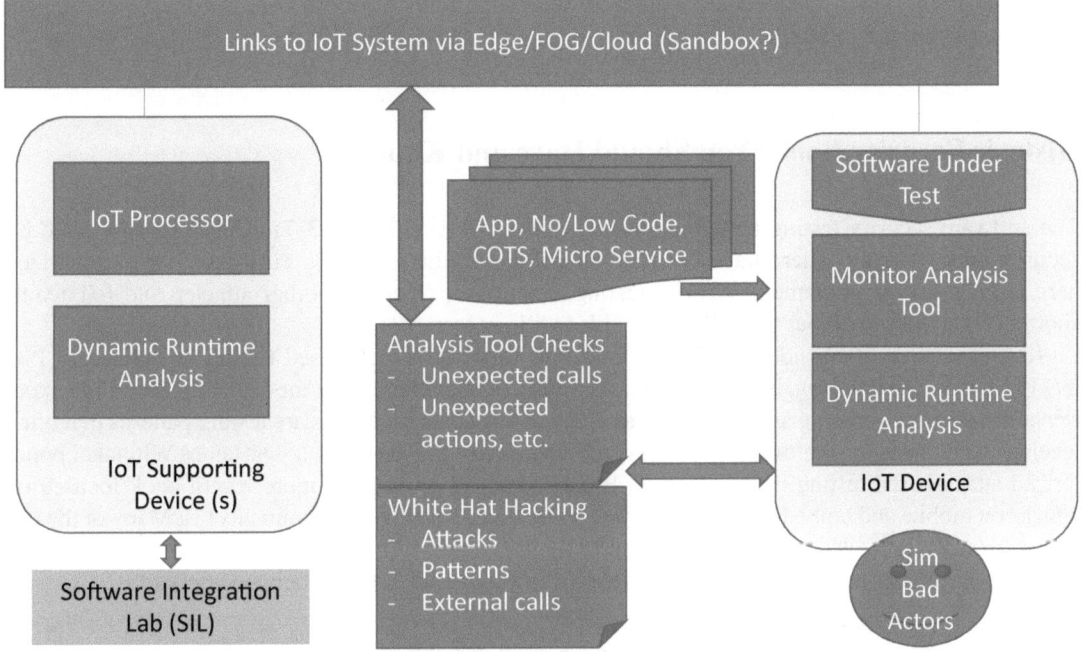

Figure 13-11. Security testing at the system level in a lab

If the file is not temporary, the tester next needs to determine if any permanent information can be accessed, abused, or corrupted. For secure IoT devices, this data should be encrypted, and here again, the tester should apply cracking encryption tools to attempt to gain access.

Finally, report how hard or easy it is to read the file (open text is bad, while encrypted is better).

Edge, Fog, and Cloud Attack: Security Testing Needed beyond a Single IoT Device

Throughout this book, I have advocated that teams should consider those elements beyond the IoT device itself after basic IoT device testing and assessment. When dealing with security vulnerabilities, these elements include the hardware and other IoT devices, edge, fog, and cloud. While ownership of these other levels of IoT security testing may not be clear, the test team should at least bring the risks forward so that stakeholders can make informed decisions. This is considered securing the larger IoT picture, but not to the whole system in the field, which is covered under the chaos engineering level (Part 3, Chapter 16). In this section, I briefly outline this attack level.

In Figure 13-11, I have created an example system test lab. It allows at least a partial assessment of the larger IoT system before going into the "real world." While not easy or cheap to set up, I have a sandbox to play security hacker-tester. It starts with an edge, fog, and/or cloud system in the sandbox to support communication. The lab configuration supports IoT devices integrated with the test lab. The SIL allows testing support, automation, test cases, analysis, and reporting. The SIL drives the IoT support devices, which are not directly under test assessment, but in the security configuration necessary; these communicate to the network. The SIL also has analysis tools to support security testing in unexpected calls, system action, and automation. The white hat hacker-testers use attacks, patterns, and framework external calls to simulate (sim) malicious users on the software under test (SUT). Finally, under test, the IoT device has monitoring, analysis, and dynamic runtime analysis tools to aid the hacker-tester. This system security sandbox SIL configuration offers a controlled test environment where many things can be tried but not released into the real world.

Historic Security Books You Should Have and Know

The software security testing activities of the infinity model (Figure 13-7) need to be supported by security attack testing. There have been books written on this type of security testing, summarized here to help testers implement security testing. There are, of course, other attacks, and IoT teams should create their own, but the following tables will get testers started.

Test teams should consider risk-based testing to focus on attack-based testing as defined in ISO 29119. In risk-based testing, higher risk areas are subject to earlier and/or more testing, including experience-based error guessing software tests, also called test attacks. Attacks are testing patterns that often leverage techniques in combinations to break the software or find errors in it. Dr. James Whittaker popularized attack-based testing in his "How to Break" series [1]. I wrote a more recent book focused on attacks on mobile and embedded software systems in their respective environments [3]. Many of the test attacks from these books have an application for security testing, as shown in Table 13-2.

James Whittaker and team [1] identified the following ideas, which I believe apply to IoT:

- Attacking IoT software dependencies
- Attacking the user (human, hardware, and software)
- Finding and attacking the backdoor(s)
- Attacking the interfaces for IoT security
- Breaking the IoT design for security testing
- Breaking the IoT code implementation for security testing
- Breaking the IoT interfaces for security testing

If some of these attack ideas sound familiar, they should as many teams use them. Also, security testers should refer to ISO/IEC 27001 Information Security Management for more helpful information.

Table 13-2. Example Software Attacks for Exploratory Testing

Software Test Attack Type	Attack Finds	Notes on the Attack
Developer-level attacks	IoT code and data structure problems that can impact security	This attack, including security issues, can find almost a quarter of errors in IoT
Control system attacks	IoT hardware and software control system errors that result in vulnerabilities	Many critical errors in IoT devices are centered on the control logic. For example, analog-to-digital (A2D) and digital-to-analog (D2A) computation problems
Communication attacks	Digital IoT communication problems open complex interfaces that become attack surfaces	Attack the software as it communicates with hardware, edge/cloud networks, other IoT systems, and other software
Time attacks	Time, performance, sequence, and scenario errors open up security attacks such as data, performance, and DoS	System software can have vulnerabilities in timing and performance factors that testing can provide valuable information on
Security hacking attacks	Software errors can expose devices to security threats	Security of devices or systems is experiencing an increase in importance and attacks
Static code analysis attacks	Hard to find errors that classic testing often misses and leaves open to security attack surfaces	Run static analysis tools and look for threat areas

Note: Details on this table and these test attacks are contained in [3]. Note: Given the size and complexity of some IoT systems, automation of some test attacks and static analysis may be needed, including the use of AI test systems.

Security Test Tool Categories

As the world becomes more and more connected, the list of tools is practically endless these days as more tools are created weekly. This section mentions a few tools by name that testers and developers may find helpful in creating better-quality IoT devices and systems. For more information, see www. appknox.com/blog/iot-security-testing-tools and https://medium.com/ brandlitic/top-iot-hacking-tools-f9355e384db0 – accessed winter 2022.

Tools for Software Security Testing

- Communication monitoring tools (e.g., Wireshark and Nmap)
  - Scan Network Devices – Software system
  - NetFlow Traffic Analyzer
  - Network Configuration Manager
- Memory dump tool – System dependent
- Web info (e.g., Shodan)
- Edge/cloud test probe/monitors
- Static code analysis tool (https://en.wikipedia.org/wiki/List_of_tools_for_ static_code_analysis – accessed spring 2022)
- Performance test and analysis tool

IoT Testing Tools for Hardware

- Printed circuit board debug tool
- Oscilloscope – To watch signals

  Other tools defined in Part 4, Chapter 20, may be used.

Software-Defined Security and AI – An IoT Future

There are five ways AI models can be used in security system testing, depending on the level of insight desired:

- System-software insights – Analyze the system and software data (requirements, design, code, reuse, teams, etc.) to discover patterns that can be used to define threats, risks, and vulnerabilities.
- Test planning recommendation – The AI discovers system, software, and data patterns. Then the AI provides recommendations on what testing should be done.
- Autonomous mitigation – The model discovers test information patterns and automatically solves problems without human authorization.
- Automated test support – Test generation, rapid test execution, and pattern detection results.
- Testing the "whole" – Find hard-to-see vulnerability patterns in security testing across IoT boundaries into the edge, fog, and cloud via connectivity and communication.

Note
For more AI information, see [12].

Many system's large data volumes will mark the world of IoT. These volumes will quickly over-come previous small and manual test efforts, even with tools and collaboration. While I recommend ongoing efforts over the lifecycle, security testing will be a challenge, and management will be tempted to save time and money by eliminating testing. With AI, modeling, and automation, the cost and time may be reduced in the future to the point that management will see some benefit. Management will want to test whether parts of the code or the IoT system are good AI automation candidates. This is so that time and money are saved and some due diligence is done. In the AI future, I recommend using AI with security models with automation to quickly "smoke test" (prove nothing is going to burn up) within short hours on the IoT device. The use of test environments (Part 4, Chapter 19) can support this kind of future to provide what management wants, which is "yes, it works and is secure." The ongoing test planning with strategies including security will lead to success. Investing in the right balance of development, testing, and AI can help guarantee satisfied customers and regulators. Hence, *IoT testers will benefit from learning AI*.

Security Finding Reporting

In other parts of this book, I have presented the need for test results documentation [9]. I repeat the need for reporting here because it is essential. In security tests, the reporting remains essential for legal and company successes. Records of what was done are essential. I was involved in one legal case concerning a software failure. The records I provided were essential to the project and the lawyers. Security testing reports should include

– Plans
– Identified threats, vulnerabilities, and risks
– Testing actions taken and other mitigation efforts
– Where IoT security was broken and how a fix was made and implemented
– Cost and schedule information

Summary

Security testing of IoT is probably worth a whole book (see [8]), but the references and materials of this chapter are a starting point for most people – but only a starting point. There is much to be learned on this subject. For many experienced testers, this chapter should be read and understood.

Remember the plan, do, check, and act model as applied to security testing. I would like to see security testers who like to attack and break things. Companies need these kinds of dedicated, skilled testers. Not everyone is devious enough to do this kind of work, but everyone can use the tools, pro-cesses, techniques, and AI to support those gifted, ethical, white hat, devious attackers. The next chapter gives another track to take with OWASP and open source ideas. I like to reuse the information on security testing and then build my own infinite model attacks. Why not join me on this IoT journey?

References

1. *How to Break Software Security: Effective Techniques for Security Testing* by Herbert Hugh Thompson and James A. Whittaker
2. ISO/IEC 20246:2017, Software and Systems Engineering – Work Product Reviews
3. *Software Test Attacks to Break Mobile and Embedded Devices* by Jon D. Hagar
4. https://en.wikipedia.org/wiki/Zero-day (computing) – accessed spring 2022
5. IEEE 982.1-2005 IEEE Standard Dictionary of Measures of the Software Aspects of Dependability, in revision as of 2022
6. www.lockheedmartin.com/en-us/capabilities/cyber/cyber-kill-chain.html – accessed spring 2022
7. www.whitehouse.gov/wp-content/uploads/2022/01/M-22-09.pdf – accessed spring 2022
8. *Practical IoT Hacking* by Fotios Chantzis, Ioannis Stais, Paulino Calderon, Evangelos Deirmentzoglou, and Beau Woods
9. "IEEE/ISO/IEC 29119, ISO/IEC/IEEE International Standard – Software and systems engineering – Software testing" – Part 3
10. https://en.wikipedia.org/wiki/Call_graph – accessed spring 2022
11. https://en.wikipedia.org/wiki/Spoofing_attack – accessed spring 2022
12. https://spectrum.ieee.org/the-state-of-ai-in-15-graphs – accessed spring 2022

Chapter 14
Security OWASP IoT Information Pointer and Logging Events

Chapter 13 presented starting point information for IoT security testers. However, the IoT security staff need ongoing information updates to stay at least current with the bad actors. The Open Web Application Security Project (OWASP) is an open source of information and a nonprofit foundation that works to improve the security of software, including IoT systems. This chapter includes a sampling of their information to get testers started using this resource. Since the OWASP community works constantly to keep information updated, this chapter should be used as a reference point to get you started, and then to find the most current information, go to

`https://wiki.owasp.org/index.php/OWASP_Internet_of_Things_Project` – accessed spring 2022

Agreement information: `https://owasp.org/www-policy/` – accessed winter 2022

The OWASP site and other security websites should be used on a regular basis by security and testing staff. This is a moving target topic area.

Intro to OWASP Top Ten Threats (As of 2022)

Top ten threats from OWASP with modifications are as follows:

1. Weak, guessable, or hardcoded passwords
2. Insecure network services
3. Insecure ecosystem interfaces
4. Lack of secure update mechanisms
5. Use of insecure or outdated components, including hardware, software, and/or system elements
6. Insufficient privacy protection
7. Insecure data transfer and storage
8. Lack of device management
9. Insecure default settings
10. Lack of physical hardening

I recommend putting these events in your security test scenario as part of an exploratory test. These events must be understood and tailored for your IoT device and system. Not every event will be applicable to every IoT system.

Tables 14-1 through 14-4 present security threat category events and specific events that when attempted by testers may expose useful security information. They get updated by OWASP as

© Jon Duncan Hagar 2022

J. D. Hagar, *IoT System Testing*, https://doi.org/10.1007/978-1-4842-8276-2_14

frequently as necessary. The OWASP team spends a lot of time on these tables, and security testers should be using these resources and even providing feedback to the OWASP team.

OWASP: This is a working draft of the recommended minimum IoT device logging events. This includes many different types of devices, including consumer IoT, enterprise IoT, and ICS/SCADA type devices.

Information security professionals should use the stages listed in Table 14-2 when conducting firmware security assessments.

Table 14-1. IoT Logging Events

Event Category	Events
Request exceptions	• Attempt to invoke unsupported HTTP method • Unexpected quantity of characters in parameter • Unexpected type of characters in parameter
Authentication exceptions	• Multiple failed passwords • High rate of login attempts • Additional POST variable • Deviation from normal geolocation
Session exceptions	• Modifying the existing cookie • Substituting another user's valid session ID or cookie • Source location changes during session
Access control exceptions	• Modifying URL argument within a GET for direct object access attempt • Modifying parameter within a POST for direct object access attempt • Forced browsing attempt
Ecosystem membership exceptions	• Traffic seen from disenrolled system • Traffic seen from unenrolled system • Failed attempt to enroll in ecosystem • Multiple attempts to enroll in ecosystem
Device access events	• Device case tampering detected • Device logic board tampering detected
Administrative mode events	• Device entered administrative mode • Device accessed using default administrative credentials
Input exceptions	• Double encoded character • Unexpected encoding used
Command injection exceptions	• Blacklist inspection for common SQL injection values • Abnormal quantity of returned records
Honey trap exceptions	• Honey trap resource requested • Honey trap data used
Reputation exceptions	• Suspicious or disallowed user source location

Table 14-2. OWASP IoT Firmware Security Verification Assessment

Stage	Description
1. Information gathering and reconnaissance	Acquire all relative technical and documentation details pertaining to the target device's firmware
2. Obtaining firmware	Attain firmware using one or more of the proposed methods listed
3. Analyzing firmware	Examine the target firmware's characteristics
4. Extracting the filesystem	Carve filesystem contents from the target firmware
5. Analyzing filesystem contents	Statically analyze extracted filesystem configuration files and binaries for vulnerabilities
6. Emulating firmware	Emulate firmware files and components
7. Dynamic analysis	Perform dynamic security testing against firmware and application interfaces
8. Runtime analysis	Analyze compiled binaries during device runtime
9. Binary exploitation	Exploit identified vulnerabilities discovered in previous stages to attain root and/or code execution

The OWASP IoT attack surface areas are listed in Table 14-3.
Table 14-4 describes the OWASP threat vulnerability project.

Table 14-3. OWASP Attack Surfaces

Attack Surface	Vulnerability
Ecosystem (general)	• Interoperability standards • Data governance • System-wide failure • Individual stakeholder risks • Implicit trust between components • Enrollment security • Decommissioning system • Lost access procedures
Device memory	• Sensitive data • Cleartext usernames • Cleartext passwords • Third-party credentials • Encryption keys
Device physical interfaces	• Firmware extraction • User CLI • Admin CLI • Privilege escalation • Reset to insecure state • Removal of storage media • Tamper resistance • Debug port • UART (Serial) • JTAG/SWD • Device ID/Serial number exposure
Device web interface	• Standard set of web application vulnerabilities, see: • OWASP Web Top 10 • OWASP ASVS • OWASP Testing guide • Credential management vulnerabilities: • Username enumeration • Weak passwords • Account lockout • Known default credentials • Insecure password recovery mechanism
Device firmware	• Sensitive data exposure (see OWASP Top 10 – A6 Sensitive Data Exposure): • Backdoor accounts • Hardcoded credentials • Encryption keys • Encryption (symmetric, asymmetric) • Sensitive information • Sensitive URL disclosure • Firmware version display and/or last update date • Vulnerable services (web, SSH, etc.) • Verify for old software versions and possible attacks (Heartbleed, Shellshock, old PHP versions, etc.) • Security-related function API exposure • Firmware downgrade possibility

(continued)

Table 14-3. (continued)

Attack Surface	Vulnerability
Device network services	• Information disclosure • User CLI (command-line interface) • Administrative CLI • Injection • Denial of service (DoS) • Unencrypted services • Poorly implemented encryption • Test/development services • Buffer overflow • UPnP • Vulnerable UDP services • Device firmware update block • Firmware loaded over insecure channel • Replay attack • Lack of payload verification • Lack of message integrity check • Credential management vulnerabilities: • Username enumeration • Weak passwords • Account lockout • Known default credentials • Insecure password recovery mechanism
Administrative interface	• Standard set of web application vulnerabilities, see: • OWASP Web Top 10 • OWASP historic risks • OWASP Testing guide • Credential management vulnerabilities: • Username enumeration • Weak passwords • Account lockout • Known default credentials • Insecure password recovery mechanism • Security/encryption options • Logging options • Two-factor authentication • Check for insecure direct object references • Inability to wipe device
Local data storage	• Unencrypted data • Data encrypted with discovered keys • Lack of data integrity checks • Use of static same encode and decode key
Cloud web interface	• Standard set of web application vulnerabilities, see: • OWASP Web Top 10 • OWASP historic risks guide • OWASP Testing guide • Credential management vulnerabilities: • Username enumeration • Weak passwords • Account lockout • Known default credentials • Insecure password recovery mechanism • Transport encryption • Two-factor authentication

(continued)

Table 14-3. (continued)

Attack Surface	Vulnerability
Third-party backend APIs	• Unencrypted PII sent • Encrypted PII sent • Device information leaked • Location leaked
Update mechanism	• Update sent without encryption • Updates not signed • Update location writable • Update verification • Update authentication • Malicious update • Missing update mechanism • No manual update mechanism
Mobile application	• Implicitly trusted by device or cloud • Username enumeration • Account lockout • Known default credentials • Weak passwords • Insecure data storage • Transport encryption • Insecure password recovery mechanism • Two-factor authentication
Vendor backend APIs	• Inherent trust of cloud or mobile application • Weak authentication • Weak access controls • Injection attacks • Hidden services
Ecosystem communication	• Health checks • Heartbeats • Ecosystem commands • Deprovisioning • Pushing updates
Network traffic	• LAN • LAN to Internet • Short range • Nonstandard • Wireless (e.g., Wi-Fi, Z-wave, XBee, Zigbee, Bluetooth, LoRa) • Protocol fuzzing
Authentication/ authorization	• Authentication/authorization-related values (session key, token, cookie, etc.) and disclosures • Reusing of session key, token, etc. • Device-to-device authentication • Device-to-mobile application authentication • Device-to-cloud system authentication • Mobile application-to-cloud system authentication • Web application-to-cloud system authentication • Lack of dynamic authentication
Privacy	• User data disclosure • User/device location disclosure • Differential privacy
Hardware (sensors)	• Sensing environment manipulation • Tampering (physically) • Damage (physical)

Table 14-4. OWASP Common Threat Vulnerability Project

Vulnerability	Attack Surface	Summary
Username enumeration	• Administrative interface • Device web interface • Cloud interface • Mobile application	• Ability to collect a set of valid usernames by interacting with the authentication mechanism
Weak passwords	• Administrative interface • Device web interface • Cloud interface • Mobile application	• Ability to set account passwords to "1234" or "123456," for example • Usage of preprogrammed default passwords
Account lockout	• Administrative interface • Device web interface • Cloud interface • Mobile application	• Ability to continue sending authentication attempts after three to five failed login attempts
Unencrypted services	• Device network services	• Network services are not properly encrypted to prevent eavesdropping or tampering by attackers
Two-factor authentication	• Administrative interface • Cloud web interface • Mobile application	• Lack of two-factor authentication mechanisms such as a security token or fingerprint scanner
Poorly implemented encryption	• Device network services	• Encryption is implemented; however, it is improperly configured or is not being properly updated (e.g., using SSL v2)
Update sent without encryption	• Update mechanism	• Updates are transmitted over the network without using TLS or encrypting the update file itself
Update location writable	• Update mechanism	• Storage location for update files is world writable potentially allowing firmware to be modified and distributed to all users
Denial of service	• Device network services	• Service can be attacked in a way that denies service to that service or the entire device
Removal of storage media	• Device physical interfaces	• Ability to physically remove the storage media from the device
No manual update mechanism	• Update mechanism	• No ability to manually force an update check for the device
Missing update mechanism	• Update mechanism	• No ability to update device
Firmware version display and/or last update date	• Device firmware	• Current firmware version is not displayed and/or the last update date is not displayed
Firmware and storage extraction	• Firmware interface • In situ dumping • Intercepting a OTA update • Downloading from the manufacturer's web page • eMMC tapping • Unsoldering the SPI Flash/ eMMC chip and reading it in an adapter	• Firmware contains a lot of useful information, like source code and binaries of running services, preset passwords, SSH keys, etc.
Manipulating the code execution flow of the device	• Code interface • Side-channel attacks like glitching	• With the help of a JTAG adapter and GDB, we can modify the execution of firmware in the device and bypass almost all software-based security controls • Side-channel attacks can also modify the execution flow or can be used to leak interesting information from the device

(continued)

Table 14-4. (continued)

Vulnerability	Attack Surface	Summary
Obtaining console access	• Serial interfaces (SPI/ UART)	• By connecting to a serial interface, we will obtain full console access to a device • Usually, security measures include custom bootloaders that prevent the attacker from entering single user mode, but that can also be bypassed
Insecure third-party components	• Software	• Out-of-date versions of busybox, OpenSSL, SSH, web servers, etc.

Summary

This chapter presented open source OWASP information. OWASP is a resource that all security staff and testers should be aware of, follow, and use constantly to stay up to date with the actions of bad actors. This chapter flowed from the general security testing of Chapter 13, which led to the concepts in this chapter. Security topics found in this chapter covered logging events, security verification assessment, IoT attack surface areas, and common threat vulnerability project information. I covered security testing as a major quality characteristic to address in testing.

Chapter 15
Internal Security Team Penetration Test Process

Chapter 13 noted that most black hat hackers and security people practice penetration (Pen) attacks at many points on a project. Here, I take this type of attack into a Pen test practice. Penetration is how the hacking world thinks, and we testers need to think like them to be most effective. This chapter deals with Pen attacks as a "special case."

The internal penetration process can access development team information that might take a black hat hacker much longer to obtain; however, they often can still get it. On an IoT project, skilled staff need to apply this particular attack, which an experienced tester can do or help others with. There can be positive communication interchange between the Dev, operations, and security test teams on risks, the priority of fixing issues, and how issues found in Pen testing can be fixed with follow-up regression testing.

Pen Test Process: A Beginning

The whole team needs to remember the statement I have often said and repeated (both in classes I have taught and in papers and books I have written), that is, "*nobody would ever do that*," because the black hat hackers are precisely the *nobody who actually would do that*.

The main goal of Pen testing is to identify vulnerabilities before black hat hackers can. *ALL IoT systems are likely to be attacked.* IoT stakeholder organizations need to understand the risks and respond to security issues as threats arise.

Why Perform Pen Test?

> **OUTSIDE TESTERS AS HACKERS**
> Microsoft has paid out around $400k in security bounties as part of its Azure Sphere Security Research Challenge. Many companies pay "bug bounties" to outside testers (think V&V), with security being a top interest. I know of white hat hackers who have bought new cars with their bounties. I choose not to hack because I have a life outside of testing, including skiing, writing books, and having other kinds of fun. *Hacking can be fun if done ethically.* Pen testing and some of the attacks in Chapters 10 and 11 can get you started.

© Jon Duncan Hagar 2022
J. D. Hagar, *IoT System Testing*, https://doi.org/10.1007/978-1-4842-8276-2_15

The attack use cases of Pen testing are the key steps that follow.

Pen testing starts with gathering information about the target before the attack. This can be done with exploratory (unscripted) testing to identify possible entry points, attempt to break in (virtually or for real), and note the findings and observations. Balancing these tests and V&V efforts takes critical thinking, and there is no "one-size-fits-all" set of security attacks or explorations for test planning. When the risks justify it, *a highly skilled tester must conduct the security attacks and testing.* The lack of highly skilled testers may be why many teams opt for purely scripted verification checking. As a result, testers miss critical errors in high-risk areas such as security.

After this, the Pen tester should find and define the Pen test surfaces to attack and document them. From the exploration information and definition of where to attack, the tester conducts attacks to show weaknesses to be exploited later or in security faults.

Next, the stakeholder's response to hacking is an essential aspect of Pen testing, specifically:

- Do stakeholders have a security policy with associated compliance (assessment and testing)?
- Does development have an openness to security risk assessment and fault elimination? (Consider it imprudent or careless if they ignore security.)
- Can the Pen testing probe the project staff's security awareness and readiness (insider threats and opening phishing emails are one way to get malware into IoT code)?
- Can testers continue Pen test attacks for vulnerabilities in the products, hardware, software, and system over the lifecycle (since bad guys never stop)?

*Projects will **never** stop every hack*, but a hacker-tester's job is to make a bad guy hacker's work harder. An IoT project will need to consider security testing over the whole life with the necessary test activities done by white hat ethical hacker-testers. Finding security issues and weaknesses should lead the whole team to look for similar risks and maybe change development practices to reduce the risks.

Pen Security Attack and Risk-Based Test Planning for Systems

Planning Pen testing and considering risks is an essential skill for testers. We need to consider all the aspects of the IoT system, not just the individual parts. The individual vulnerabilities and parts are an important start. However, look at the following information. Figure 15-1 is an example of a test person working on an IoT system of systems. In the figure, we see all the IoT items this person is using. We know that many stakeholders have an interest in IoT system data. The questions to ask include

- Who needs the data?
- Who doesn't need the data?
- What systems interact with the data, and where does the interaction occur?
- How many and what kind of black hat hackers might need to be considered that the project is defending against? (Are they after personal data, a piece of data, or to prove they can get into a device or system?)

The user is the lady in Figure 15-2. She has a smart glucose monitor and an IoT insulin pump implant. These things keep her alive and her doctors informed. The connections can be made using her smartphone and watch. Just in this little example, we have a lot of attack surfaces to consider in the security Pen. This is terrible news for her yet good news for bad guys.

Faults that make an attack surface to build Pen test on include

- Missing commands or missed processes
- Comm issues and integration

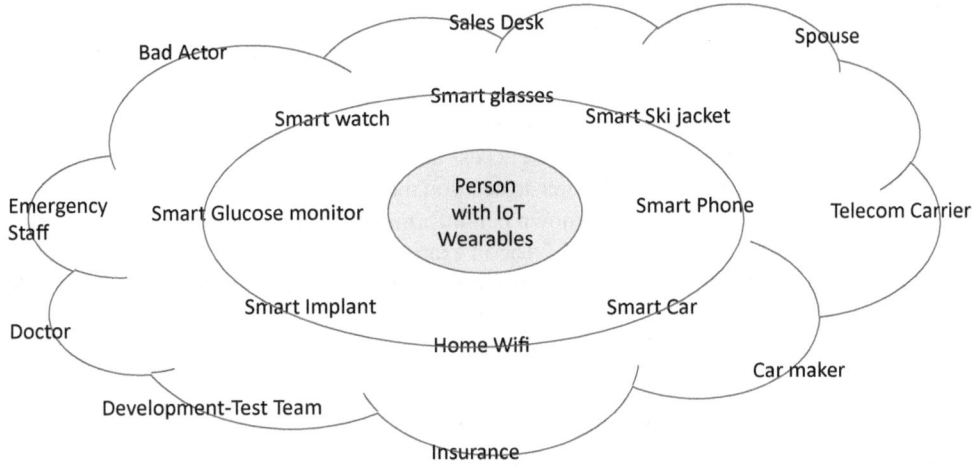

Figure 15-1. Hypothetical person with many IoT systems

IoT Navigation Security Use Case Example

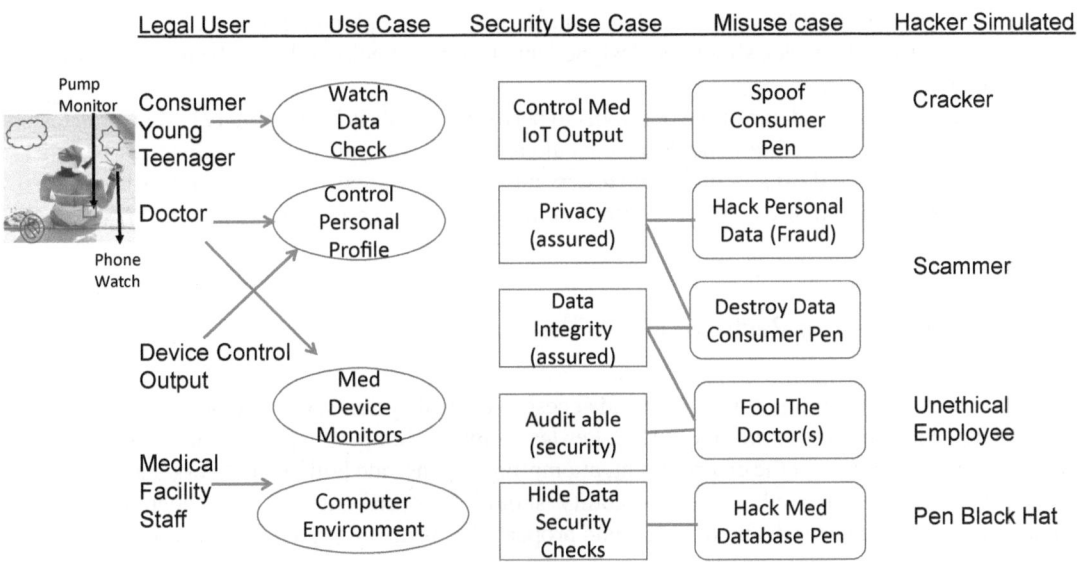

Figure 15-2. Penetration use case example for an integrated IoT/medical system

- Attacks from each cloud user access point (from outside of the cloud)
- Attacks at each connected device (the inside circle)
- The lady herself
- Fault detection and error processing

We, as testers, need good security Pen test planning and a long list of risks. Your job as a tester is to define a test plan, including the risks, what security attacks might be recommended, and why (because management will ask).

Full-Scale Penetration of the System

Using the information planning and from the earlier exploratory stages, now the hacker-tester tries to gain access to the system with focused attacks [1]. Access may be by a hacker-user or via inserted software (e.g., virus malware to keyloggers). The access and information flow should be maintained and expanded over time. For example, as the hacker-tester, you may gain more privileged access. The targets in Figure 15-1 will be varied based on the known pen information. Having a security use case design is part of this strategy plan. Consider Figure 15-2 for an example of focused use cases for a detailed plan.

As a tester, ask yourself, "does the lady or her family want hacking done to her or other bad things in this use case?" Testers need to perform the "standard" test efforts (due diligence), and then they need to play the "bad guys" (right side) by doing

- Spoofing Pen.
- Hack personal data attacks (identity theft, etc.).
- Pen vulnerabilities that can be used to "destroy consumer data" (this could result in lifesaving data lost).
- Try to figure out how to "fool" or highjack the doctors since they are deciding how much medication to use (here, you are playing an insider "unethical employee").
- In the end, use nasty tactics by attempting to hack the medical device database (can hurt many people, so great caution is needed).

Each use case should at least be considered in risk and security planning. Use cases that stakeholders consider worthy of attacks, should be designed into test pen attacks and then the team goes to work conducting hacking test.

Tools are often used, and one of the best tools to perform exploits is Metasploit [2]. Many Pen testers commonly use this tool. Additionally, testers should apply the concepts of Part 3, Chapters 10 and 11, at this test planning and design point. Some testers may be really good at this kind of test-hacking.

Given the preceding figures, can you think of other Pen tests to apply, how and why?

Simplified Process to Perform a Pen Test

Pen testing aims to determine if unauthorized access to critical systems, environments, and data types can be achieved. In the process of Figure 15-3, Pen tests should attempt to exploit security vulnerabilities and weaknesses throughout the environment, attempting to penetrate both the network and application levels. If access is achieved, the vulnerability should be corrected and the penetration testing reperformed until the test is deemed "safer" and properly limits unauthorized access or other malicious activities.

The effort starts with the test plan allocating time and staff to a Pen test process. The hacker-tester staff gathers information about the kinds of risk, processes, and software of the IoT system. Next, a network and IoT device architecture survey should be started and refined. In this set of activities, the tester identifies communication ports and services by scanning the system internally and externally. A vulnerability assessment (scan) is done from these sets of information, and a social engineering survey is produced. These provide information about firewalls and how the hacker-testers might evade detection. These steps are repeated as needed, and knowledge of the IoT system improves.

Password cracking – This is meant to find user passwords. This can be done by accessing the user's stored or transmitted data or brute force, where the hacker tries to "guess" the password often using tools (e.g., pkcrack). A variation of password cracking is where only one or two passwords are tried before going to another user account to avoid getting "locked" out of the system.

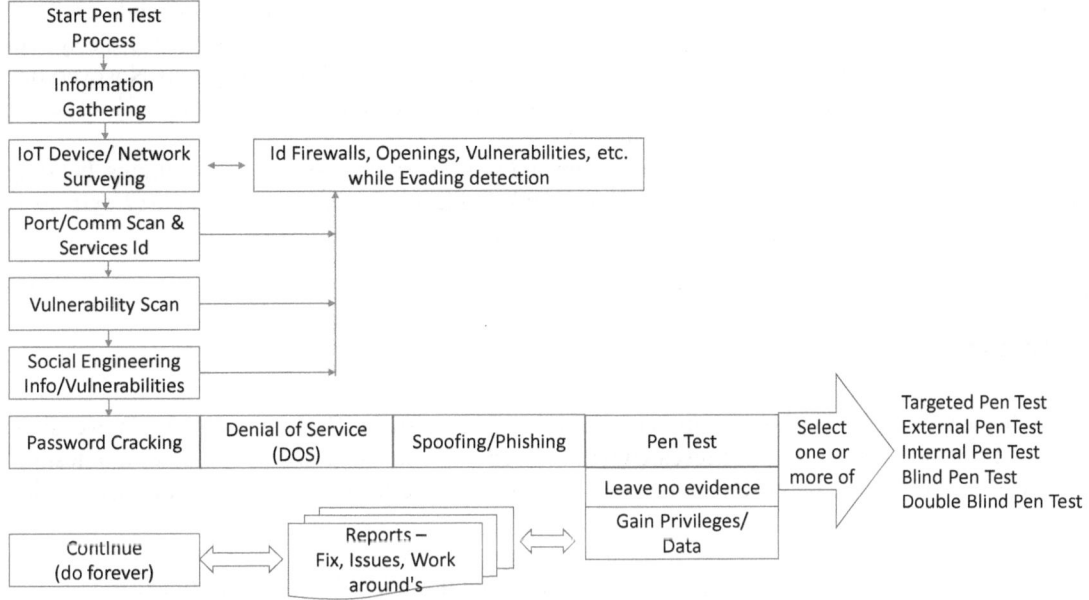

Figure 15-3. Pen test process steps

Reference: Jon Hagar Class

Denial of service (DoS) attack – This is meant to shut down an IoT device or system, making them inaccessible to its intended users, human or nonhuman. This can be done by flooding messages on network traffic used by an IoT device. For example, sometimes, in a DoS attack, an opening of an attack surface that the hacker may see is a security access system message or warning.

Spoofing/phishing – This is meant to gain user account access by the fraudulent practice of sending emails or other user information (file attachments) purporting to be from reputable sources to encourage individuals on the inside to reveal personal security information, such as passwords, accounts, or other data. Watch out for things like "click on this link to get..." or "this is your best friend; please send me a system access password; I am lost in China."

The following is a list of Pen test attack techniques from Figure 15-3 that can help an individual or organization achieve Pen test processes:

Targeted Pen test – This test targets specific features, often new ones, or software functions with risk. The hack is performed by the organization's development/IT team and the Pen testers. This combination is done so that everyone can see the test being carried out.

External Pen test – This type of Pen test targets a company's externally visible clouds, edges, or IoT devices. The objective is to determine if the attacking party can get in and what they can access once they get in.

Internal Pen test – This test simulates a hack behind a firewall by a tester with standard access privileges and assesses how much damage an employee could cause.

Blind Pen test – The IV&V team simulates the actions and procedures of an actual attacker since the information given to IV&V is the same as a real hacker would start with. The IV&V can be a team or a single person.

Double-blind Pen test – An IV&V blind Pen test expanded to assess IoT operations security monitoring, incident identification, and response process.

Once the hacker-tester has information that yields security threats, one should show the Pen test has been completed. Again, you may get the statement, "nobody would ever do that." "Yeah, right," says the bad guy hacker, "just keep telling yourself that, and I will keep hacking."

The Pen test team may be internal or external. The testers are trying to show vulnerabilities. They need a devious thought pattern, but their efforts should not destroy the project by hurting the system or leaving things like malware in the code or access to the code of the IoT device/system. Not everyone is suited to this job by personality or knowledge. Hence, subcontracting to outside, well-respected hacker-testing companies will often be better.

A word of warning at this point, hacker-testers and projects should be doing these things, but only with the knowledge of management or creditable stakeholders. *Real hacking of many IoT systems is illegal.* You do NOT want to lose your job or go to jail over this kind of testing.

Summary

This chapter is a dive into the security testing world of Pen testing. There are many tools, websites, and books on this subject, so more research will be needed if you are tasked to do this kind of testing. I advocate that the best defense (staying safe from IoT device hacking) is a good offense (having tester-hackers attack your IoT device).

The next chapter introduces the IoT test environment and labs that support security testing as well as functional and nonfunctional quality assessment, which flows into Part 4 of this book.

References

1. https://en.wikipedia.org/wiki/Penetration_test – accessed spring 2022
2. *Metasploit: The Penetration Tester's Guide* 1st Edition by David Kennedy, Jim O'Gorman, Devon Kearns, Mati Aharoni, No Starch Press, 2011

Chapter 16
IoT Test Environment Introduction

In the previous two chapters, I considered security testing for IoT systems. To support security testing, teams need test environments, labs, and tooling. This chapter addresses lab planning and requirements, preferred test environments for full IoT integration, field testing, cost and scheduling a test lab, and test lab tool considerations.

Additionally, in this chapter, I introduce the topics of the test environment, which are covered in Part 4. Depending on the nature and size of the IoT device, the test environment will vary. Test environment options introduced in this chapter include

Option 1 – Desktop lab
Option 2 – Simple lab
Option 3 – System/software integration lab (SIL)

- Security test lab

Option 4 – Full SIL
Option 5 – Real-world SIL chaos engineering

The book referenced in [1] addresses setting up the test environment extensively. I will not repeat all of that information here, but I will tailor the ideas of test environments to IoT and security.

Test Lab Lifecycle

I view the creation and maintenance of a test environment as a project inside a project. The environment can be short and easy (Option 1) or massively complex and risky (Option 5). Let's begin with a mini-test environment production lifecycle as shown in Figure 16-1.

This lifecycle picture in Figure 16-1 is a variation of information from Deming [2] and the National Institute of Standards and Technology (NIST) software security cycle [3], found in Chapter 13. The lab operations and requirements (plan phase) start during the test proposal. The test lab environment functions with the software test architecture (STA) and various supporting configuration processes (do phase). For example, as the IoT device changes in hardware, software, and operations, the test environment must change configuration and usage. During the testing (check phase), the tester assesses the quality and runs security attacks in the test system integration lab (SIL). The SIL (sandbox for security) environments must be maintained. Finally, the results and data produced by the test

© Jon Duncan Hagar 2022
J. D. Hagar, *IoT System Testing*, https://doi.org/10.1007/978-1-4842-8276-2_16

environment drive project decisions (act phase) with information on IoT device maturity, risk retirement, or actual issues that may need to be fixed. This repeated cycle of test environment usage continues over the entire project lifecycle.

> **SIL MAINTENANCE OVER A LONG DEVELOPMENT CYCLE**
> Test environments that function longer on a critical IoT system will need maintenance and equipment recalibration because things wear out over time. In one case, certifications on support equipment for the IoT hardware expired during the test schedule, and there was no time to stop testing. The solution, with some risk, was to continue using the equipment until the test was completed. The risk was that "if something was not right on the calibration, the completed tests would be invalid." The test manager wrote a memo documenting this situation and risk. Management and quality control engineering okayed the deviation. Luckily for the test manager, the equipment was recalibrated and passed certification at the end of the testing cycle. The testing and schedule were saved.

Test Lab Refresher

In IoT software system testing, having a good test environment lab is paramount. The lab is tailored to the system and IoT device so that each lab will be different. Critical points from lab development in [1] include

- Labs provide the environment to do comprehensive testing/V&V.
- Labs have a lifecycle the same as any other engineered product:
  - Requirements
  - Cost and schedule
  - Design
  - Implementation
  - Testing the test lab
  - Support from trained and skilled staff

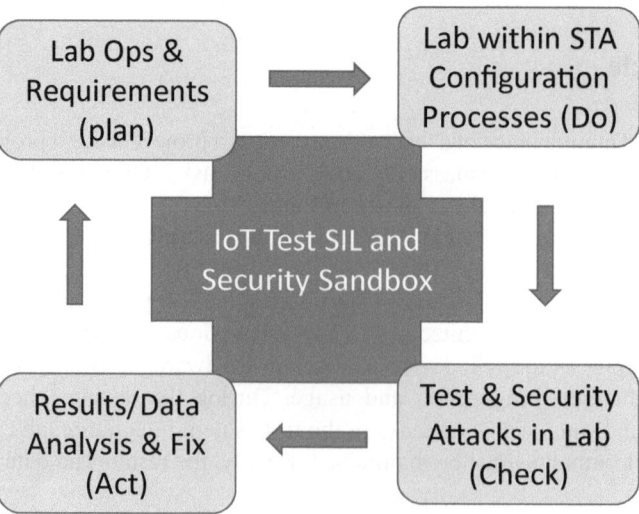

Figure 16-1. The IoT SIL/lab activities (repeats throughout the lifecycle)

Reference: Jon Hagar Class

- Ops of the environment
- Management and reporting
- Full lifecycle

- Simulation is a key early part of labs.
- Prototype lab versions support early testing.
- Developers-security-operations and testers use labs.

  - Features and tools in the lab will change.
  - Rapid reconfiguration may be needed.

- Labs support product testing and integration.
- Labs support pre- to post-release testing/V&V.
- Labs may need to integrate with field testing.
- Labs may need to be certified for use.
- Labs during Ops and maintenance (O&M) need to be maintained and updated.

  - Labs are not without lifecycle challenges and issues.

- Automation of lab activities, but automation must be used correctly.

  - Management, simulations, execution, analysis, reporting

- Lab tools vary by IoT device.

  - Dev, test, security, Ops, and others

- Labs must support data capture and analysis, including AI.
- Labs support the assessment of qualities:

  - Security, performance, safety, reliability, etc.

- Labs are developed and run by independent specialists.
- Labs require proper funding for maintenance and upgrades of hardware and software.
- Labs have specific retirement and disposal concerns.

For IoT, creating a SIL or a lab starts with planning requirements, continues during the lifecycle, and ends with the lab's retirement. Development of these facilities is conducted by a multidisciplined team of engineers and management.

IoT Lab/SIL Planning and Requirements

The test environment or SIL must support some or all of the following test/V&V efforts:

1. Software testing

   1. Structural testing
   2. Functional testing
   3. Quality testing

2. Integration testing

   1. Communications
   2. Other IoT devices or systems or systems of systems
   3. Other systems
   4. Hardware
   5. Software

3. System testing

    1. As much of the system(s) as can be put in a lab
    2. Interfaces to the outside IoT world (chaos engineering)

4. Provide the needed tools

    1. See the following test tool clause

5. Miscellaneous support
    1. Planning, risk, cost, schedule
    2. Growth
    3. Parts (hardware and software) aging
    4. Reporting
    5. Specialized testing
    6. User testing

This list may not include every item and consideration for the many variations of test labs, but it can give a lab design team a good start.

Preferred Test Environment with Full Integration of a Complex IoT System

A test environment, after its creation, integrates with the IoT device under test. There are parallel development efforts conducted between the development and test teams, as shown in Figure 16-2. Close coordination, work, and communication between these teams are a must. Teams often need to use prototypes to support the IoT project (see the Test and Proof of Concept Pre-Contract study). In Figure 16-2, the development team is doing engineering following the flow of the figure's left side. The development team creates the design, code, prototypes, models,

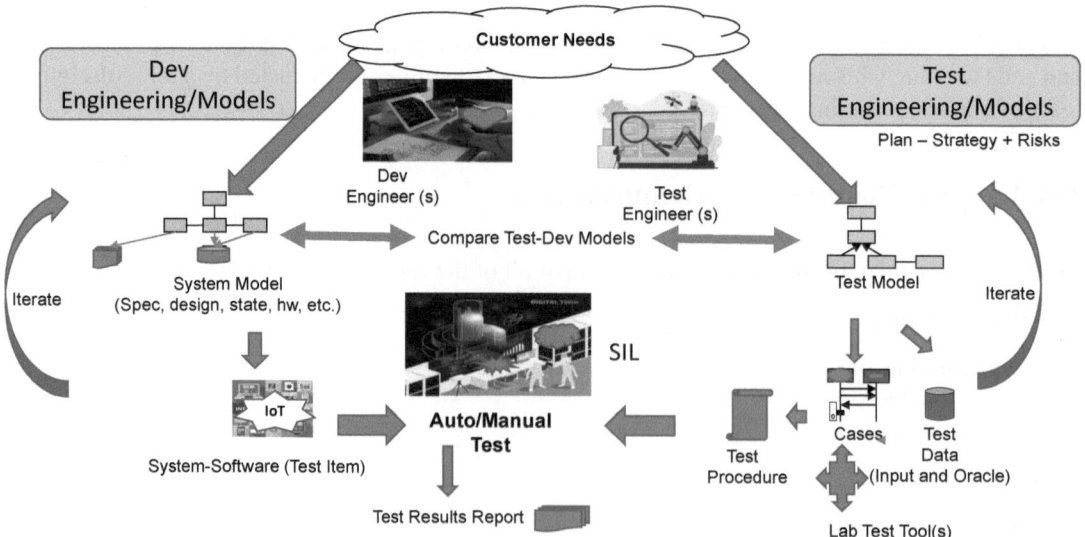

Figure 16-2. Example development vs. SIL test development steps

data, and documentation. These are created iteratively to one degree or another, whether agile or more traditional.

In parallel, the test engineering team is also creating test systems, hardware, software, and models. The two teams can compare the development and test models or information at various project stages. This helps to find IoT quality and integration issues early on. In such an effort, testers are a valuable parallel support team to the development team since both provide valuable design information to each other. The test environment is readied as the IoT device, hardware, and software mature in this parallel effort. This means test models, data, cases, lab tools, and usage procedures should be constructed. The parallel development and test efforts help both IoT teams shift test design efforts to the "left" on the schedule of the lifecycle. When the IoT products and the test environments are ready, they support rapid test execution. As the test environment is designed, automation of testing should be considered. In optimal cases, both the IoT device and test environment(s) are used and tested in parallel, so the testing cycle is almost complete when IoT products are ready. This is so the IoT system, test results, and reporting can be given to the stakeholders. These ideas support Agile (Part 3, Chapter 12).

There are, of course, variations on this preferred example as many factors such as supplier lifecycles could impact this example.

TEST AND PROOF OF CONCEPT PRE-CONTRACT
In one case, the development and test effort started even before management had approved the production of a device. This early agile development cycle produced hardware, software, and a straightforward test environment. Testers executed tests that worked but also showed 80 percent CPU usage, with only 5 percent of the software being done. This meant there were going to be performance problems for the software. So, when the project was approved, development and test focused on making the software as efficient as possible, and performance testing was done constantly. Ultimately, the IoT device was created and tested, meeting timing performance margins. The early Dev-testing agile sprint was very successful. This agile Dev-test collaboration is shown in Figure 16-3.

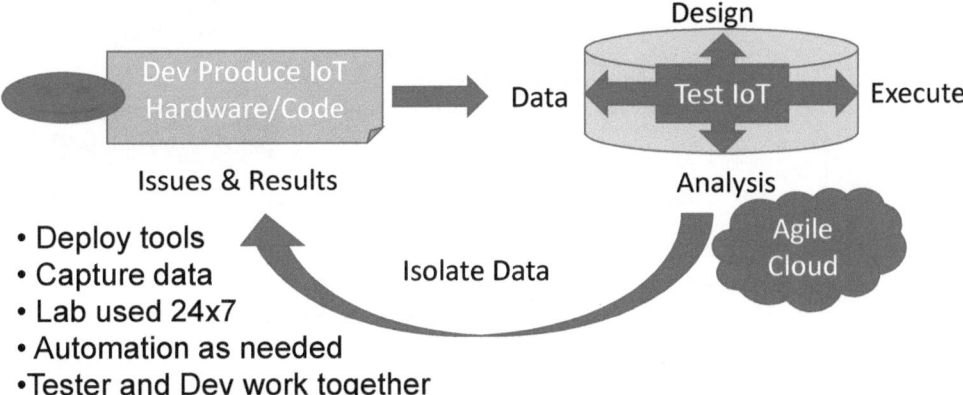

Figure 16-3. Dev-test agile IoT collaboration in real time and post-assessments

Reference: Jon Hagar Class

The collaboration between groups benefits IoT in any lifecycle and project. This example shows that testers and hardware, software, and system engineers get better IoT products with parallel development and interactive feedback cycles. Test tools, automation, and data capture should be ongoing engineering. The massive amounts of data possible from IoT will require data analytics and isolation of important information from the cloud where test results are recorded. The testing speeds mean that post-test isolation is also needed because not all IoT information and issues are immediately apparent.

The Field Test Environment, Analytics, and SIL Working Together

The IoT test data analysis extends from the lab into the real world. For those IoT systems that must enter test and analysis in the real world, the IoT with development, operations, testing, security, and data analysis must continuously work together. The development team completes testing on each product release and updates in the real-world test environment with testers and security staff, as shown in Figure 16-4.

As shown in Figure 16-4, the amounts and speed of test data are likely to increase. This leads to ongoing testing/analysis, chaos engineering, use of the cloud, realtime data feedback, and AI monitors. These activities make IoT systems even larger and more complex. Testing and analysis outside the lab should continue as the IoT device moves into the real world. Operations will share data with testers to keep that interface active. This data sharing with testers may require showing stakeholders the benefits (price, free things, features, etc.). Information shared can include user data, tester views of the data, media, measurements, and other information. The amount of data created by hundreds or millions of users will lead to AI and smart data analysis (Part 4, Chapter 22). Yes, such data will be of interest to the marketing staff and management, but the development and test teams need to be equal in their usage of real-world data. Some of the data will end up being used in real time analysis thus changing how the IoT system is used and maybe where it "fails." Also, more of the data will need to be used in long-term analysis, looking for trends, threads, issues, and other helpful information to improve IoT systems of the future.

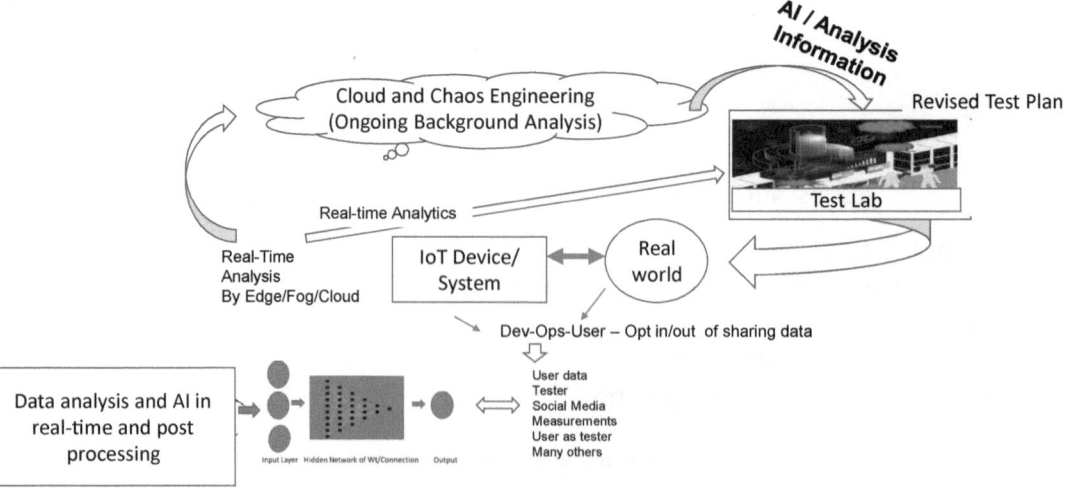

Figure 16-4. DevOps-Sec-test and analysis never stops in the real world of complex IoT systems

Reference: Jon Hagar Class Composite

This ongoing engineering may seem costly, but companies have found gold mining in data usage. The marketing/salespeople may lead the data gathering, but all engineering staff need to be first in line to get and use the data. For example, new product development, upgrades, and issue fixes will keep stakeholders happy and sales going. Once the product is in the real world, the user may have to opt in to allow the sharing of their data. Companies will find ways to make this sharing of data happen. As data is shared, a constant flow of information happens. Yes, users will likely be given the option *not* to share usage data, but again companies will find ways to encourage data sharing, in whole or part. The data may flow from the device/system, the edge, fog, or cloud. Some of the data will flow directly back to the development team in real time. Other data may take more time and processing in the background of the cloud. This AI/analysis information should flow into the company. Testers should look at and for

- Improving use cases
- Issues/bugs
- Missed test cases to bring into the test lab
- Test lab improvements
- Better input data streams for the test lab to use
- Changes to risks or risk list
- Test designs that can be improved

The staff may be involved in other projects and test activities. Continuous product and engineering improvement at the right "good enough" option will be the hallmark of a successful organization.

Deep Dive on Test Environment Cost and Schedule Introduction

In Part 1, there is information on cost and schedule in IoT testing/V&V. In this section, we take a step back to address a question many IoT test teams ask, "how do we plan, estimate, and schedule our IoT test environment?"

There are no absolute formulas or fixed percentages, but I am asked where to start when I produce a bottom-up, detailed cost and schedule estimate. In bottom-up, I create a detailed list of each item of cost in the SIL and how long it might take to make it or buy it. I have spent months doing this on some projects. However, I did start with a rough target based on a percentage of test and Dev team efforts. Such a very rough set of percentages can be seen in Figure 16-5. It is only one of many hypothetical allocations, but it gives a team a start.

I started with a little bit of development efforts and how they might allocate percentages of effort. I would note that there were changes to the numbers and allocation almost daily on a real estimation effort, so I had to change test numbers.

Next, I had a list of efforts and tasks (items under the triangle). The exact nature of these things and details are not important for this example. I then allocated percentages of V&V/test effort to each effort task. Some percentages were more significant because, from experience, my teams spent a lot of effort on them. Using percentages allowed me to change my actual numbers, those things management wanted, as development changed their numbers. I had this information in a spreadsheet, and I had to make a few changes and regenerate my spreadsheet.

Finally, I knew creating a lab and SIL within the STA was a considerable effort. Depicted in Figure 16-5, I had SIL creation percentages for each major effort task. For an approximation, I start with the cost of the SIL at 30 percent of the development. Some will say this number is high and want to lower it, but if I am talking about IoT device systems (option 4 or 5), the SIL/STA will be prominent and have many parts (hardware, software, Ops, etc.) and tools. Again, as a percentage, I could recalculate my SIL costs and schedule quickly as actual numbers changed. There is nothing magical or absolute about these percentages. Your percentages will be different based on your SIL/STA context. The ideas here are the percentages and the allocations. The next chapter has more information on STA. This approach is a top-down estimate, providing a sanity check of a more detailed bottom-up estimate.

Estimates for your project context may not look like mine. I create line-item cost, task, and schedule for each detailed option item in a bottom-up estimate not presented here. Items would include individual hardware, software, tools, people, support systems, training, maintenance, testing of the test items, etc.

Test Tool Introduction for IoT

There are many kinds of test support tools for an IoT SIL environment. The exact tool vendors and names are IoT addressing device and project specific. I will not be addressing specific tools in most cases for this book, but readers should be aware of tool categories for the IoT test environment. The following are tool *categories* that should be considered for IoT SIL. Figure 16-6 illustrates an example IoT test tool environment.

Such a toolset would be a combination of tools. The capabilities of tool types in Figure 16-6 include

- Planning, risk, and team management
- Cost and schedule support
- "Test cloud, architecture, and SIL" environment with a supporting test database to record all test results
- Analyzers for results and data retention for the long term.
- The AI/data analytics system can cover both SIL test data and data from the outside "real world."
- "Processed data" is provided to testers for future testing, management, and other stakeholders.

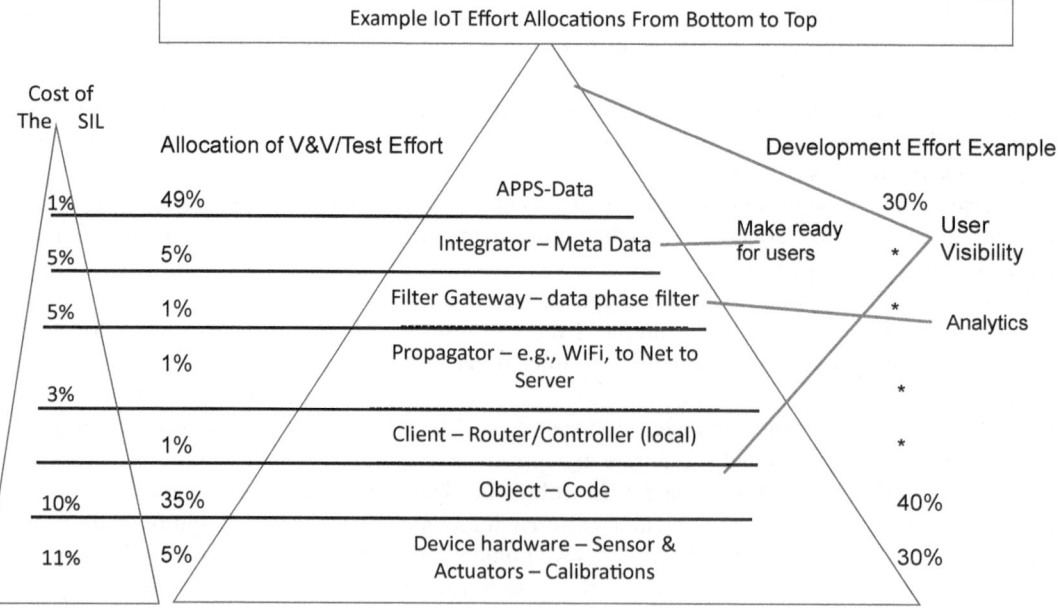

Figure 16-5. Sample cost allocation

Reference: Jon Hagar Class

- Center testbed and IoT system integration lab (SIL), which is made up of
- "Keyword/tools," test automation framework tools
- Local SIL "test database" recording system
- Support "hardware and software"
- "Recorders and scopes" supporting IoT hardware (e.g., probes, timing, etc.)
- AI/data analytics tools in the upper right of Figure 16-6 (For details see Part 4, Chapter 22)
- Communication and collaboration by providing analyzed information to the team (lower left of the figure)
- Documentation – Plans, procedures, inputs, outputs, data streams, analyses, views, issues, etc.
- Real-world support, if applicable (lower middle of Figure 16-6 (For details see Part 4, Chapter 22)

This example is iterative and repeated over the project lifecycle. This STA environment is only one example of many possible, but it satisfies many IoT system team needs. Communication and collaboration in real time during testing and post-test efforts over the complete lifecycle are of the highest importance in tooling. Communication would include the following stakeholders:

- Test team
- Managers (program and organization)
- Customers and other stakeholders (governments, regulators, etc.)
- Dev
- Ops
- Support staff (QA, QC, CM, SCM, AI, data analytics, etc.)
- Lab devices (signals) and remote labs (field, support, subcontractors, etc.)
- Data analysis – Memory coverage, data memory usage, ports, timing performance, stressing, load, etc.

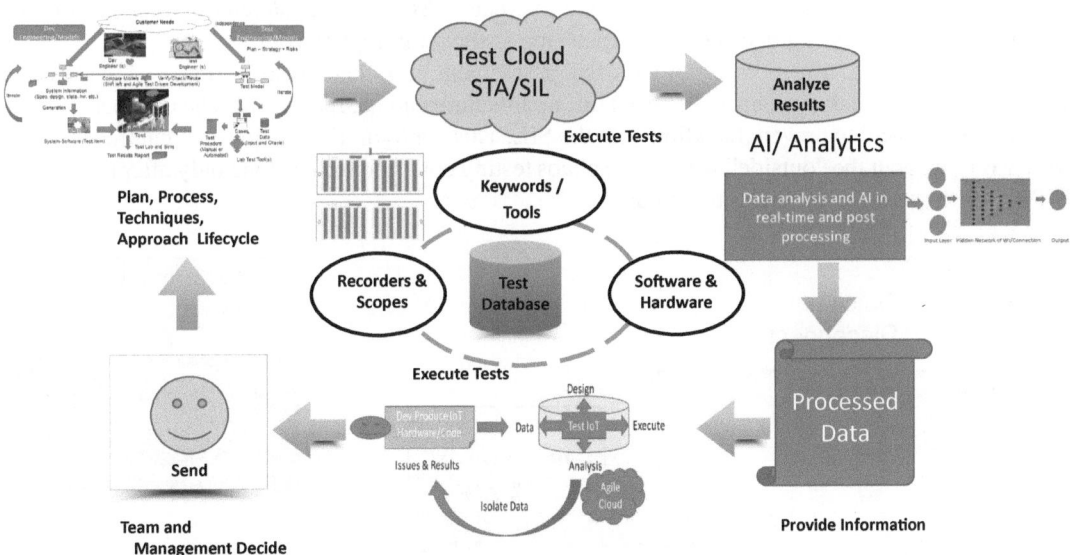

Figure 16-6. Example tool-driven STA/SIL

Reference: Jon Hagar Class Composite

AI and data analysis tools can include

- Test coverage in dynamic testing (structural and functional) identifying covered and uncovered testing
- Quality coverage (especially performance, memory usage, usability, security, validating requirements, etc.)
- AI and data analytics (Dev, Ops, test, quality, sales, management, etc.)
- Graphic and numeric displays

Technique and approach tooling includes

- Test automation and execution for the test lab and architecture (Part 4)
- MBT tooling (modeling) and execution (Part 3, Chapter 11)
- Emulators and simulators (Part 4, Chapter 19)
- Mind mapping (Part 3, Chapter 10)
- Fuzzing (Part 3, Chapter 11)
- Combinatorial test tools (Part 3, Chapter 11)

Hardware support (commercial)

- Examples – Scopes, signal generators, software-defined radios, signal capture equipment
- Specialized – Hardware emulators, in-circuit emulators, simulators (custom or OTS)

Test Hardware Setup for SIL-Chaos Engineering Support with ZIF Connectors

There are sets of tools, hardware, and software to create a SIL testbed that can interface to the real world to support chaos engineering. A sample setup is shown in Figure 16-7.

This example SIL test toolset supports IoT assessment and chaos testing (Part 4, Chapter 19). There is a needed quick disconnect between this architecture and the real world. The connection to the SIL is into a firewall to prevent backflow into the SIL. This configuration is optional, but the designers were worried about the "outside" world. The chaos testing configuration is done only after the SIL and IoT black box testing is sufficiently mature.

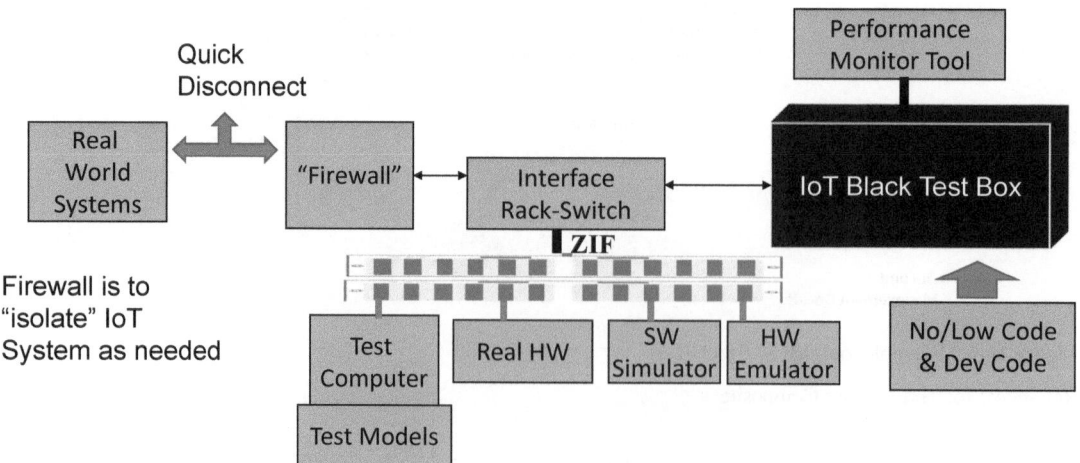

Figure 16-7. Example ZIF (zero insertion force) connectors of a chaos supporting SIL testbed [4]

Reference: Jon Hagar Class

Inside the SIL structure, tools, hardware, and software support rapid, agile configurations over the IoT system test lifecycle. This was a development design choice to support testing and analysis of critical resources. The testing starts in the IoT black box, which houses the IoT computer processing hardware, but not the IoT support hardware. The software under test is loaded into the black box (e.g., OTS, no-code, microservices, low-code, operating system, and development code). Each of these is mapped into the computer's memory under test. The performance monitoring tool, a computer and special software, can check internal black box performance timing and memory usage.

Of interest in this example is the patch "interface rack-switch" panel. This hardware design selection of ZIF connectors was made to allow fast switching of hardware and software over multiple time configurations without damaging equipment. This SIL design feature allows fast and early configuration of IoT devices and supports the software. For many IoT testbeds, this setup is recommended.

There can be a variety of tools connected via the patch panel. This configuration starts with a simulation "test computer" that runs simulations, MBT keyword automation, test setup monitoring tools, and test models. "Test models" address test design cases, test suites, test procedures, and results documentation.

Next, the patch panel allows IoT hardware and software to be quickly configured or reconfigured. In the early schedules of agile testing, where the team does not have the IoT hardware, the ability to have simulated or emulated (Part 4, Chapter 19) hardware allows the test of the IoT black box software. As the real hardware arrives or a new version of the hardware is configured, the test SIL setup allows rapid reconfiguration without damaging equipment.

Summary

The test environment or lab is essential for IoT systems where we must test both hardware and software to have complete testing/V&V. This chapter introduced this complex subject covering only the top levels. The following chapters cover the SIL and test environment architecture in more detail.

References

1. *Software Test Attacks to Break Mobile and Embedded Devices* by Jon D. Hagar, CRC Press, 2013
2. Deming – https://en.wikipedia.org/wiki/PDCA#:~:text=PDCA%20(plan%E2%80%93do%E2%80%93check,study%E2%80%93act%20(PDSA) – accessed 2022
3. NIST security cycle – https://csrc.nist.gov/publications/detail/sp/800-218/final
4. http://principlesofchaos.org/ – accessed spring 2022

Figure Reference

https://img.freepik.com/free-vector/laptop-software-assisting-testing-process-tiny-people-testers-automated-testing-automotive-executed-test-software-auto-tester-concept_335657-2437.jpg?w=826; https://img.freepik.com/free-photo/close-up-image-programer-working-his-desk-office_1098-18707.jpg?w=826; www.freepik.com/free-vector/illustration-digital-twins-testing-simulation_20878564.htm#query=Software%20test%20lab&position=0&from_view=search

Part 4
IoT Architectures, Environments, and Integrated Independent Testing

After introducing architecture and test support environments in Chapter 16, this part expands on architecture and explains why architecture is critical for IoT more than other test environments. Since the real world of IoT is infinite, this book cannot address every possibility of software test environments and architectures. This part presents some general test environment classification, but readers will have to extrapolate to the local environments they need for IoT testing.

The supporting parts of IoT software test architecture may be new to some testers from the web or IT world. The architecture supports include specialized software integration test (SIL) labs, test automation tools, and independence in testing, often done by independent verification and validation (IV&V) groups. The final chapter begins by considering data analytics, including test information and support from AI. The last chapter, while an introduction analytics, leads testers to start the journey into AI and analytics, which are rapidly growing support areas for testing.

Chapter 17
Architectures Critical to Project Success

IoT is part of the cyber-physical world of systems that are often complex and involve risk. As systems and software have grown in usage, the need for architectures has increased. Examples of these systems include the ideas of "smart" cities and even "smart" houses where numerous IoT devices, computers, and edge and cloud systems interact with each other, often in unexpected ways, both good and bad (failures).

Cyber-Physical Systems

NIST (National Institute of Standards and Technology, US government) defines "cyber-physical systems (CPSs)" or "smart" systems as "co-engineered interacting networks of physical and computational components." The huge scope of CPS technologies depicted in Figure 17-1 includes

- The IoT
- The Industrial Internet of Things (IIoT)
- Smart cities (made up of IoT and IIoT)
- Smart grids (IoT and IIoT)
- "Smart" anything (cars, buildings, homes, manufacturing, hospitals, appliances, personal devices, and others)

Many of these are large complex systems or systems of systems. Since CPSs function in the real world with software and a wide array of networks, testers will need to be doing test engineering outside the small software labspace. This part of the book focuses on IoT environments and architectures to support testing, which expands into hardware, physical interaction, operations, and varied environments found in the universe of systems.

The industry struggles with classifying this topology which extends into test environments and software test architectures. The NIST CPS is one view of IoT/IIoT, and another view is depicted in Figure 17-2. The figure's classification ranges from the familiar software structural code testing in a software-only test lab to full field testing in the real world, including a variety of network connections. Knowing when to stop testing is a familiar problem to testers. This problem grows as the normal boundaries of software testing disappear and testers move into real-world and systems of systems testing.

Many development organizations will likely determine a test strategy and plan to limit or eliminate many linked aspects of these testing topologies because of cost and ownership. This limitation is acceptable as long as the information about the IoT devices defines the topology limitations that the planned IoT testing addresses. The limit may be done in product specifications, fine print, or user

© Jon Duncan Hagar 2022

J. D. Hagar, *IoT System Testing*, https://doi.org/10.1007/978-1-4842-8276-2_17

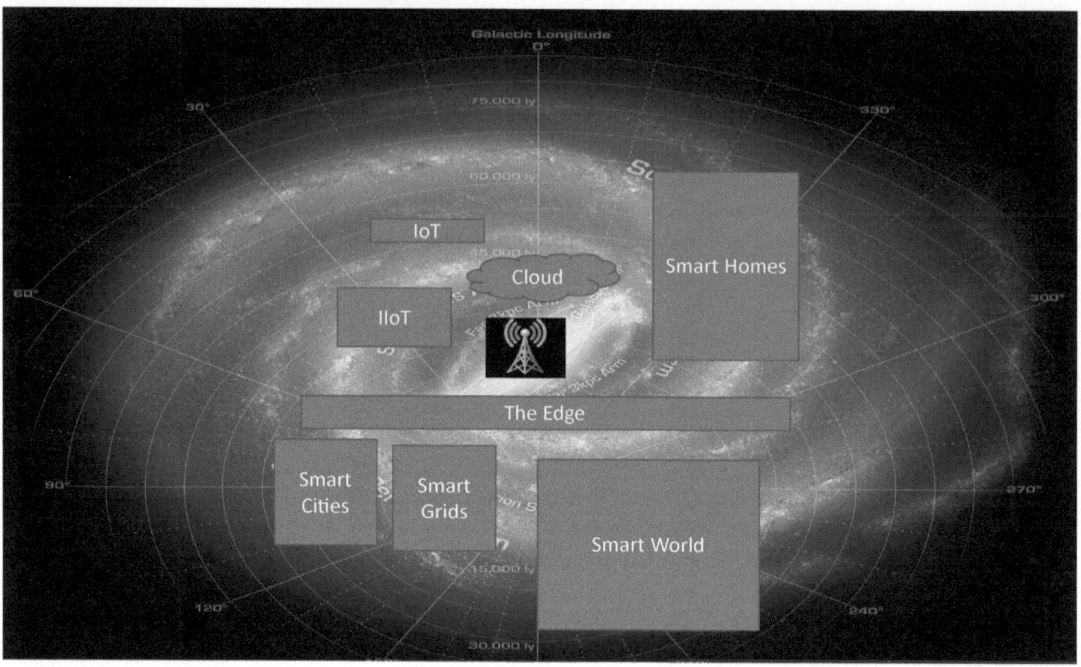

Figure 17-1. IoT technology universe

Reference: NASA image and Jon Hagar

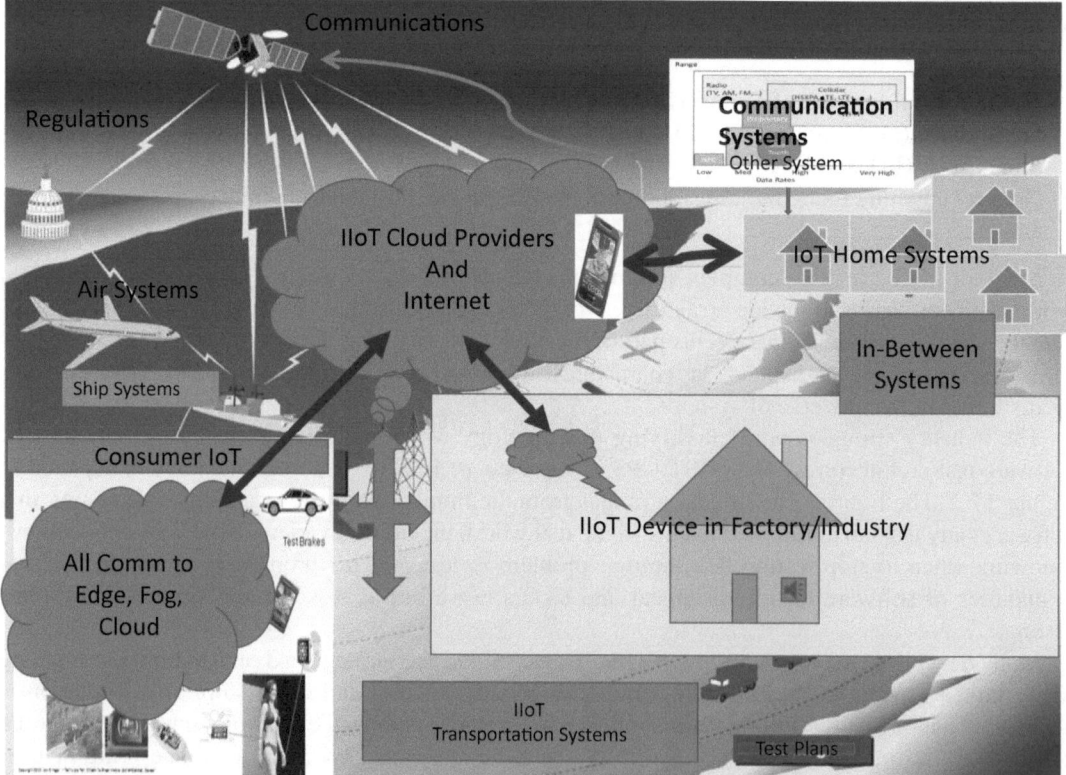

Figure 17-2. IoT of everything

Reference: Part 1 image

IoT Product/Project Maturity Steps View

Figure 17-3. Example IoT product and organization maturity framework

Reference: Jon Hagar

guides, but disclosing such information will likely become common as IoT/IIoT matures. For example, the industry sees this already where some wearable devices disclaim being a medical or safety system. I suspect that such disclaimers will be evaluated and then required by regulators.

Such disclaimers will also allow users and stakeholders to decide where to establish their testing/V&V activities. For example, it has been common practice for years in the large corporate IT world to have internal IT groups test and evaluate IT systems, hardware, and software before pushing them into the corporate user base. This practice becomes more important as security and reliability quality factors increase for IT-IoT groups. One can expect companies, government organizations, and extensive user communities to expand such IT-IoT evaluation groups into active test/V&V efforts.

An example IoT product maturity framework is shown in Figure 17-3. It ranges from the newbie company of Part 1, where they are just looking to stay alive with proof of concept, to level 4 products, where income revenue and cost savings for an organization occur. The likely situation is that the framework test environment will change and grow with a project as it grows and thrives. With success, products and projects will move upward through the levels.

Figure 17-3 continues the mapping and classifications of this book's Parts 1 and 2. However, here I focus on critical elements of test environments and architecture.

Environments and Architectures for IoT

IoT testers and companies currently tend to localize testing considerations to the IoT device and maybe the edge. This shallow focus can leave areas of large-scale IoT systems (e.g., system of systems, factory, city, or country) undertested. Given that IoT devices are predicted to be everywhere, in everything, and used at many different architectural scales, a problem that test organizations will face is how to define a software test architecture with associated test environments that are sufficient and comprehensive to support various factor levels of testing (see Part 1, Chapter 4).

As I have noted earlier, this situation leaves many risks untested since teams do not fully understand the whole picture of device usage, the Internet, and the larger scope found beyond the local test plan of the device and its ill-informed test strategy. Further, since the testing industry does not fully understand test architecture [1], engineers will struggle to define IoT environments thoroughly enough to do their jobs. The nature of unintended consequences [2] from lack of test knowledge about proper IoT world testing will lead to an increased likelihood of failures and challenges in security, reliability, interoperability, safety, functionality, use, and others [3]. These challenges are critical to ultimate stakeholder-user satisfaction.

One might expect the phrases "software test architecture" (STA), "IoT test architectures," and "IoT test environment" to be defined in IEEE Standards Association or even SEVOCAB. While not without controversy on universal correctness, these references do not define these phrases or the concepts surrounding them very well, at all. And standards such as ISO and IEEE offer little help on STA, currently. A general Internet search on software architecture yields over 5000 hits, but the top sites do not offer much help and only represent a single author's viewpoint. While individual authors have valid information, I will use the following definitions in this book.

Architecture Definition of Terms for This Book

To start refining the definition of the phrase "software architecture" or "software test architecture," I find it is helpful to take a step back and consider the historical usage of the word "architecture" from other sources. The oldest historical usage is from civil engineering and building construction. The Wikipedia website [4] offers the following definitions:

> *Architecture (civil engineering) – is both the engineering processes and the products of planning, designing, and constructing buildings and other physical structures.*
> *Test Environment – physical and logical support structures where testing is done.*

The test environment aims to allow human testers to exercise new and changed code via either automated checks or non-automated techniques. After the developer completes the coding, integrates, and configures it, the items are moved to test environments. Different types of testing suggest different test environments, some or all of which can be done in parallel testing. For example, automated user interface tests may occur across several virtual operating systems and displays (real or virtual). Performance tests may require a normalized physical baseline hardware configuration to compare performance test results over time. Availability or durability testing may depend on failure simulators in virtual hardware and networks, all of which can be done concurrently.

Tests may be serial or parallel, depending on the sophistication of the test environment and the type of development test lifecycle model. A significant goal for Agile and other high-productivity software development practices is reducing the time from software design to delivery in production. For Agile and many IoT DevOps systems, highly automated and parallelized test environments are essential to rapid software development.

Historical Reference: Architecture in Engineering and Literature

For me, building architecture can result in engineering functional and elegant structures. As a mathematician, I look for both in a solution space. The construction industry dates back thousands of years. If enough of these large systems and systems of systems use the STA concept and refine it with a definition, its usage can be shared, and a consensus reached. This history should be our starting point for defining the phrase, software test architecture, as software testers.

Schools offer specialized training for students to become architects supporting the building and construction industry. In architecture, there are hand-offs between architects and other building engineering disciplines (e.g., civil, structural, electrical, mechanical, heat/air-conditioning, etc.). Additionally, as the architecture progresses into the construction phases, it is common for the people doing the actual work to find and correct architectural "issues" in the implementation efforts. Still, many times they must consult with the original architect.

Can you see how these concepts and ideas might help software testing? The following list presents a definition from individual test practitioners and Professor Dr. Nishi Yasuharu from the University of Electro-Communications, Japan, taken from conference materials [5].

- STA is a big picture of test design.

  - Test engineers have to grasp the big picture of test design because test cases increase by over 100,000 cases and get much more complicated.
  - Test techniques and coverages cannot prevent large lacks of test cases though they can prevent small lacks of test cases.
  - Quality of test design depends more on total balance than the priority of each test case.

- Test architecture is just architecture of test design.

  - In software testing domains, people confuse big pictures of test design and big pictures of test process or test management.
  - In software development, software architecture is not described in project plans though test architecture is described in test plans.
  - What kinds of tests you design should be prior to ordering test cases.

- Test architecture consists of "test viewpoints" and relationships between them.

I like parts of Professor Yasuharu's definition for STA as his usage is closer to the definition I envision. Still, as testers evolve the IoT concept (reusing ideas from the construction industry), there is room for improvement, as the boxed definitions address my STA usage and related concepts.

Software test architecture (STA) *is the process(es) and the product(s) of planning, designing, and constructing software tests done with supporting test structures.*

Note
For IoT, STA includes hardware of the IoT hardware and software systems, interfaces, and Internet connections, and it may include external hardware and world systems (things connected to the IoT device).

Supporting test structures include test tools, sub-environments, planning, documentation, tooling, views/viewpoints, and analytics (Part 4).

For IoT systems, an essential test architectural support structure is the environment. The website in [6] gives the following definition of the computing environment.

(Computing) Environment (to support testing) – The overall structure and facilities wherein a user, computer, or program operates.

(Software) User – Typically, humans are the only users who interact with the software system. But, in IoT, the software user is expanded to include hardware, interfaces, and other systems.

Summary

This chapter introduced Part 4, covering overview concepts and definitions. Since IoT has new topics in testing, such as software test architecture (STA), many concepts are still evolving. They are essential in the big picture of IoT systems of systems. There is a history that STA and definitions are built on, but as IoT and projects evolve, test teams need to stay current with the concepts and definitions started in this chapter. They are expanded in later Part 4 chapters. The next chapter addresses the general ideals of software architectures.

References

1. J.D. Hagar, "Defining the phrase software test architecture emerging idea," in 2017 IEEE International Conference on Software Testing, Verification and Validation Workshops (ICSTW), 2017, pp. 313–316
2. https://en.wikipedia.org/wiki/Unintended_consequences – accessed winter 2022
3. "IEEE/ISO/IEC 29119, ISO/IEC/IEEE International Standard – Software and systems engineering – Software testing" – Parts 1 to 5 series
4. https://en.wikipedia.org/wiki/Architecture – accessed winter 2022
5. S. Masuda, Y. Nishi, and K. Suzuki, "Complex software testing analysis using international standards." Porto, Portugal: IEEE, 2020, pp. 241–246
6. SEVOCAB – https://pascal.computer.org/sev_display/index.action – accessed winter 2022

Chapter 18
Overview of IoT Software Architectures: Products and Testing Support

After introducing the architecture concepts of Chapter 17, this chapter provides an overview of general software and test architecture concepts. While this book has testing as a primary focus, I believe IoT testers (and developers) must understand and work in testing with models of IoT device hardware, system, and software architectures. IoT projects cannot build and test in a vacuum by just focusing on software. Projects need a variety of support architectures for knowledge and collaboration of the teams, as shown in Figure 18-1. Typical architectures include support for the development team, test/security for testing, operations for when the system goes into use, and the product's architecture with hardware, software, and system defined.

This chapter looks at the architecture support elements of development, test, security, operations, and the IoT product (hardware, software, and system).

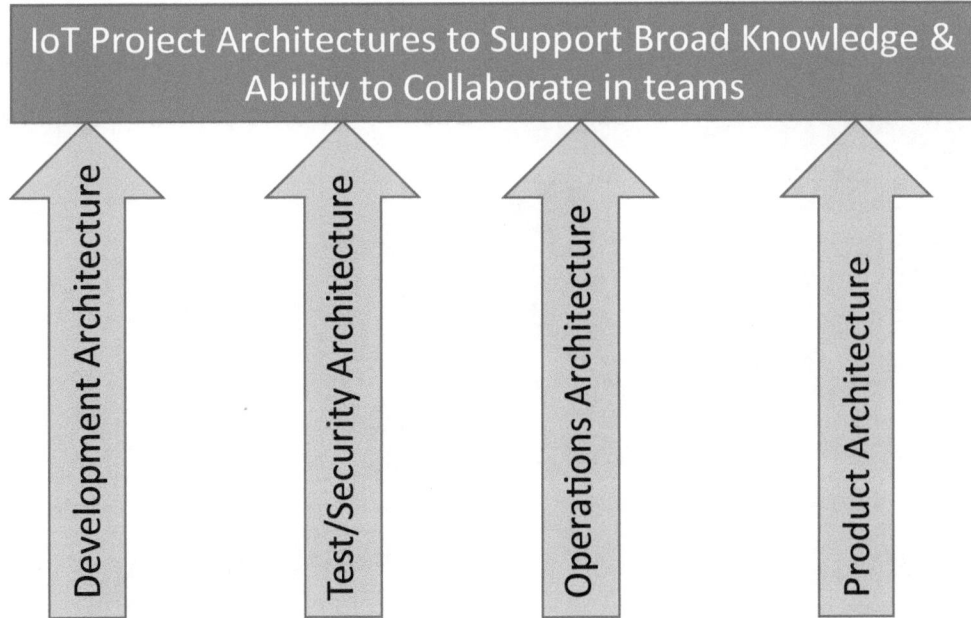

Figure 18-1. Architecture support elements

J. D. Hagar, *IoT System Testing*, https://doi.org/10.1007/978-1-4842-8276-2_18

A Quick Look at IoT Architectures

Figure 18-2 presents an example model of how testers might view an IoT architecture. In this example, testers have an architecture of a hypothetical system for controlling traffic. The IoT app uses traffic lights to control traffic flow (when to change, how long to stay on, adjustments in real time, etc.). Below this layer is one of the data analytics components, possibly located on an edge device. After this, there is an access layer to interface going to cars at the edge, buses at the fog, a central traffic control hub at the city cloud, and finally cell phones that have global connectivity. Next is a task scheduling of functions within the OS. There is also a user interface, though access layer restrictions may limit it. Because of the risk of traffic chaos, if something goes wrong, the security layer exists and finally needs support functions such as I/O, time, reporting, etc. An actual architecture will probably look different and be much more complex, particularly if one tries to model things like the car interface, the traffic control system, and any phone apps as well as any other attack surface that might be applicable.

A generic IoT architecture hierarchy structure is shown in Figure 18-3.

In Figure 18-3, a general communication and control architecture is depicted. It starts at the bottom with possibly a human user. The human interfaces with the IoT device via sensors, actuators, and computational elements. At this point, the human user may not be aware of the other aspects of the

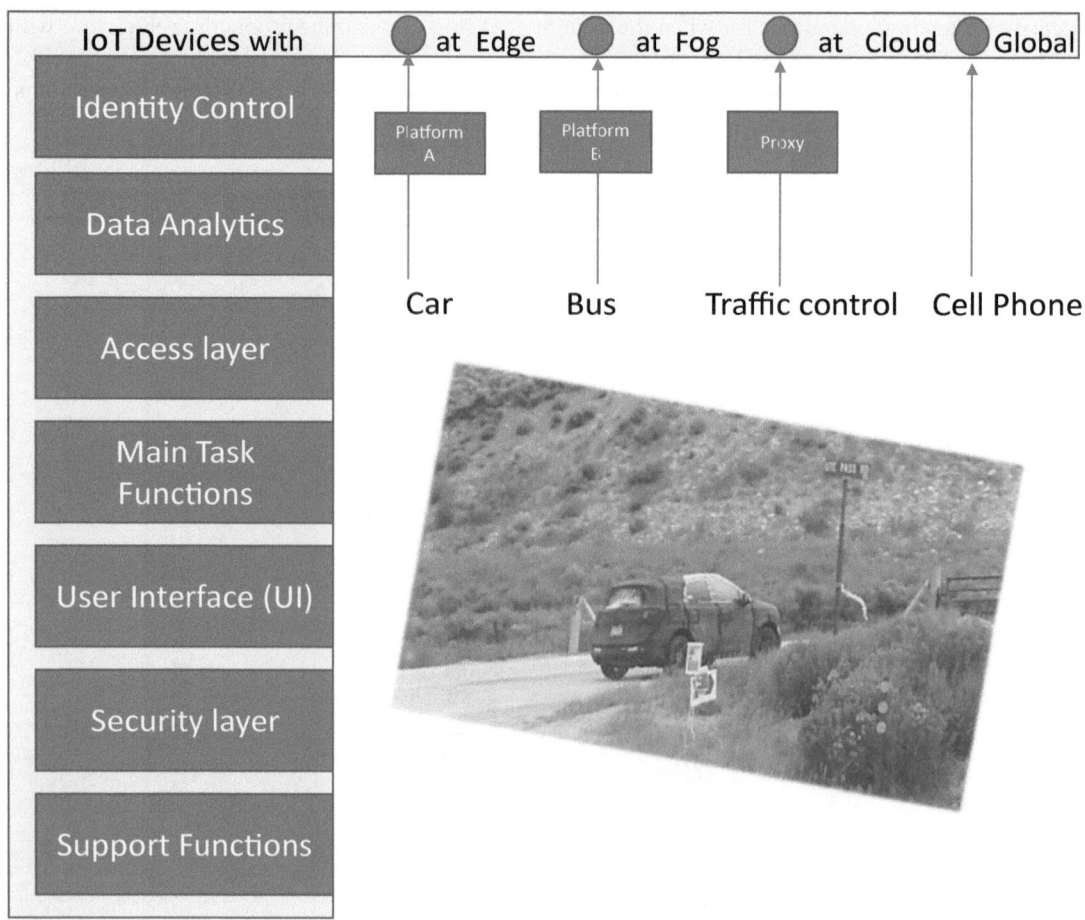

Figure 18-2. IoT system traffic control architecture ecosystem

IoT systems. Many of the architecture layers exist past the device and are functioning today; these are an edge layer and a fog layer, which goes into the cloud. Many users may be aware of their data and information going into the cloud but may not know what this means or how it works. Finally, the large worldview is where the cloud and networks share the IoT device information. This figure raises a question: "Where do we do IoT testing, and who owns what layers or levels above a device?" Just testing a single architecture level may miss usage issues that transcend any particular level. Users will not care which level is responsible but will care when a failure happens that impacts them.

Some software projects, made with IoT, focus on developing and testing the new software and hardware architecture elements. The belief is that existing systems, hardware, and software have been "proven" by use. However, with this understanding, the significant picture of architecture and integration may leave problems which cross these architectural boundaries. These problems go undiscovered until the device goes into large-scale use. This can be an ugly surprise for stakeholders. However, at least at an integration level, testing these levels and the quality assessment should happen for the new and reused IoT elements.

REUSING SOFTWARE IN A NEW ARCHITECTURE
For example, the reuse of Ariane rocket software on a newer, faster rocket led to a failure because the code that was left running after liftoff had a measurement overflow that crashed the flight computer. The code feature existed in the old rocket to make launch day recycle on the pad faster, and it worked fine for the first Ariane rocket. However, the faster acceleration of the new system caused an overflow that was not thoroughly tested because of a belief that the old software was proven by actual system usage. Trusting reuse led to a failure, and this "blind trust" will likely happen in IoT [1].

This book does not address the deep details of specific IoT/embedded systems, hardware, and software architectures. I expect readers to create unique test architectures and environments as part of the book's testing activities (Parts 1 and 2).

In IoT, architecture product elements include unique hardware and software to produce an overall system. These elements are common to most cyber systems and thus need assessment testing. Further supporting IoT products, documentation is needed for software, hardware, and the system, specifically interfacing and integration areas.

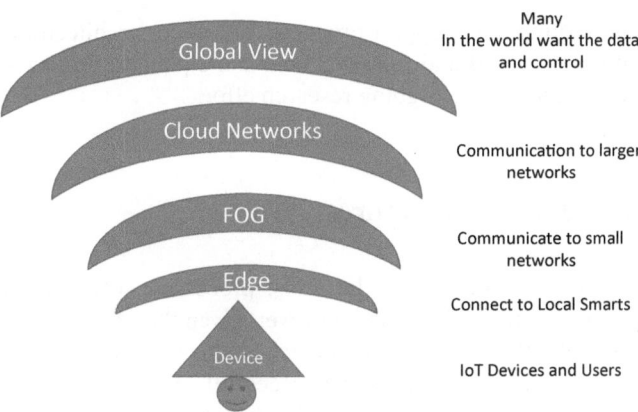

Figure 18-3. Generic IoT communication/control architecture

The views in this IoT test architecture are made up of users, development, testers, management, operations, and other stakeholders. These views include levels of testing, use cases, risks, and other information that a tester provides. Testing from the user's view is essential. Still, testers must also address the information associated with many other views, including management's metric reports, stakeholder costs, operations limitations, and what development may not have provided (e.g., structural testing).

Past the generic architecture picture, there are a variety of IoT architectures recognized in the industry. Currently (as of 2022), the widely used general IoT reference architectures include

- Internet of Things – Architecture (IoT-A) 2013 – The IoT-A reference model and architecture were developed through a European Union (EU) Lighthouse project. IoT-A is generic and supports creating concrete architectures applicable across various domains.
- IEEE P2413 – Standard for an Architectural Framework for the Internet of Things (IoT) – This ongoing IEEE standardization project aims to identify commonalities across IoT industrial and consumer domains.
- Industrial Internet Reference Architecture (IIRA) 2014 – IIRA was explicitly developed for industrial IoT applications by the Industrial Internet Consortium (AT&T, Cisco, General Electric, IBM, and Intel).

From these, architecture can have the following capabilities:

- Managing devices and their data
- Connectivity and communication
- Analytics and applications
- Quality and security factors
- User/stakeholder support

A focus of many traditional test teams is on functionality testing. This focus is a first verification checking step that is necessary but not sufficient. Reference architectures also describe mechanisms to address nonfunctional requirements such as flexibility, reliability, quality of service, IEEE 982.1, interoperability, and integration. These areas need to be tested with functional and nonfunctional quality evaluations.

Some current industry IoT platform–centric reference architectures (vendor specific [2]) include

- IBM IoT Reference Architecture
- Intel IoT Platform Reference Architecture
- Microsoft Azure IoT Architecture
- Amazon Web Services (AWS) Pragma Architecture

Finally, environment, tooling, and support processes are major architectural elements. However, having IoT tools and processes used and followed does not equate to project success. IoT software tooling and support factors must be an ongoing research effort.

Overview of IoT Support Architectures

This short section will provide pointers to the tools, processes, software, and support elements necessary for developers (Dev) to do their job. However, given that this, in large part, is common to historic technical projects, I will not go into great detail in this section, instead leaving that for other books. My experience in the software world is that successful development staff often spend as much time in the "test labs" learning about the device's hardware and software as the testers do. These efforts should happen early in the lifecycle (e.g., planning and development decision analysis).

However, with support from testers, the development staff must come to understand what this book calls the IoT test environment and architecture.

The following list should be a development team's starting point of requirements when considering a general software development architecture. Testers should be familiar with these concepts to understand and help the development staff. The IoT project team should consider these essential IoT supporting capabilities:

- Control and automation of lifecycle steps
- Modeling and simulation
- Device technical management
- CM/SCM with a version description document (VDD) for each "release"
- Data communication protocol tooling
- People factor support (ease of use, familiarity, flexibility, etc.)
- Data storage (cloud or other storage)
- Data and information analytics (AI, ML, etc.)
- Rapid application development and deployment tooling – continuous integration (CI) and continuous test (CT) and continuous deployment (CD)
- Integration tooling up to the whole (system, hardware, software, operations, etc.)
- Security and privacy tooling
- Reporting and communication tooling
- Management (human)
- Resource, cost, and scheduling project planning tools to manage the solutions that you build/test
- STA and test environment

As a starting point, here are current famous examples of general-purpose end-to-end IoT platforms that provide some of these features and are suitable for a wide range of IoT applications (as of publication date and names subject to change):

- IBM Watson IoT – Built on top of IBM's Bluemix PaaS platform, IBM Watson IoT supports realtime analytics and cognitive computing.
- Amazon Web Services (AWS) IoT – AWS IoT is an expandable platform that includes a toolkit, access authentication and authorization, system registry, an IoT gateway that communicates with devices using MQTT, WebSockets or HTTP, and a rules engine.
- Microsoft Azure IoT Suite – The Microsoft Azure IoT Suite is a comprehensive IoT platform that supports device management using the Azure IoT Hub, standard communication protocols including MQTT, AMQP, and HTTP secure storage, rules, and data analytics.
- ThingWorx – ThingWorx is an IoT platform that supports model-driven rapid application development. Features include device management, modeling, standard protocols including MQTT, AMQP, XMPP, CoAP, WebSockets and data analytics, and various toolkits.
- Kaa – Kaa is a free, open source IoT platform published under an Apache 2.0 license, supporting local self-hosting. Kaa includes REST APIs and SDKs for Java, C++, and C, including device management, data collection, CM/SCM, notifications, load balancing, and data analysis [3].

Mind Maps of IoT Environments

Figure 18-4 introduces an example IoT device environment that testing must support. The mind map is one of many possible environment taxonomies, but there is no universal standard for IoT test environments.

Start in the center with "IoT environments" to read this map, including software, hardware, system, and operations. Testers of IoT devices should think of testing the device within a whole environment. The map branches into three main divisions: industrial IoT (IIOT), consumer, and what this book calls the "middle" sector. Many testers may be familiar with IIoT and consumer IoT devices [4] as these are in common usage today. The middle IoT sectors are those systems that combine aspects of industrial and consumer. The middle environments cover personal consumers, industry, and government users mixed to varying degrees as well as cross consumer and industrial boundaries.

The consumer IoT sector shows that personal devices can be worn, provide information to be aware of, and improve life. The consumer sector also includes devices found in people's homes to control their homes, provide entertainment, and even help manage food inventory. The final parts of the consumer sector are systems that monitor the body or exercise, provide reminders to the individual, and possibly provide medications (e.g., smart pills). Many of these aspects have concerns about safety, security, money, and individual consumer's life quality.

IIoT systems have already come into use and are growing. IIoT represents potential savings and information gathering, benefiting organizations such as companies and governments. IIoT has sub-classification entities that include constructs, factories, and energy. Constructs represent a significant resource used in "large, industrial, or government" settings. Examples of constructs would include buildings, infrastructure, and even cities. Using IIoT, these resources can be built, managed, and optimized for resources such as quality, people, cost, or schedules. Factories and production facilities have detailed factors such as automation, robotics, inventory control, internal factor/facility transportation routing, waste management, and general management, to name a few. Finally, a significant factor in the industrial world is energy using terms such as usage, manage, control, and distribution.

The left side of the IoT map is the largest. The middle sector comprises transportation, communication networks, public security and safety, retail, and healthcare. Each of these environments overlaps industrial and consumer to some degree. There are many subsector categories and possibilities in these environments. Each node in the mind map represents an environment that testers may need to address and consider for an IoT device. The middle IoT sector of the map has quality factors such as safety, security, financial, functionality, usability, and reliability.

This book advocates that in IoT, a comprehensive test environment and architecture must be considered by testers as well as in part addressing the lower levels of the device, edge, and app cloud. The development team and tester must consider the abundant test architecture elements outlined in the environment map. Testers must focus on the software system or system of systems, addressing the integrated cyber-physical system in the "real world," reflected in the test environment. IoT adopters

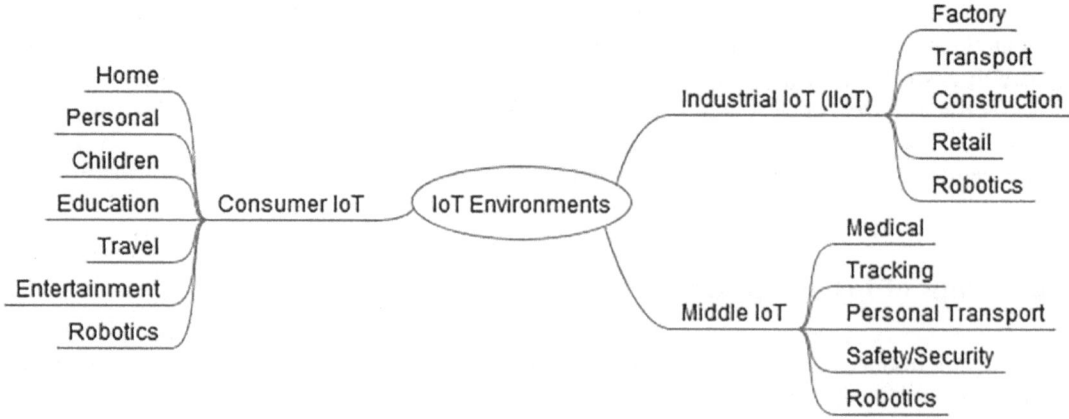

Figure 18-4. Example IoT device environment mind map

may be tempted to take the minimal path of simplistic low-level software testing (e.g., test lines of code), ignoring the more significant architectural and environmental issues. However, this is likely to delay finding many problems until a device is in actual use. Finding problems in the field can substantially increase the costs [5].

Individual vendors may be able to "test" much of the right side of the mind map, but parts of the left side of the map will be challenging to address with testing done strictly by vendors. This situation raises the question of who owns the testing of the large-scale integrated IoT environment (e.g., the system of systems, a whole city, a factory, or a country).

The later chapters and subsections explain each classification in more detail and provide example devices and locations. This topology classification is not the only mapping possible. It is not complete (it does not include all possible IoT/IIoT environments), but it is one way to organize environments and architectures. Many readers will have other specific topologies that they should document in their test plans. Different topologies are fine, and universal classifications may become possible over time as IoT/IIoT environments evolve.

TESTING FROM MOUNTAIN TOPS
Car companies use a high-altitude test company located in the mountains of Colorado, USA. Why? Even with the manufacturer's test environments, tools, and maybe even independent verification and validation (IV&V), some unique test environments in the real world are hard to duplicate. Here are just a few examples: snow, ice, a mix of snow and ice, mud, high mountains (14,000 feet/4500 meters), dirt roads that can be driven at 55 mph/120 kph in cold (–40 c), rain, and other "special conditions" not found at sea level. The cost to produce a simulation environment for these conditions would be enormous. Is there a best test environment? No. Each environment has advantages for testing, as well as issues. What in Colorado is so different? The testing company likes high altitude, mountain roads, snow, dirt, ice, rough roads, etc. See Figure 18-5 for a picture of an IoT car being tested on a mountain dirt road.

IoT Software Test Architecture (STA) Introduction

This section introduces STAs, which are expanded in later sections and chapters. I believe STAs are very important for many IoT test efforts, particularly for larger and more complex systems. Readers should start with this introduction and then proceed to other sections of interest exploring STAs.

Figure 18-5. IoT software test environment car (note: covered in drapes)

Picture Copyright 2009 Laura M. Hagar

STAs are a specialization of general architectures, including hardware, systems, and software. ISO 42010:2011 "Systems and Software Engineering – Architecture Description" is a general reference to architecture relationships shown in the boxes of Figure 18-6. ISO 42010 defines an architecture in terms of a description, aspect(s), stakeholders, and an entity of interest. The boxes show relationships between ISO 42010 and STA, not a flow of information or logic. The entity of interest implies environments where the entity operates under one or more contexts. Contexts are defined by stakeholders who have a perspective on system use. Finally, stakeholder perspectives define concerns related to the stakeholder and the architecture aspects. These architectural concepts from ISO 42010 are very high-level elements of architecture.

Figure 18-6 transforms these generic elements into an STA-specific model example. The ISO 42010 Architecture Description box transforms into the Test Plan & Strategy, which could follow ISO 29119-2 as an example. The ISO 42010 Entity of Interest box becomes the system/software under test (SUT). The 42010 Environment box maps to the Test Environment; the Stakeholder box maps to the tester staff, development (Dev) team, and users. The Stakeholder Perspective box continues as defined by ISO 42010. The box of Architecture Aspect maps to test models as defined in 29119-3. Next, the Concerns box of ISO 42010 becomes the Risks and Needs of the testing planning effort. Finally, the context box of ISO 42010 is mapped into views/viewpoints for the STA. Each of these STA elements is defined in this chapter.

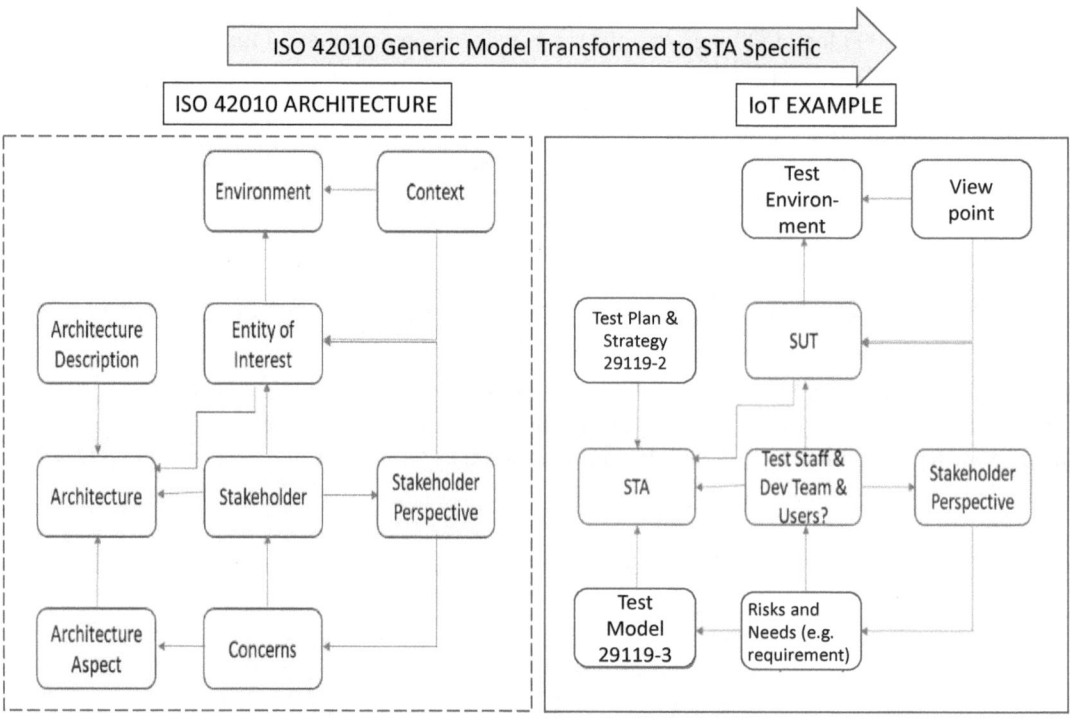

Figure 18-6. ISO 42010 translated into an STA IoT example model

A specific STA model for actual test projects would be built slowly over time and repeated iterations. The STA may start with outlines and concepts in the test planning efforts. However, as project stakeholder needs (requirements) expand, you can also expect the STA to grow in complexity. Generally, reference STA details that are included as part of test planning should be

- Device onboarding and testing time phases
- Risk-based testing using the STA
- Updating device firmware and software as it evolves and changes in the project lifecycle
- Addressing and testing new configurations of software, hardware, and operations system usage
- Testing larger-scale remote operations like disabling, enabling, or decommissioning devices
- Security and privacy, as well as other qualities necessary for stakeholders
- Performance of system in time and resource use
- Reliability and trustworthiness
- Functional qualities essential to the specific IoT device (project unique analysis)

The test plan, strategy, and STA cannot address everything in the entire lifecycle at once. Plans will be refined and modified over the lifecycle. A more detailed example of this generic STA model is shown in Figure 18-7, which presents an example STA classification for IoT for a car speaker system. This example model is only one possible classification system and maps to the STA of Figure 18-6 but includes many common critical points for producing and testing IoT software systems. This STA should reflect aspects of the product's system architecture that stakeholders experience. In this expanding example, testers now see the views/viewpoints being expanded (see the sections that follow for more information). From a stakeholder perspective, testers can now consider user views, models, simulation analysis, test automation, combinatorial testing, exploratory testing, and system tests.

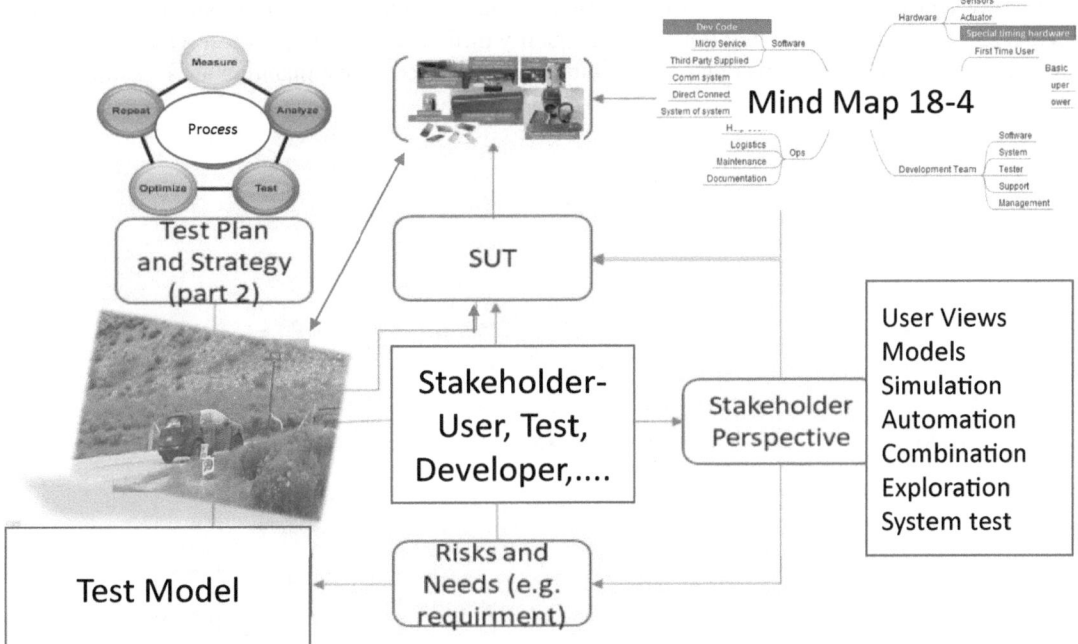

Figure 18-7. Example high-level STA refinement of Figure 18-6

Note: The view/viewpoint mind map of this figure is explained as follows.

IoT Software Test Architecture (STA) Details

To explain the STA, I have produced a mind map. The map has significant elements contributing to the STA. I had suggestions on the subpart names of the mind map from Professors Nishi (already identified) and Santoshi as well as the paper found in [6]. Each of the significant elements of the Figure 18-8 map is introduced in the following subsections.

Major Element: Test Plan (a.k.a. STAp)

STAp is the project test plan, but when viewed from an STA perspective, as in this mind map, testers see the test plan as an association within the STA, including the other elements of the STA. Test plans are explained in detail in Part 1 of this book as well as in references at [5].

Major Element: System/Software Under Test (SUT)

The SUT is what the STA is created to test. It is indirectly part of the STA. Items under test include the software items to be fielded or delivered to stakeholders. Items under test include the software, hardware, system, and documentation. Each can be built into a special test item configuration to support testing. For example, hardware may be built to have probe points for test access. Or the software may have special code to support testing in the lab. Typically, these special SUT configurations are not delivered to end stakeholders, but in some cases, "test configurations" may be delivered to end users. However, there can be a noteworthy or high-value test support configuration. There is a risk of special test configurations having functional or quality differences that impact test results or usage in the field. A balance between risk and benefit must be considered in test planning an STA of SUT.

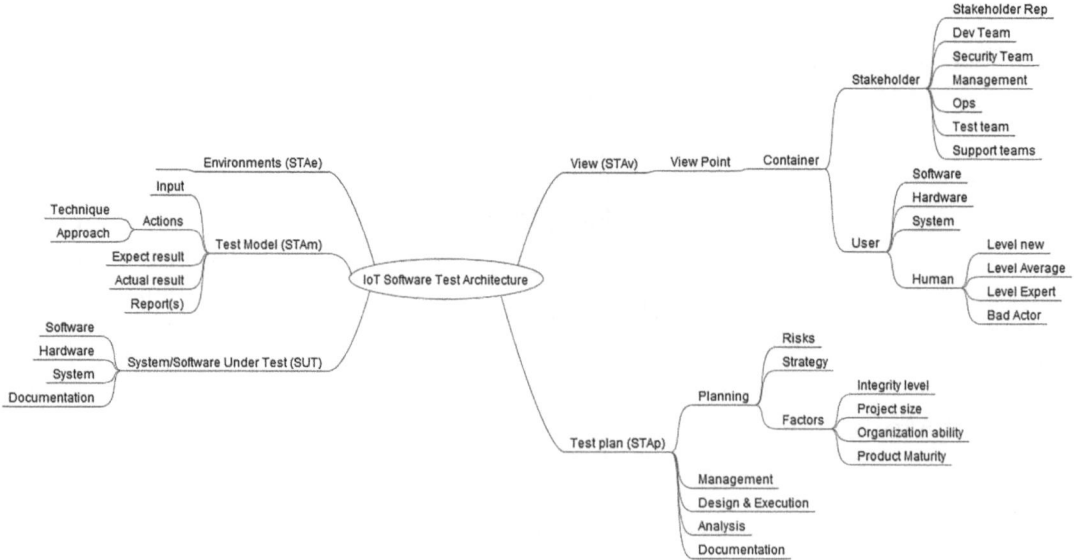

Figure 18-8. IoT test architecture (STA) mind map [6]

While this is a book on software testing, in IoT, the complete system, including hardware, is part of the SUT that teams need to consider. The software test planning may not address hardware and system element testing, leaving those for separate test plans, while on smaller projects, these elements may be included in a test plan.

To evaluate the software, the system and hardware for IoT must be part of the SUT, including

- Hardware – Different versions, configurations, vendors, etc.
- Software – Different versions, configurations, vendors, etc.
- System – Different versions, configurations, vendors, etc.
- Documentation – While many people don't think of documentation as an item under test, it should be. It may be a part of what the tester evaluates in many cases. For example, verifying documentation should include

  - User guide documentation
  - Regulatory, legal, and device limitation statements
  - Unique configurations of hardware, software, system, and languages
  - Support group targeted design documentation for development, test, management, support, hardware, systems, customers, etc.

Note
Documentation can be internal to the system (softcopy) or external (paper hardcopy); both would be considered part of the release documentation accompanying the overall IoT product(s).

Many teams think of the SUT as a single thing at any point in time during the development life-cycle. My experience is that the SUT may have unique configurations and software builds to support groups like agile development, testing, security, and performance. See the story that follows for an example.

STAp BY HARDWARE
A unique hardware build was needed for an IoT device to do a performance test. The IoT device had hardware "triggers" for fault detection and correction, which caused correction logic in the software to be executed. To create these faults in the device, sensors would cause damage to the hardware, which was a significant expense, since the hardware device could no longer be used in testing. To solve this problem, the development team created a unique software build where the hardware interrupt (fault) would be tripped, done by the software at a preset time. This approach allowed the assessment of recovery logic functions and timing performance testing. This unique SUT build was a compromise. While not totally realistic, it saved on the IoT device expense.

Major Element: Test Environment (STAe)

My experience during testing IoT and embedded devices is that having the correct test environment(s) is a critical success element. The book devotes Part 4 to explaining test environments with examples of different levels. Also, the system-software integration lab (SIL) is defined in Part 4, Chapter 19. STA SIL test environment is introduced with an overview in Figure 18-9.

The test environment includes development, test, and the real world; these and associated sub-activities are covered in Part 4, Chapter 19. This lifecycle figure of the test environment starts with test management working with development-operations and support teams to generate a test plan (see Part 2) and the test products (hardware, software, system) to control. These elements flow into the test design and model processes. After these elements, test generation, setup, and execution can be done. These lifecycle activities provide the data to evaluate in analytics, the results of which are documented in a test report.

Major Element: Test Model (STAm)

The test model has many configurations and aspects. The test model should illustrate the design and techniques of the testing. IoT test designs and techniques are defined in Part 3.

Major Architecture Element: Views, Viewpoints, and Containers (STAv)

This section is not without debate among the modeling and architecting people, many of whom do not consider test modeling and planning to include views/viewpoints. The world of standards, particularly test standards, has not reached a consensus on these topics. I present my experience-based practices with due consideration of other authors and standards.

Definition: View – A view is a conceptual representation of one or more structural aspects (logical, physical, processes, development) of an architecture that illustrates how the architecture addresses concerns held by stakeholders.

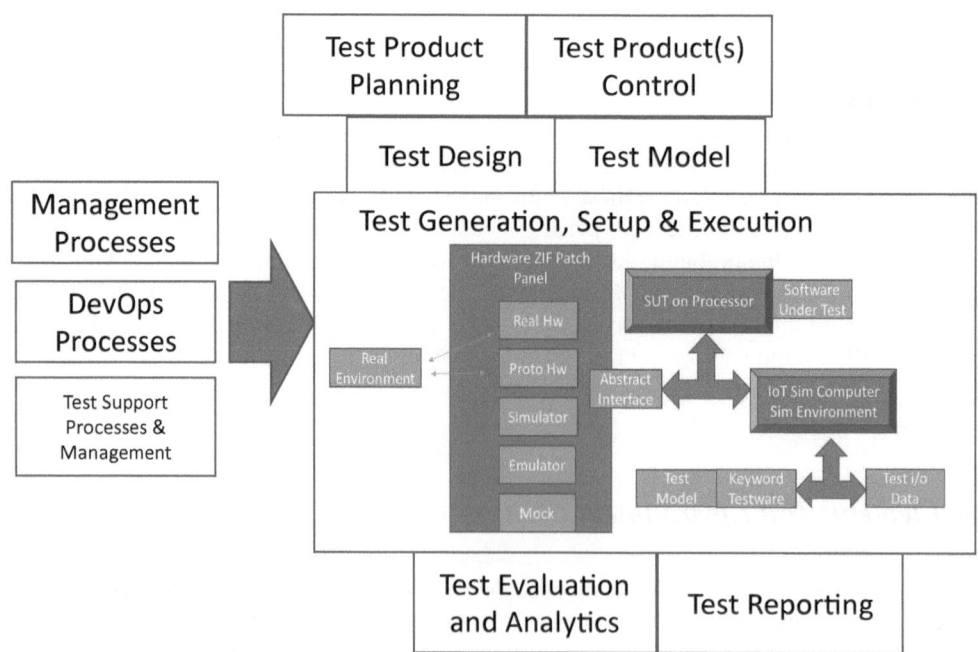

Figure 18-9. STA test environment (SIL) lifecycle overview

> **Note**
> The view represents as much of the system as necessary to address the perspective of a related set of concerns. A subset of the system generalizes the model or representation and has boundaries of what is included in the characterized representation (i.e., not everything is included in the model/view).

Many modelers and teams will try to have one complete model of the architecture crammed with views, concerns, containers, and viewpoints. However, this is not possible in all but the most straightforward cases. No set of views and architectural elements can completely address all perspectives and concerns of a modern system without becoming so large and complex as to be unusable.

Advice: Divide and conquer in an architectural model is a best practice.

For example, test planners should have as many models, views, and viewpoints organized into containers to aid testing and stakeholder understanding but remember KISS (keep it simple, sir). The divisions and containers allow information hiding.

Definition: Concern – Concern can be the needs, stories, requirements, risks, and data of the view/viewpoint.

Definition: Viewpoint – A viewpoint is a grouping of patterns, information, and conventions for constructing one type of view in abstraction to focus on the grouping.

> **Note**
> A viewpoint defines stakeholder concerns that are reflected in the view. The information can include concerns, guidelines, principles, patterns, and models for constructing the views.

Example views and viewpoints are given in Figure 18-10. The views shown include software, hardware, users, the development team, the operations team, and other systems. After views, individual viewpoint examples are presented. These are just samples, and many more could be possible.

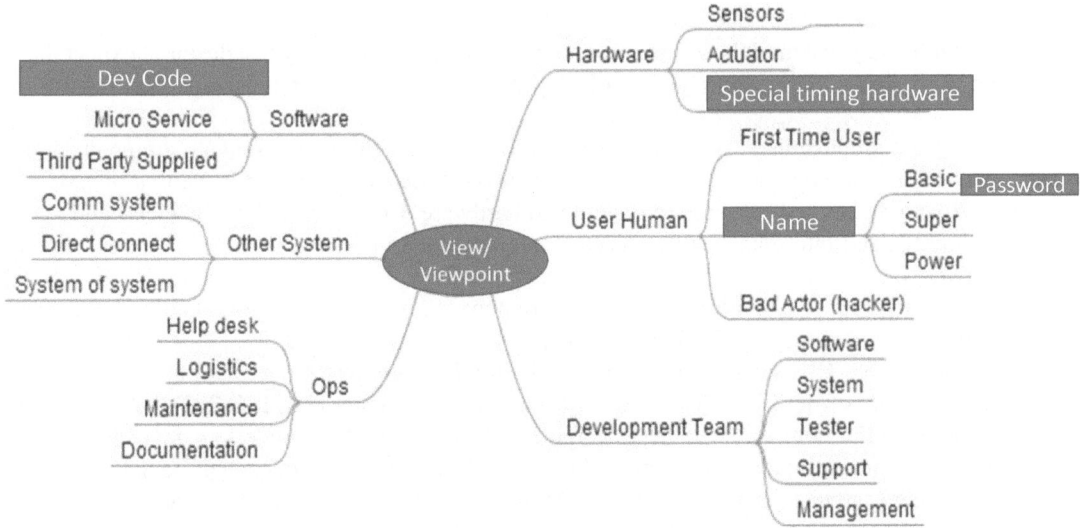

Figure 18-10. View/viewpoint mind map [6]

Definition: Container – A container is a conceptual collection of related views and viewpoints, done for summary convenience and to support information "hiding" in the model. For example a container might be: The development team container holds the system staff, programmers, hardware engineers, testers, engineering support, and managers.

A container is used to address the issue of architectural model complexity and the advice of divide and conquer given earlier. A single massive architecture model cannot address every view, viewpoint, and concern without becoming too large to understand. For example, testers need to address the complexity of the STA into manageable pieces. The pieces must be understandable and of use to the team and stakeholders. One can expect to iterate, edit, and modify these architectural components.

The first step in creating these architectural components starts with the following (more or less in this order of production):

– Identify the problem from test planning.
– Identify significant elements (functional, logical, physical, qualities, etc.) in the architecture.
– Rapidly create a representation using concepts such as a mind map, figure, or table of how these elements interact with

  • The world
  • Other parts of the system
  • Other elements of the system
  • The user
  • The stakeholders

– What information from these elements are input, managed, stored, and analyzed into model elements?
– Allocate a mapping of these elements to hardware, software, system, or users.
– Identify concerns.
– Identify how elements interact with development, testing, operations, security, users, etc.
– Identify environments (development, test, operations, security, training, management, etc.).
– Start identifying and allocating the preceding items to views and viewpoints.

I like to use mind maps, model pictures, tools, tables, and whiteboards to support this work. I get other members of the project to review, comment, and input. I am not bothered that my first attempt at the preceding analysis is just a starting point that is likely to change 300 percent before they start to be accepted and used. Testers need to have thick skins.

I end up with many test models, views, viewpoints, and containers in my test design. I expect these to change throughout the testing and project. Table 18-1 gives examples of view/viewpoint categories.

Many software architectures focus on views that model the system's internal structures, data elements, interactions, and operations. Table 18-1 considers these topics and then focuses on external elements, expanding beyond software to system and hardware. Containers allow the views and viewpoints to be organized into manageable pieces. A recommended practice is for software systems to follow a divide and conquer method as part of the test architecting process. Views, viewpoints, and organizing containers result in a test architecture framework. For software systems, these combinations feed into test planning (e.g., ISO 29119-2 [5]).

Additionally, Table 18-1 presents concepts on the model kind, test model, and test model viewpoint. A model kind is from ISO 42010 and is traced to the "test model" of ISO 29119. A test model describes the essential component of test design and techniques. Test models in ISO 29119 have inputs, step-by-step techniques, and outputs. Also, in ISO 29119, test models can be defined in the Unified Modeling Language test profile (UTP). At the lowest level of software test design, the viewpoint concept relates to the test input (conditions) and techniques (processing of testing conditions) that generate outputs (results). Thus, a model can have various test conditions following the test

Table 18-1. Architecture to STA Mapping [6]

View	Viewpoint	Descriptions	Modeled in	Examples
Project process	Project overall usage	Describes the relationships, dependencies, and interactions between the system and the stakeholders, users, other systems, and external entities with which it interacts	SysML, BPM	Examples: business process flow, interface diagram, and modeling
Logical, functional	Software, hardware	Describes the system's functional elements, their responsibilities, interfaces, and primary interactions	SysML, UML	Requirements, stories, needs, etc.
Physical	Hardware, physical usage	Describes the system's physical elements, their responsibilities, interfaces, and primary interactions. Hardware is the core of this element	Circuit diagrams, interface diagrams, architecture models	Interfaces, electric diagrams, communication channels, scale models, drawings
Logical, functional	Quality characteristics	Quality factors including reliability, safety, security, maintainability, performance, and concurrency	SysML, UML	ISO quality standard
Logical	Information use, information flow (within IT)	Describes the way that the architecture stores, manipulates, manages, and distributes information and data	High-level view of static data structure, object models, and information flow	Database, data, information
System physical and logical	System with full hardware and software implemented	Describes the system elements, their responsibilities, interfaces, and primary interaction. Example model with SysML	SysML	System as a whole, other system relationships, elements, interfaces, communication channels, etc.
Development process	Requirements analysis, coding, and support team	Describes the architecture that supports the development processes. Captures communication in the team of stakeholders involved in building, maintaining, and upgrading the system	SysML, BPM	Create hardware and software including Dev testing and support efforts
Development process	Testing	Plans, models, and designs the test activities within a project including developing test environments (a project inside of a project)	UML, UTP	Create test plans, architectures, environments, design, test, and other documents
Development process	Deployment and maintenance	Dev, security, test, QA, management, and other stakeholders	BPM	Deploy and maintain the system/software
Project process	Data analytics and sales	Describes the environment into which the software system will be deployed, including timing, logistics, and lack of upgrades in the system, technical needs, and others. Address the hardware, software processing, network interconnections, and other logistic impacts	SysML, BPM	Reviews data from system-software usage to support analytics, sales, and process improvement

(continued)

Table 18-1.   (continued)

View	Viewpoint	Descriptions	Modeled in	Examples
Project process	Operations and analytics	Describes how the system will be operated, administered, and supported during use. Factors to include: data processing, storage, data analytics, AI, help desks, error reporting, feedback to operations/maintenance, staffing management, and lifecycle	SysML, BPM	Reviews data from system-software usage to support testing, maintenance, and system upgrades

Table Definitions: SysML = Systems Modeling Language, BPM = business process modeling, UML = Unified Modeling Language.

viewpoints as described in [6]. For example, the test model input conditions and results for a legitimate software system user will be different from the test inputs and outputs of a bad actor in security testing. Both are test viewpoints with different test model conditions that must be exercised in testing.

Summary

This chapter provided an overview of general software architecture concepts focusing on STA, including the STAv and STAe. IoT testers (and developers) must understand and work in testing with models and architectures. Being just a code-only manual "tester" limits the career and ability to do a good job. Testers should be systems and big picture STA thinkers. The next chapter addresses STAe of this chapter's mind map with details on the system-software integration lab (SIL) environments and supporting test automation.

References

1. www.esa.int/Newsroom/Press_Releases/Ariane_501_-_Presentation_of_Inquiry_Board_report#:~:text=On%204%20June%201996%20the,path%2C%20broke%20up%20and%20exploded – accessed winter 2022
2. www.ibm.com/developerworks/library/iot-lp201-iot-architectures/index.html – accessed winter 2022
3. www.ibm.com/developerworks/library/iot-lp101-why-use-iot-platform/index.html – accessed winter 2022
4. https://en.wikipedia.org/wiki/Internet_of_things – accessed winter 2022
5. "IEEE/ISO/IEC 29119, ISO/IEC/IEEE International Standard – Software and systems engineering – Software testing" – Parts 1 to 5 series
6. J.D. Hagar, "Software Architecture Elements Applied to Software Test: View, Viewpoints and Containers," in 2022 IEEE International Conference on Software Testing, Verification and Validation Workshops (ICSTW), 2022

Chapter 19
IoT STA System: Software Integration Lab (SIL) Environments

Chapter 18 provided an overview of software test architecture concepts that lead to test environments. This chapter addresses specific examples of test environments that are one of the main elements of STA and IoT testing. For IoT, these environments are critical and numerous. The examples in this chapter cover various example levels that IoT teams may find their device system can be tested, but there are *many* possible variations.

The IoT team needs to consider the levels by evaluating their test plan, risks, stakeholder needs, company goals, integrity levels, and IoT factors. IoT environments include hardware, software, system, and operational considerations.

I recommend the team conduct a complete decision analysis (Part 2, Chapter 9) with management and possibly other stakeholders to determine the number and types of environments. Each environment will cost time and money, but not assessing the IoT system properly can cost even more.

The STA environment will need the IoT hardware, software, and system to be tested. Test tooling, simulation modeling, and automation will be significant cost drivers. The test concepts of this chapter will be a starting point in the decision analysis: plans, risks, cost, schedules, stakeholders, regulations, and test team skills. The concepts of this chapter can be applied to any of the test factor levels in Figure 19-1.

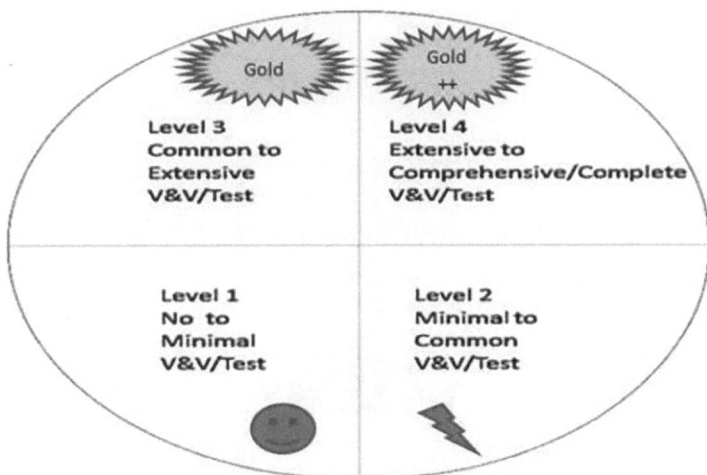

Figure 19-1. IoT factor levels

J. D. Hagar, *IoT System Testing*, https://doi.org/10.1007/978-1-4842-8276-2_19

Environment: Development Team Testing and Integration Support

The most basic testing level is structural-based verification testing conducted on a software simulator or hardware system, such as an emulator. At this level, the development testing, Agile or traditional, ensures that the code implements design information and software standards. This testing is usually done at a module/object level, with small code segments being executed in isolation from the rest of the system. The comparison and review of results at this low verification level can be labor intensive unless automation tools are used.

The development team testing support allows structural testing per ISO 29119 and Part 3 of this book. As part of these outlines, structural testing goes from lines of code to data flow. This testing can start at the lowest code element level (module, object, etc.) and continue through integration of the software. The testing in this environment is illustrated in Figure 19-2 and can be done by developers, testers, or a mix.

In Figure 19-2 development test environment, the software under test (SUT) is driven by drivers or software that invokes the SUT. The SUT may need stub or mock object code elements if the SUT makes calls on other software. The SUT is driven by test cases and procedures to produce test results. The correctness of test results needs to be determined. This effort can be undertaken using the drivers or by reviewing the stored data results. Finally, the SUT needs to execute on either hardware, an emulator, or a simulator (software that simulates the hardware; see simulation sections later). Many development test environments are commercial. Thus, there can be numerous configurations of development test environments, each with advantages and disadvantages.

As the SUT progresses in its lifecycle, integration of the software occurs. Software integration can be bottom-up, top-down, objects/classes, or mixed. Most development-operations and agile teams expect continuous integration (CI) with frequent (constant) builds, nightly complete builds, and constant ongoing tests (CD). This leads to building the software system to the system level, including integrating the off-the-shelf (OTS) elements. As the software is ready for continuous delivery (CD), it is good practice to have tool support in the development test environment.

Supporting tools can include compilers, project lifecycle management, software configuration management (SCM) with build scheduler, reports, static code analysis (SCA), and driver/stub code execution support and analysis. There are, of course, other possible supporting tools.

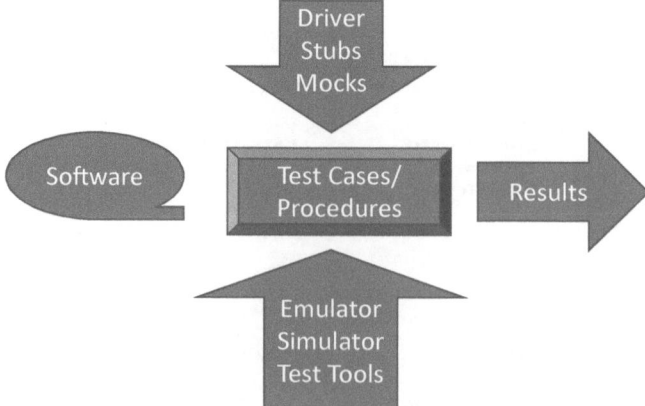

Figure 19-2. Development test environment

Reference: Jon Hagar

> **Note**
> An SCA tool and line of code testing found a structure that should *not* have been compiled and integrated. The compiler was open source and not the best. The miscompiled code left a security hole that could be exploited if it had not been found by the SCA tool and code coverage testing. The low-level SCA testing and analysis found that a new compiler should have been employed.

Integration into or with the software relates to integration with the hardware, which relates to the integration of the overall system. Whether your team has done continuous integration, test and deployment, Agile, development-operations, or just "plain" integration, IoT integration should be an ongoing planned task over the lifecycle. The larger and more complex the IoT device or system is, the integration should be a team focus. Some teams may distribute the focus across different skilled groups and people. This approach can work, but with increasing size and complexity, having a person to lead integration supported by a multidisciplined team makes success more likely. A software integration team can be a scrum, ad hoc, or called out on an organization chart. However, the integration team should have management's charter as well as authority and assignment. This authority should continue from the software level upward.

> **Note**
> One new development staff member, who had never worked in teams professionally, inserted simple code from a freeware OTS element. This approach was to provide a feature he needed in his code. The device was medically related, and when the nightly build from the SCM tool ran, it flagged the new code/OTS as *not fully approved*. The senior programmer reviewed the OTS code and found a possible security flaw. The new programmer had a learning experience.

Environment: Hardware Team Testing and Integration

While this book is focused on software IoT testing, I provide a passing overview of hardware integration based on experiences because IoT devices have considerable functionality from hardware.

Most hardware integration is bottom-up. The hardware team needs a new board. There are many tests of the design before committing the expense of actually building a "burned in" board. First, they design a board and then redesign it several times. Then, they build a prototype board with significant parts, such as actuators or sensors. Once they believe they have a working circuit board, then, they do a "smoke test" running prototype assembly code to ensure the board does what it is supposed to do. Next, they build a "final" sensor and actuator. Then they hook the board to a sensor and apply power (another literal smoke test). Then, they hook the actuator in with another smoke test. At this point, they may call the software development team and say, "can system-software testers hook up our board-sensor-actuator to current production software to see what happens?" It is not quite as straightforward or as fast as the preceding steps sound, but these steps give you an idea of why CI, CT, and CD for the system team are necessary.

Here are some of the key items needed for a hardware-software integration lab, including

– Scopes
– Signal generators
– Hardware stimulator
– Hardware simulators

- Wire wrap prototype I/O boards
- Inspection devices – cameras, probes, etc.
- Recorders for electrical, heat, humidity, time
- Support for software configurations and abilities
- Plans, schedules, and costs of the system
- Team abilities and skills

An example set of hardware test support devices might look like those in Figure 19-3.

FAST IOT HARDWARE PROTOTYPING

In one hardware lab, the team produced wire-wrapped prototype board assemblies for early testing. These prototype boards had a layer of circuits, sensors, and actuators in a box or on a board. A different version of the device was produced over several years. One day, the devices started failing when they went into integration test evaluation. Nobody was sure why, but they asked the lady who had just retired from the soldering line to come back to put a box together for a day. She put the box together and then waved her hand between each board level in the box using an inspection port with a cotton glove. The support engineer asked, "What the heck are you doing, because that is not in the assembly language instructions." "Well," she said, "I am checking for solder leaks by looking to see if they pick on my glove and if there are, I fix them."

This story is an example of a person doing informal agile testing that is not documented in a test procedure but has found a consistent way to problem-solve with hardware.

Note

Wire wrapping has been around since 1946. It's a fast, reliable, simple, solderless way of creating complex, one-of-a-kind circuits and systems. And it's easy to learn. Printed circuit boards are for everything but the simplest of circuits.

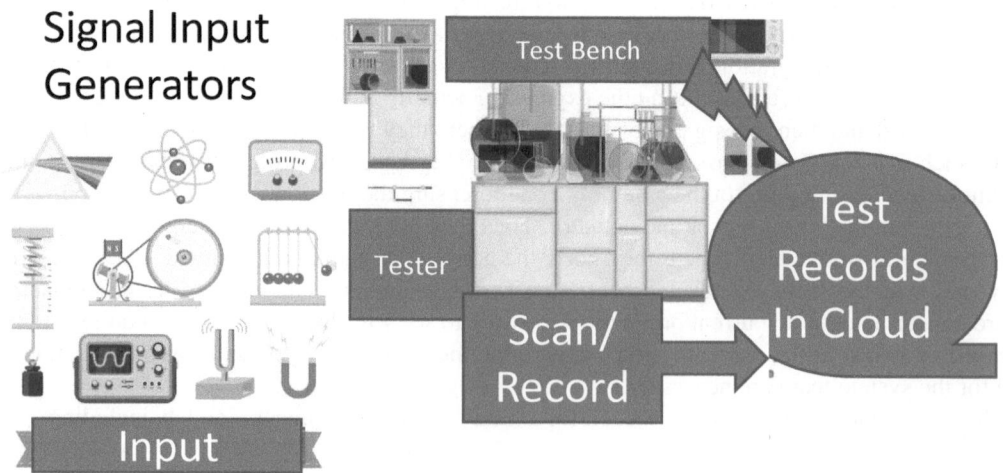

Figure 19-3. Sampling of hardware equipment test bench setup

Testing, be it hardware or software, is everyone's job:

- Where should hardware testing and integration happen?
- At the vendor?
- In China, if there are a series of hardware vendors?
- When the hardware gets to your IoT test team?

Well, the answer is "yes" to all of these questions. Testing the hardware and integration should happen at all stages of the project lifecycle. The ultimate responsibility for IoT testing will follow the final project hardware and software testers.

Environment: Full Hardware-Software-System Integration

I have been recommending levels of testing environments and to build them up slowly. I advise companies and teams not to wait, to plan iteratively and integrate early, and to use testing all along. This full hardware-software system integration and the testing environments have many configurations. Next, I will outline STA environments from simple to complex.

A Simple Integration STA SIL

Figure 19-4 is maybe the most straightforward, full hardware-software-system integration and test environment. This complete and yet simple hardware-software-system test environment can support small IoT device testers trying to address the IoT software and hardware including actuators and sensors. There is internal memory data storage. There is a connection between the users, testers, and bad actors (hackers). Finally, there is a connection to "other systems" on the hardware. Since the device is IoT, part of this connection will be the Internet. On the integrated software side, the elements that are integrated are the operating system (OS), off-the-shelf (OTS) software, commercial software, firmware (burned in the hardware), and the "official" software under test (SUT) code. This IoT hardware, software, connections, and functions must be integrated and tested as a system. Errors can come from many places in the IoT system. Even though testers are tasked with testing the "official" code, they cannot overlook the other software, firmware, and even hardware in the IoT system. These items are all part of the software testing problem, which management may "forget" or not understand. The testing within an integrated SIL may seem simple at first, but testers will find it is not as easy as the example sidebar story illustrates.

LACKING HARDWARE CM

On one hardware-software-based device, the software was undergoing revision. A new hardware sensor would roll in and out of the lab during the evaluation process. Revision CM part numbers were checked for each test configuration's hardware and software. On running a test, the tester noticed that a failure occurred, and the team set about trying to isolate the failure to the software, since it was the only project CM managed item that had changed. After weeks of work, the team found that the hardware had changed, but the outside contract vendor had failed to change the hardware CM tag. The vendor was not using proper hardware CM, and it impacted the software testing. Their reason was that the hardware design was the same, and all that changed was some new transistors that were "functionally equal." This equivalence was not valid. The vendor's lack of proper CM wasted many labor hours of testing time tracing a hardware integration failure.

These parts of a system form the configuration of the device. The software manages the device's data and talks to other systems via communication lines, as well as controlling the actuators, reading the sensor data, and communicating with the Internet. This IoT SIL environment addresses the device as it will be used. It may be sitting on a desk or a test fixture, yet this IoT SIL configuration is nothing fancy. In this example, the tester will play the "users" (normal) and "bad user" (hacker). The tester will run testing – playing both roles – while stimulating and monitoring the software, hardware, and other system inputs and outputs. This testing would likely be manual with minimal use of tooling. This simple example might work for an IoT stereo speaker controller for a smart house, toys, and simple stand-alone IoT systems.

The example that follows is the most basic test environment with the human doing the "modeling" of the wind generator control device found in Figure 19-5. The testers must concern themselves with how to do user input (on, off, view data), input wind (speed), and measure output (power). Simple if you can be out in the field with real wind, but what if there is no wind? How do you measure power output and wind speed? What if the system is a series of systems interconnected? These efforts lead to the next IoT STA SIL.

A More Advanced IoT STA SIL with Rapid Integration Reconfiguration

I address IoT integration separate from test since my experience is that as the complexity of a system grows, integration challenges become a significant task area requiring their own environments and support. Often, testing does not just plug and play for either hardware or software integration. Projects have integration issues with off-the-shelf (OTS) items and developed hardware and software, which grow exponentially as the number of elements and interfaces grows. As an IoT system grows in complexity, so does its SIL.

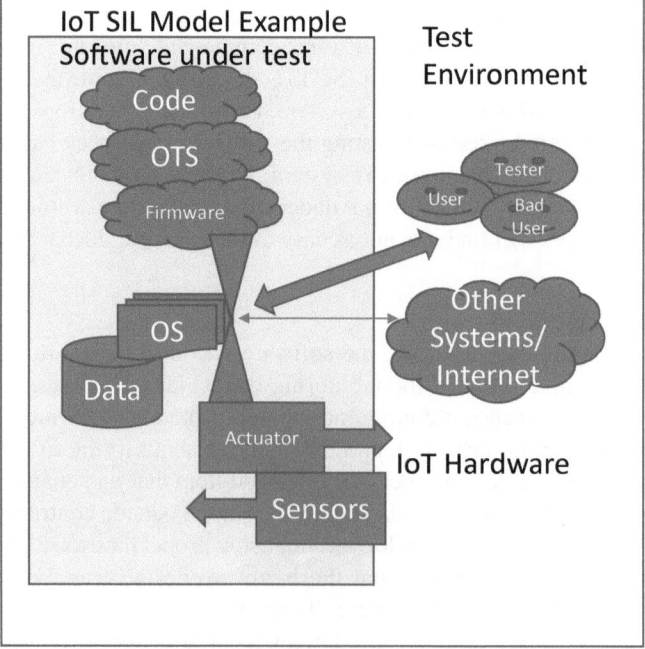

Figure 19-4. Simple example IoT SIL integration

Reference: Jon Hagar

On one project, everything was firmly hardwired in the simplest SIL configuration since this configuration was using the actual IoT device. However, I introduced more ability to quickly change hardware and software configuration versions in the second expanded hardware-software-system integration and test environment. Test integration challenges involved the planning, scheduling, and integration interfaces, each of which must be considered in the SIL integration environments; an expanded example is shown in Figure 19-6. The software configurations can quickly be loaded into the processor (blue box).

Figure 19-6 SIL environment expands on the simple model to support more complex IoT device testing. This SIL configuration uses racks to hold different pieces of hardware. In this configuration, the SIL has a "Hardware Patch Panel" with zero insertion force (ZIF) connectors that allow different versions of hardware to be in the rack and reconfiguration to happen quickly without risking damaging wires. (ZIF connectors are used to prevent wire damage.) I have seen SIL labs connected with tens and more devices via a ZIF patch panel. The patch panel can also connect different IoT device version

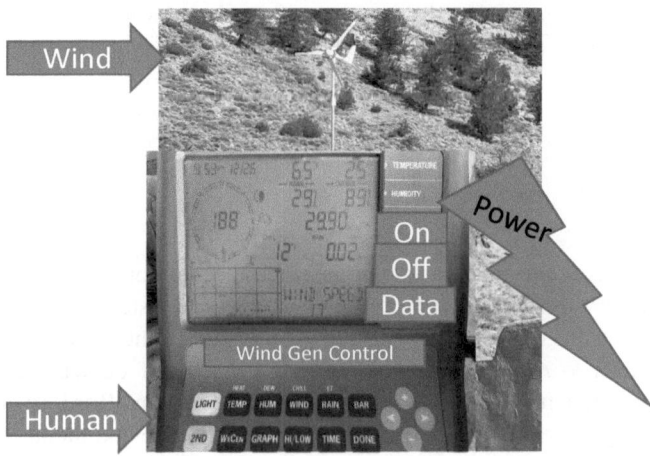

Figure 19-5. Basic test environment

Reference: Jon Hagar

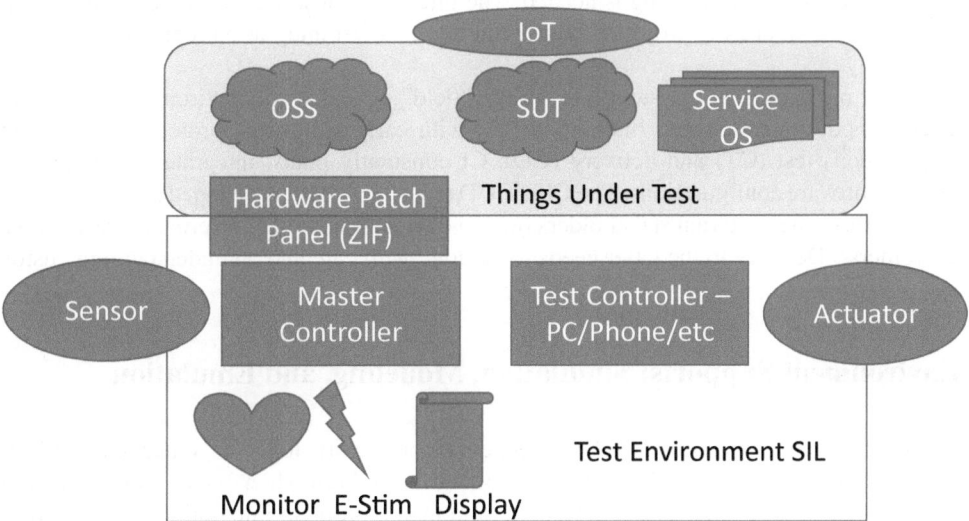

Figure 19-6. Next level of SIL complexity

Reference: Jon Hagar

configurations (think different cell phone versions). Also, the ZIF patch panel can support different test support devices, such as scopes, timing devices, models, communication lines, etc. Potentially, there may be multiple IoT SUT device configurations under test. These different configurations can be "switched" in and out of the integrated test configuration. Given this diversity of configurations, the ability of the SIL to quickly change configurations becomes an advantage for rapid testing.

Of course, this is a temporary test SUT integration configuration. Once some integration and testing are done and the configuration of integration items stabilize, a final actual software and hardware integration can occur. Care must be taken that the hardware and software interfaces of the ZIF are correct (see the sidebar: Miswired "Smart" Box).

MISWIRED "SMART" BOX
The test team received a new smart hardware support box version in one lab. The box was months late, and the team rushed it into the SIL, plugging it into a ZIF. However, the ZIF plug would not work because the new hardware integration plug was wired upside down and backward according to the interface specification. The vendor said, "Return it, and we will rewire it," but that was likely to take weeks. So, the test team quickly rewired the ZIF connector plug so that a new box could be integrated within the SIL and the IoT device under test. Later, when a corrected box was delivered, only a new ZIF connector had to be changed while weeks of testing time had not been lost.

The tooling, such as ZIF, equipment racks, scopes, signal generators, analyzers, etc., are significant advantages for integration efforts. Not every software engineer or tester can run a special test tool, but if the test environment has the right people who can use such tools, much integration and test time can be saved. *It takes a team to run an integration lab*, but if the IoT device has complexity, it is an excellent plan to have as much support and tooling as your team can afford. Even if you have to rent or borrow integration equipment and support staff, it may save time and money vs. chasing "ghosts" with staff who may not have the right equipment or knowledge.

IoT planning works best if test teams include integration efforts. The more OSS, OTS, software, OS services, hardware interfaces, etc., a project has, the more integration planning must be done and maintained. Also, the more prototypes and evaluation copies of integration elements included in the configuration, the more test planning is needed. The integration test team I worked with met weekly and often as needed on one big project. The story indicates a fast integration replan by choosing their options carefully.

You can test piece parts at a system level in the "field" or real-world environment, but this is not advantageous as doing continuous integration (CI) with early builds and prototypes, which in turn lead to continuous test (CT) and delivery (CD). CI constantly plans, integrates, tests, and repeats hardware and software configurations. The CI/CT/CD team needs to stay engaged and agile. Deliveries in CD may be to a tester, internal stakeholders (e.g., a development team, external representatives), or even "customers." Delivery to the latter needs to be done with care and an understanding customer.

Test Environment Supports: Simulation, Modeling, and Emulation

Availability limits can happen due to scheduling, cost, complexity, logistics, other real-world issues, safety, or many other restrictions. So, how do you integrate and test when one or more of the required elements are not available? In a software test environment, a problem integration efforts have is how to test when one or more of these elements are not available or need "special" stimulation to drive the testing.

After getting some experience, most testers find that software errors leading to failures *still occur* even after they have done a good job of testing because of restrictions in the test environment. The restricted environment can occur for a variety of reasons, including

- IoT hardware and software configurations are different from what was tested.
- Users, including bad guys, do things that the software test plan, risks, and test models did not cover.
- IoT device input/output ranges related to external sources are not what was tested.
- States and conditions that the software or system is in are different than expected.
- The availability limitations of the test environment.
- Security threat indications from OTS software such as

  - The battery life drops or high-power use is caused by system changes or events.
  - The time the device was active over what was expected.
  - Unexpected network connects/disconnects.

Modeling, simulation, and emulation support for the testing system can allow engineers to solve some of these problems before testing gets to the real world, even though the simulation and modeling are not perfect compared to real environments. These imperfect restrictions occur because the real world is different from the test environment.

> **Note**
> Model-based testing and keyword automation are not what this section is talking about.

The test environment should mirror the "real world" as closely as possible. Figure 19-7 expands the simple test environment to include simulation, computing support, and recording capabilities.

Figure 19-7 has a "test lab simulation" of the wind as a tester-controlled simulation that inputs to the IoT device. I can now simulate the wind via a generated current as if it was from the windmill hardware, which is input into the IoT device under test in my example system. The "external test computer" can support the testing simulation, test lab simulation of wind monitors for things like time, and expect output signals (e.g., from security threats). I can generate test simulation inputs to select "buttons" on the device. I have also added a "recording hardware" device to record electricity and other generated system outputs. The external test computer system drives all this testing using the test lab simulation, test cases, and recording hardware. I can now generate many more tests for the IoT device of interest, but it is still not the real world.

Murphy's law states that "anything that can go wrong will," which reminds us that the real world is filled with unexpected stress cases (not tested) that can put the software or hardware into unknown states, have bizarre inputs from users, and have time effects that impact the software. In testing, testers want to control the real world to eliminate Murphy. Not controlling the testing environment and suffering Murphy can lead to system faults or even failures that may be repeated. Some programmers call these "features" or say test "garbage in/garbage out" (GIGO) as no one will ever do that. But then, they will want to ignore such failed tests. Testers want to eliminate these kinds of situations or at least be able to explain why a test should *not* be ignored. The GIGO excuse is one the bad guy hackers love to find.

Simulation models can be simple or complex. Simple: Generated data is often a "predetermined" stream of inputs to the test system. Other simulation models take information to drive calculated input values within a closed-loop test system with feedback between outputs from the IoT device, which are used to determine new input values.

In such a test environment, simulation models can be hardware, mock objects, wire-wrapped hardware, users, other systems, etc. These can be done before the final elements are available and aid in testing interfaces and proof of concept. The complex simulation has more development issues (cost, schedule, production, testing, configuration control, etc.). An example is shown in Figure 19-8.

In Figure 19-8 simulation model, testers have a computer that "feeds" inputs, which takes the wind generator's place. Testers can set different parameters and conditions (wind speed, direction, temperature, fluctuations, etc.). Testers can also feed these into a series of wind generator systems and then into the cloud, where the data is stored. This approach allows an extensive assessment of a wind

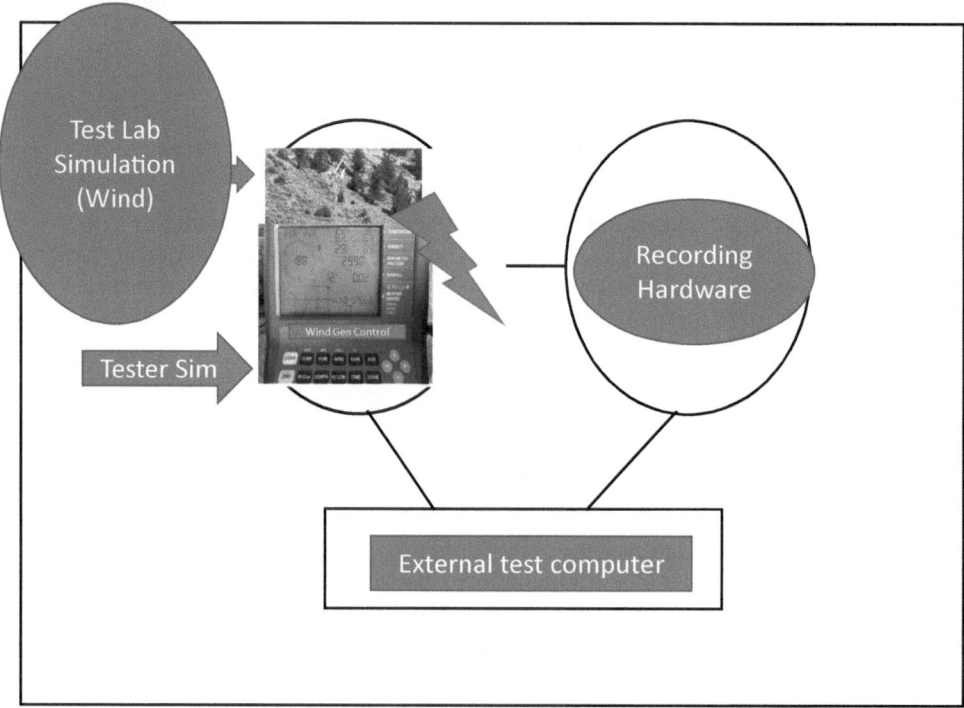

Figure 19-7. Test environment with simulation support

Reference: Jon Hagar

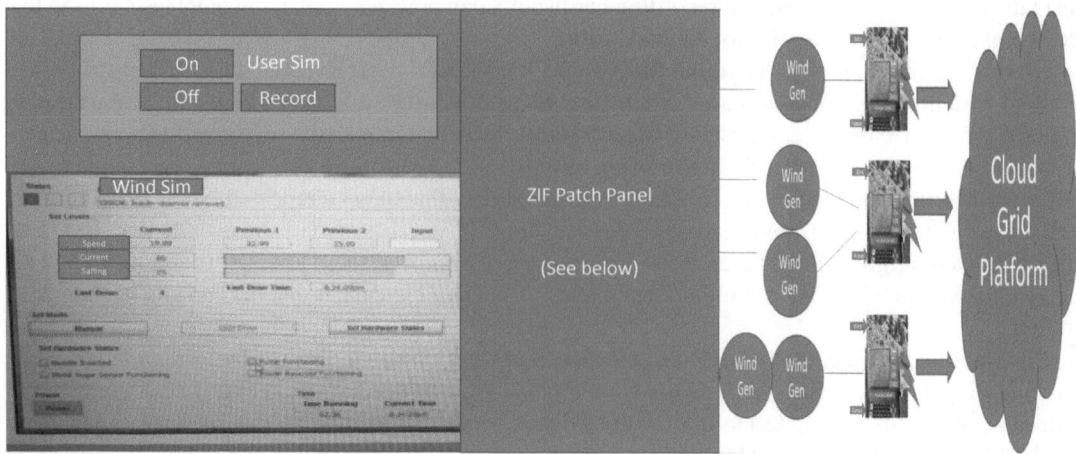

Figure 19-8. Simulations driving testing of many IoT devices connected to the cloud

Reference: Jon Hagar

generator system. The test team building such a SIL system should be able to answer the following questions:

- How do testers test the SIL and supporting simulations (test the test system)?
- Do the simulations accurately reflect the sensors and actuators?
- How much control and fidelity does the SIL architecture need?
- How much vendor information can testers get on any of the IoT elements, and should testers trust this information?

Note

Our initial simulation planning was a "best guess" in one lab. This meant the input and output of the simulations were based on engineering understanding and requirements. Still, during development, testers needed to take actual measurements of the system's inputs and outputs once in the field. Actual measurements improved our simulation test model for later IoT system versions.

Simulation-based tools provide success criteria or analysis capability that allows engineers to judge the success of the SUT without relying entirely on human judgment. A large number of simulations are based on requirements or design information. These simulations output data into the SUT and record expected outputs. These outputs can then be compared to results generated by the tested software. Some of the simulations are based on individual equations or logic sequences, while other tools simulate aspects of the entire system.

Simulation combined with modeling and emulation can allow early hardware and/or software versions. This supports early agile and development-operations testing providing feedback to engineering. Such a device is shown in Figure 19-9.

This figure is an example of a prototype emulation with wire-wrapped circuit boards of an IoT robot. The prototype was produced to assess the toy's concept. The prototype has crash sensors to stop forward movement, a motor to move wheels in all directions, and an emulation board running prototype software into circuits that are just "wires" wrapped around sensor and actuator motors. The system has a Wi-Fi Internet communication channel to send and receive data. A basic test of the toy's concept was the goal – however, this prototype grants real-world experience.

Figure 19-9. Example emulator and wire wrap boards for an IoT toy

Reference: Jon Hagar

Hardware emulators imitate hardware behaviors while the hardware and software are still being designed. Emulation allows testing before the final hardware is built, "burned," and finalized. Typically, this is done with special-purpose prototype hardware that can do in-circuit emulation. The emulator may have limitations but can be included in a test environment with other hardware and software for assessment as part of CI/CT/CD lifecycles.

Mock objects, devices, emulators, wire wraps, prototype boards, etc., are not as pretty or rugged as final designs yet. Still, they allow testing to assess coverage, concept assertions, integration with peripherals, power, visualization of the system, software/app testing, and system prototype validation.

Full System IoT STA SIL with Simulation and Modeling

At the system level, testers can test software in SILs with actual hardware in the loop and as much of the "real world" as possible. An extensive realtime, continuous simulation modeling and feedback system of computers can test the software in a realistic environment. Realistic is defined here as the software being tested as a "black box" with the same interfaces, inputs, and outputs as an actual IoT system. To test a realtime IoT software system, testers surround the computer with a first level of electrically equivalent hardware interfaces. Testers input signals into this testbed to simulate the performance of the system and hardware interfaces. The inputs stimulate the IoT SUT, which responds with the computed outputs. These outputs are read from the hardware of the SIL into a workstation. This "advanced SIL" workstation software computes appropriate new inputs and then feeds back into the SUT. This arrangement forms a closed-loop simulation environment that allows the SUT to be exercised realistically. Of course, some IoT hardware and connected systems must be simulated if the interactions with the real world cannot be supported within a SIL. In addition, unusual situations, threats, and system or hardware error conditions can be input into the SUT without negatively impacting the real IoT systems. The IoT SUT runs in actual real time. Therefore, there is no speed-up or slowdown of the system. Such a configuration example is shown in Figure 19-10.

Given the dynamic and complex nature of such an STA SIL, a system of support works will be needed. Many IoT SIL environments will start small and evolve into the complex SIL shown in Figure 19-10. In this example figure, testers start with the real environment or as much of it as is technically possible while still allowing testing.

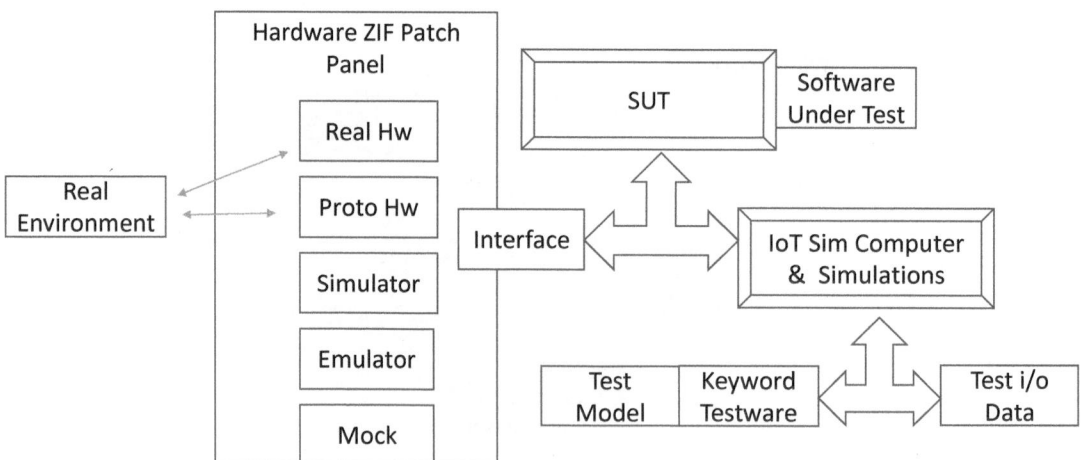

Figure 19-10. STA SIL with simulation, emulation, and test models

Reference: Jon Hagar

The first major support element is the rack and hardware ZIF patch panel. This support element can contain real IoT hardware; prototype IoT hardware; test support simulators of systems, hardware, etc.; hardware emulators; and mock devices. The ZIF patch panel allows versions of all these test support elements to be quickly configured in the test SIL. I have used such a system where the team reconfigured a test set up within a few minutes and ran a new test that might have taken hours simply by using the ZIF connectors. These connections go into an abstract interface and then the SUT computer. This interface does not impact functionality or characteristics, such as timing. However, care must be taken in the planning, executing, and testing of this kind of SIL due to the many complexities.

This closed-loop test system has a support computer system that runs simulations, test models, keyword "testware," and input data to drive the SUT. These software-driven support elements allow testing beyond anything the ZIF connector rack systems do not provide. This closed-loop test system runs in real time (faster than the SUT) and allows high test automation levels. The tester writes or models keywords, creates input support data and simulations, and runs the tests. Tests can run for minutes, hours, or even days to assess qualities like performance or reliability. Finally, all test results and data are recorded in the cloud for post-test analysis. Post-test analysis can be fundamental in IoT because some software errors "hide" (the error is not immediately obvious) in the output data. One system I worked on generated Terabytes of output data, which needed to be reviewed by the test team.

This SIL can include stubs, mock wire wrap hardware, simulations, and emulations, which will allow early testing before all hardware, software, and other systems are available. The planning, scheduling, and costs of such a SIL configuration will be nontrivial. However, the basic IoT real-world SIL test system may need to be expanded into a complete IoT system, as shown in Figure 19-11.

This next level in the SIL "real-world" continuum is where testers build on the last level and introduce more elements, such as other IoT systems, integrations, the cloud, SIL features, and capability. In this example, testers have multiple IoT devices, many "process support" SIL components, edge systems, and connections to the cloud Internet. This configuration moves our IoT testing into a system that may be necessary for some test groups.

The testing and support staff increase over time. In this configuration, the STA SIL deals with multiple devices and systems. Likewise, the "support process elements" grow, as shown in Figure 19-12, to include tooling such as CM/SCM, an issue and status reporting system, test managers, requirements management, modeling, and support for testing, such as combinatorial, keyword, and model-based testing (MBT). The support process automation and integration of these tools often take special test staff skilled in the programming and integration of tools.

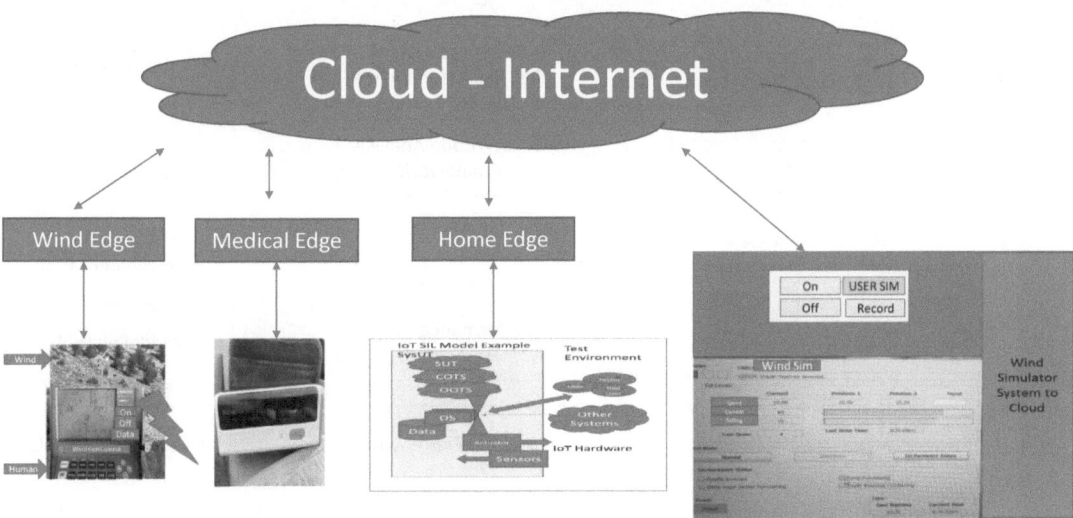

Figure 19-11. STA SIL systems of systems testing

Reference: Jon Hagar

Testers create highly automated models and keyword systems to run test procedures that drive and respond to a "sim/emulation framework" system. This test system allows the automation of tests that may run over multiple systems for minutes, hours, days, or even longer. Given such levels of automation and long runs, the SIL needs systems that support frameworks, such as "quality/code analysis," performance testing, security testing, automated test execution frameworks, and the noted test process support areas. The testers become system test "programmers," as well as users of these test systems. Again, this setup requires a lot of work, a lot of coordination, and a lot of programming, first to create it and then maintain it.

These STA SIL environments are examples, but many variations exist. The STA SIL test planning must consider and address what is viable, needed, and achievable given resources and risks.

Environment: Real-World Full System Software Test

We now enter an area of testing that is very much in evolution. As IoT, embedded, IT, the Internet, old systems, etc., all "plug" together, as shown in Figure 19-13, who owns the assessment of the system/ parts, and how do humans accomplish it, or should testers do it? Many constraints factor into the boundaries, and projects must understand and respect those boundaries.

The answer is that some level of testing and assessment will be needed. It may be done by all of us as user-stakeholders, though I suspect significant failures of IoT-Internet system of systems may lead to calls for "government" action. Companies may do system of systems testing on their large projects. For example, see Figure 19-14 for a utility company's view of testing an IoT wind generator system online with their historical systems.

Testing for any complex piece of software in a system of systems will never exercise all possible use cases, processes, user inputs, and/or hardware configurations. This fact is accepted by most of us in software engineering. However, this belief can lead some organizations to avoid doing anything but minimal test efforts. Regulations and governments are likely to step in. But, as software users become

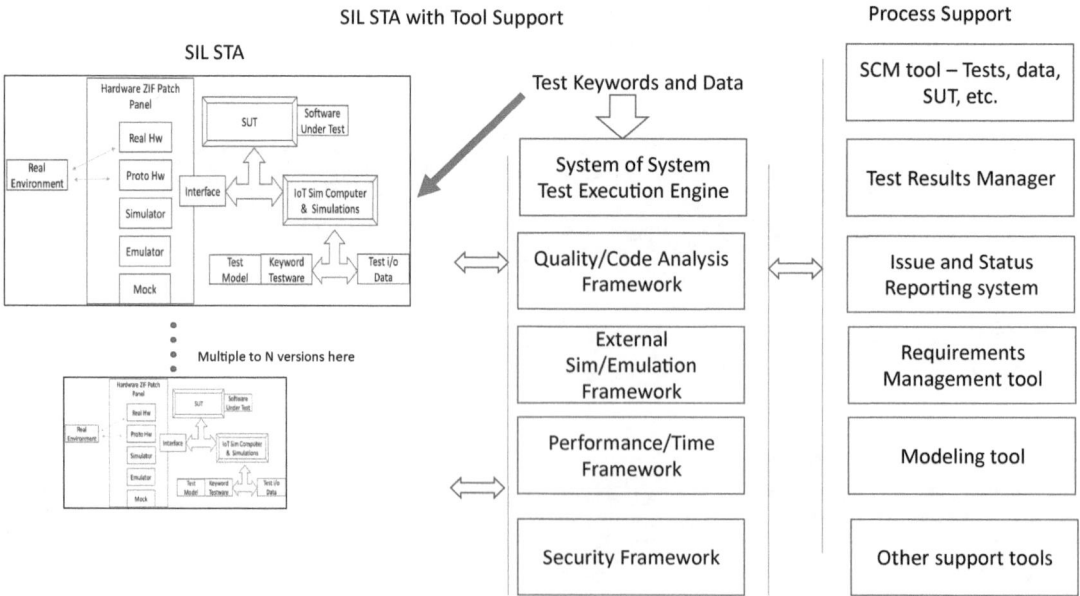

Figure 19-12. STA SIL with tool support for a system of systems

Reference: Jon Hagar

savvier and the competition continues, most companies cannot afford to have software that fails, so levels of STA SIL must become part of test planning and regulations.

Consider Figure 19-15 of the whole world of IoT/IIoT/IT, systems, software, consumers, communication, transport, home, and government regulations. Do testers just let this happen and then have a massive crash-and-burn someday? I would like to think that an IV&V/test organization might be able to help. This seems to be an opportunity for some companies with the right skilled employees.

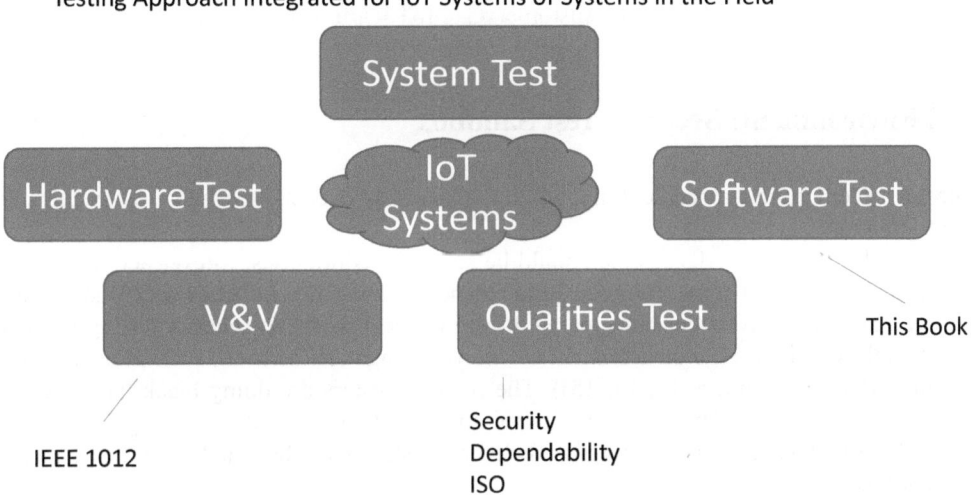

Figure 19-13. Full system test after integration

Reference: Jon Hagar

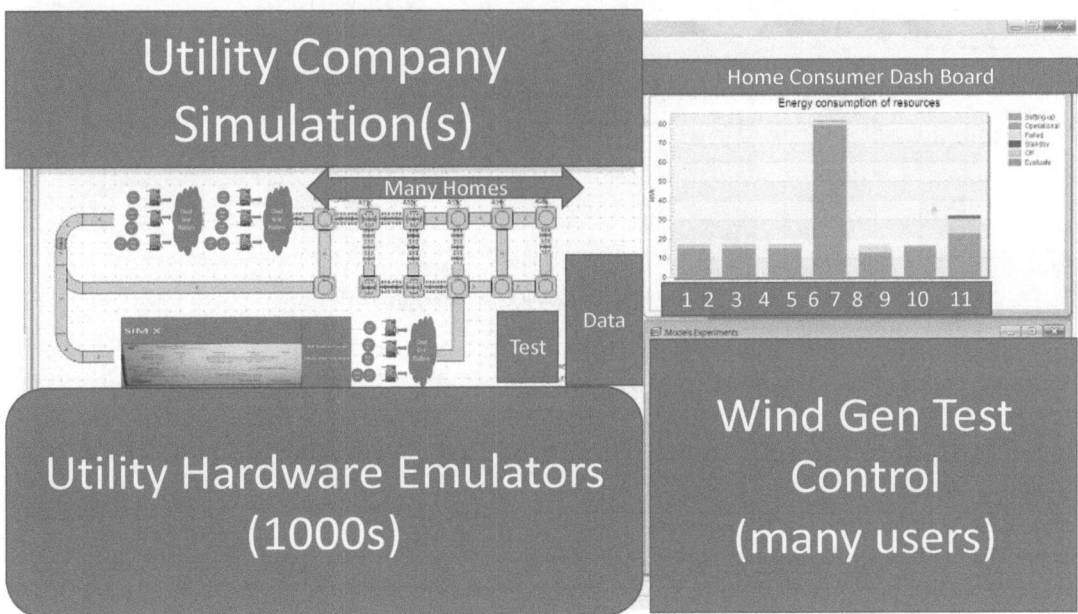

Figure 19-14. Utility company view of "testing" their systems with IoT wind generator systems

Reference: Jon Hagar

LESSONS LEARNED FROM AVIONICS SYSTEM TO IOT
Under regulations and close government oversight, air travel has become statistically safer. Aircraft manufacturers have vast and expensive test labs. For example, one is called an "iron bird" (aviation). This is a ground-based test device used to prototype and integrate aviation systems. Still, like the STA SIL environment listed earlier, it has all the systems, including simulations and many types of testing. The iron bird can be reconfigured to another aircraft quickly so testing can continue. The aviation industry also has extensive real-world testing, where a test aircraft with unique instrumentation can fly around. And finally, each fielded airplane gets tested before delivery to a customer. Can IoT learn from this story?

Special Environment: Security Test Sandbox

This special test environment is essential to do right (for safety reasons) for many IoT systems. Please pay close attention.

First, like all testing, security testing should be seen as a continuous, forever activity for IoT. To support security testing, at some point, an isolated special environment is advised. Testing an entire system in a realistic environment but separated from the real world so security testing can occur is ideal. A "sandbox" isolates the environment so "nasty" things can be tried, such as attacks, viruses, and phishing (Part 3, Chapters 13, 14, 15)). The security testers are doing black hat–type hacking before the "bad guys" do it. The testers don't want the tools they use and the testing they do released into the real world. Cybersecurity testing needs to be constantly on the watch, and the safest place is the test sandbox.

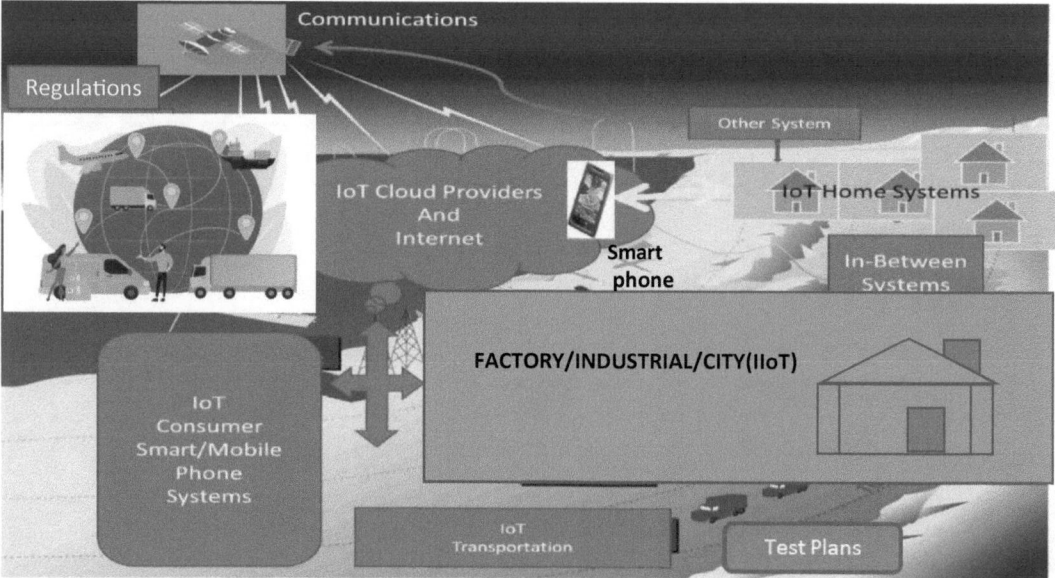

Figure 19-15.   The whole IoT/Internet world

Reference: Jon Hagar

SOLARWINDS AND ORION

A cybersecurity firm named FireEye discovered that a sophisticated attack had compromised its software. SolarWinds, an IT firm that FireEye used in its software, was the victim of a supply chain attack that gave hackers access to potentially thousands of targets (other companies that were customers of FireEye), including FireEye. "The SolarWinds hack was and continues to be one of the biggest espionage campaigns recently discovered," said Thomas Rid, a professor of strategic studies at Johns Hopkins University [1].

The sandbox is a SIL environment where testers run code, observe, analyze, and code in a safe, isolated (not connected to the real world) environment on a network that mimics end IoT user operating environments. This environment is designed to prevent threats from getting into the real world. The tester often uses it on IoT software that is untested or on untrusted code (e.g., OTS, malware, etc.).

Sandbox testing replicates the IoT functionality needed to test the system accurately. The goal of the sandbox is to assess IoT security while not impacting live systems. When establishing the STA SIL, regulatory issues, government, customer, and international regulations and laws may come into play.

There are three primary IoT STA SIL sandbox configurations:

- Virtualization/isolation – This method uses a virtual machine or isolated hardware configuration to run software in a nonconnected environment.
- IoT simulation/emulation – The sandbox simulates software and emulates the hardware of the IoT systems but may not use all of the system hardware.
- Complete system emulation – The sandbox is a complete IoT hardware configuration to test the behavior and impact of suspect software elements.

While complete system emulation is the best, it is expensive. IoT simulation/emulation is a middle-of-the-road approach, but it is not inexpensive. Virtualization is cost-effective but lacks certain levels of realism and may not be acceptable to some regulators.

The IoT security testing workflow can be outlined as follows:

1. Determine security risks and testing plans [2].
2. Identify software to be assessed.
3. Set up the sandbox SIL.
4. Load software into the sandbox.
5. Conduct security test attacks.
6. Document and report test attack results, taking follow-up actions as needed.

Here are a few good security practices to aid the security testing workflow:

1. Assess the security of the fielded IoT devices over time and repeat security testing as needed.
2. Use black hat attack techniques [3].
3. Use STA SIL setting, tools (look for new ones constantly), and configurations for the best advantage.
4. Use only authorized software and file updates coming from SCM. Configuration management (CM) people, testers, and project staff need to follow CM/SCM project practices.
5. Monitor security constantly in operations testing.
6. Keep the sandbox, threat intelligence, and attacks constantly updated (bad guys are a moving target, have an IoT security tester in operations).

7. Limit access to the devices, systems, or networks under test, the test processes, and the facilities where testing is happening. Good security practices are not maintained by lackadaisical attitudes or bringing "friends" in to secure environments.

IoT Chaos Engineering "Live" in the Real World

This approach builds on system of systems testing concepts. STA can include "real-world" testing on an actual deployed system. This testing is done because a SIL environment does not have all the real IoT system's parts, elements, and interactions, which is why it is called chaos engineering.

Chaos engineering is the discipline of experimenting on a *live* software system in production to build confidence in the system's capability to withstand turbulent or unexpected conditions and behaviors [4]. Chaos engineering is very risky since it is testing and assessment using a *live* production environment. A problem in the IoT is that chaos engineering of a system can impact the entire production system and users, hence the name chaos.

Testing the IoT system or system of systems is controlled as much as possible but performed within the real-world production environment. Therefore, total control is impossible. Chaos engineering assesses dependability, resilience, security, functionality, and performance qualities. Chaos engineering testing assesses IoT device qualities against

- Failures in the IoT device and software in the real world
- IoT supporting software, hardware, and software structure failures
- IoT connection communication failures
- Integration issues

Chaos engineering typically implements a perturbation model during the testing. The perturbations are meant to mimic rare or catastrophic events that can happen in realistic production because of the size and complexity of the real-world system [6]. IoT perturbations can include

- Failure of communication system channels
- Slow traffic (performance)
- Failure of other IoT devices (dependability)
- Attacks on the systems (denial of service (DoS) and security)
- Software error in the IoT device under test (loss of functionality)
- Cascading errors (resilience of the devices connected)
- Return to service timing (a device or system is not available for some time)
- Resource limitations (servers full, too many devices, other unavailable devices, etc.)
- Human failures (monkey testing, stuck input channels, etc.)

Level of SIL Environment vs. Project Factors

The nature and level of the IoT test environment have relationships to the project factors of Part 1, which are

#1 Product integrity – IEEE 1012
#2 Product maturity and risk-based development-operations
#3 Organizational ability
#4 Project size – People, cost, time, large complex software systems

None of the following are absolutes; they are just starting guidelines. The test team will still need critical thinking in test planning of the SIL, perhaps following Figure 19-16 as a start.

The factors of Part 1 that are summarized in Figure 19-16 can be understood as

#1 Product integrity – IEEE 1012 presents processes associated with determining integrity levels. This can be used as a starting point. In lieu of IEEE 1012, as the assigned integrity level increases (see Chapter 21), so do the STA and SIL needs. This can be seen as a one-to-one correlation.

#2 Product maturity and risk-based development-operations – As products mature or risks change, the SIL environment often matures and increases in factors, such as size, cost, complexity, and schedule. The correlation between these factors is not solid and subject to variation. Some IoT products mature, so the testing efforts and SIL usage can decrease. However, if the usage of the product increases via more sales and types of users, then the risks grow, testing becomes more challenging, and the SIL must have more capability.

#3 Organizational ability – Organizations with more ability tend to have more capable and more extensive test labs and facilities. They have more resources and a history of reuse. Hence, they also undertake large IoT projects and systems.

#4 Project size – This correlation is probably the most obvious and straightforward. Small IoT needs to call for a small SIL. Extensive IoT systems need more complex and a more significant SIL.

There is no one-size-fits-all mapping on tooling, automation, STA, and SIL environments. However, these are the factors to consider when test planning the STA and SIL.

Bringing STA to a Large-Scale Software Test Architecture/Environment

There are many IoT sectors where STA should be considered during early test planning and analysis. An example of multiple test plans, test efforts, test/V&V teams, and contractors is shown in Figure 19-17.

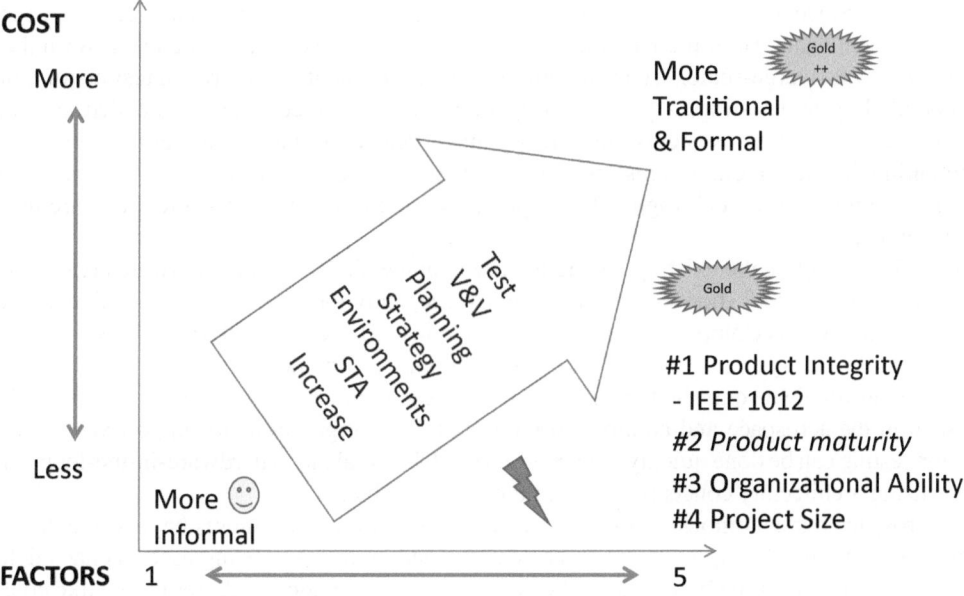

Figure 19-16. Cost vs. IoT project factors

Reference: Jon Hagar

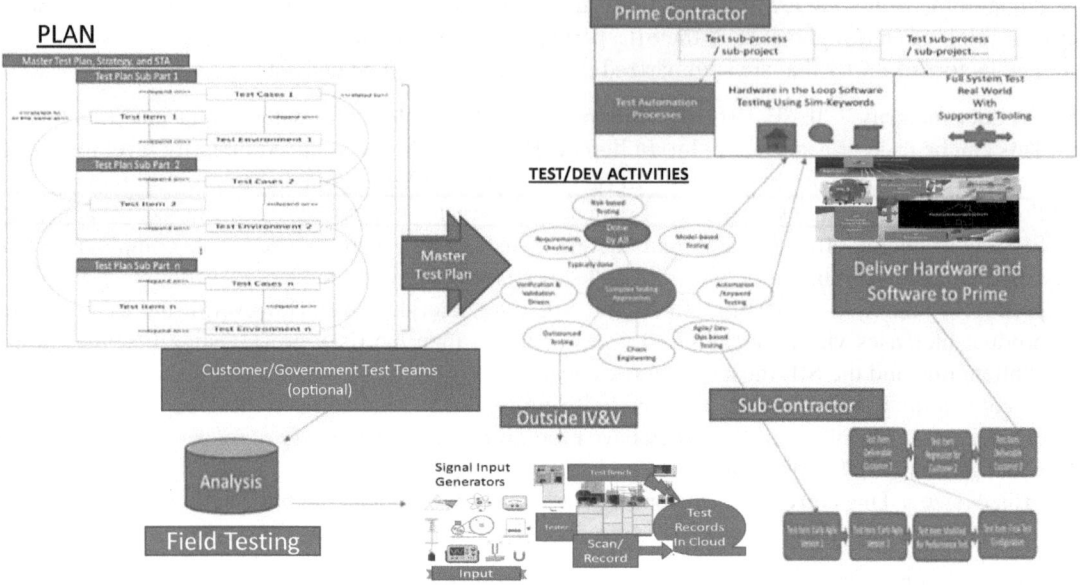

Figure 19-17. Example of a complex system of systems evaluation with the prime contractor, subcontractors, customer, and IV&V testing

Note: See Appendix D for a more comprehensive view.

Reference: Jon Hagar

The test efforts start with multiple test plans and include a master test plan, as shown in Figure 19-17. The test plans drive a test lifecycle in the center. The prime contractor creates products and has a set of test environments to assess the IoT system. There are also subcontractors for hardware and software. After testing, these elements are delivered to the prime contractor as defined in the master and lower-level test plans. Since this is a large system, there is an outside IV&V contract and customer (government) assessment team. All of these elements work together. These elements and teams form the STA.

The industry has large-scale test architectures and environments with proven test history, including embedded system testing, a good starting point for IoT since many embedded systems are evolving into IoT. These systems historically have scaled to larger implementations. Testers implementing IoT test architectures and environments should look to existing success stories to avoid "reinventing the wheel." Figure 19-18 presents a list of example historical test architectures and environments.

IoT needs to build on the existing testing history in embedded and cloud software areas. The table on the left of Figure 19-18 shows examples of aerospace, automotive, government, and tech industry. For example, the Airbus Company has an Iron Bird facility [5], a configurable test lab with all of their airplane's hardware, systems, and software. Simulations and models control the test architecture system to do testing in a controlled environment. This test architecture/environment is very similar to those seen in the aerospace and automotive industries. Reconfigurations to support verification and validation testing can be done quickly from a prototype lab to real lab to hardware-in-the-loop software test labs using architecture concepts shown in the figure's table.

The aerospace and automotive industries typically take test environments beyond the "iron bird" concept and more into testing in the field. The car covered in drapes is being used in chaos field testing. This example is a car being tested on the back roads of Colorado, USA, for things like cold, high altitude, and poor driving conditions. Testers know that the system is under test because of the drapes, and there is an instrumentation "chase" vehicle usually traveling behind the car.

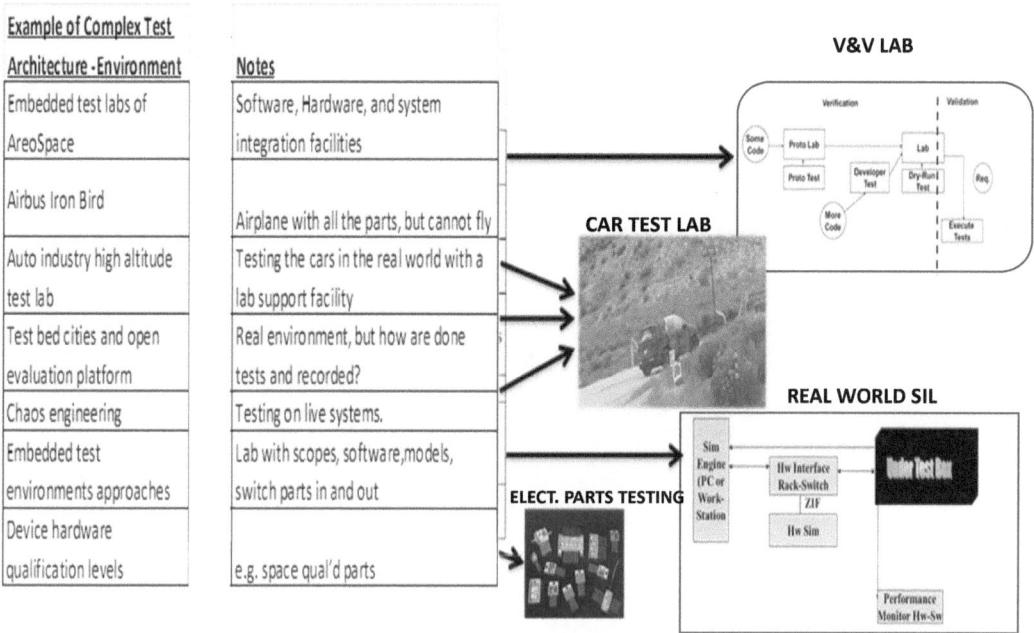

Figure 19-18. Examples of in-use complex test architectures and environments

Summary

This chapter provided examples of test environments that are key for IoT testing. These environments are critical and as numerous as IoT systems themselves. The examples in this chapter covered various example levels that IoT teams should consider in their test planning and design. The next chapter outlines test tool categories which support the SIL environments.

References

1. www.cnbc.com/2021/01/26/the-massive-solarwinds-hack-and-the-future-of-cyber-espionage-.html – accessed 2022
2. "IEEE/ISO/IEC 29119, ISO/IEC/IEEE International Standard – Software and systems engineering – Software testing" – Parts 1 to 5 series
3. *How to Break Software Security* by James A. Whittaker and Hugh Thompson, Addison-Wesley Professional, 2004
4. "Principles of Chaos Engineering." principlesofchaos.org – accessed 2022
5. https://en.wikipedia.org/wiki/Iron_bird_(aviation) – accessed spring 2022
6. Zhang, Long; Morin, Brice; Baudry, Benoit; Monperrus, Martin (2021). "Maximizing Error Injection Realism for Chaos Engineering with System Calls." IEEE Transactions on Dependable and Secure Computing: 1. *arXiv:2006.04444*

Figure Reference

1. https://img.freepik.com/free-vector/chemical-laboratory-furniture-bench-equipment-front-view-images-set-with-cabinets-tubes-liquids_1284-16307.jpg?w=740 and https://img.freepik.com/free-vector/physics-classroom-equipment_3446-34.jpg?w=740

Chapter 20
Tools for the Software System Integration Lab (SIL)

Chapter 19 examined the importance of IoT system-software integration lab (SIL) environments, including products, support, and configurations for testing. Since IoT systems have unique hardware and software, a SIL tool environment must also be customized. The range of tool customization is extensive. This chapter looks at supporting tool configurations. However, readers should recognize many options, from off-the-shelf to customized tools. Categories of tools will be considered; however, most tools will not be recommended by name.

Test Lab Needs

Test labs typically have specially trained staff and operating procedures to aid the testing. Test labs can include specialized equipment and tools that aid in engineering. Software tools include planning, management, modeling, execution, and analysis, as well as those that report on test results. Hardware support tools can include specialized computers with test points, probes, oscilloscopes, input devices, as well as data gathering tools and recorders.

While significant cost, schedule, and care may go into creating a lab, I have found that most laboratories and tools have limitations, which impact how the software is tested. Limitations in test tools may mean errors are missed and information gleaned is incomplete. For example, a lab cannot possibly have all the real-world users. A test environment may only have one or two users in the test loop. So, testers must simulate users using tools, but how realistic are these limited users? Also, test labs do not have every piece of hardware or interface(s), which can vary from system to system.

Given the limitations of labs or SILs, here are some standard tools on a project to assist the testers. Many of these tools are mandatory whether on a project or in the lab environments. Standard tools an IoT test team should have and use on a project:

– Management of lifecycle, cost, and schedule (Agile and DevOps)
– Software configuration management (SCM), status reporting, and tracking (these are critical tools for any project)
– Requirements management for verification, tracking, and closure (another critical tool)
– Data analytics with AI for designing and assessing test inputs and outputs (see Part 4, Chapter 22)
– Model-based test tool(s) for planning, design, and oracle generation

    ➤ Keyword tools, execution, monitor, and interface to models
    ➤ Performance modeling and load test tools

© Jon Duncan Hagar 2022
J. D. Hagar, *IoT System Testing*, https://doi.org/10.1007/978-1-4842-8276-2_20

- Quality assessment tools including security, performance, safety, reliability, etc.

  ➢ Security tooling to support sandbox testing (see Part 3, Chapters 13 and 15), network sniffers, fuzzing, spoofing, etc.

- Hardware support tools, including simulation and emulation systems
- Hardware test tools for electronics analysis, interfaces, and physical testing
- System support and analysis tools
- Test reporting tools including planning, risks, results, and issues
- Data visualization and presentation tools (part of a database system)
- Static code/model analysis tools
- Inspection/peer review support tools
- Integration modeling and planning tools
- Math-based technique tools supporting combinatorial, statistical, or design of experiments (DOE)
- Edge, cloud, and user lab test tooling with a simulation of the edge/cloud
- Team and crowd communication tools
- Security testing tools for Pen, analysis, cracking, and threat testing
- AI and embedded IoT tools
- Additional lifecycle and project management tools

Many of these are general test support tools that the tester can find in general references. In the following, I explore a couple of the key (for me) IoT support tool areas.

Modeling and Requirements Management Tools

IoT testers and analysts have successfully used modeling tools to do the following things in test labs:

- Allow automation of test processes using

  - Closed-loop simulation
  - MBT
  - Keyword testing [1]

- Allow generation of IoT test data inputs

  - Run hundreds of Monte Carlo simulations of code (e.g., control systems, guidance, navigation, etc.) to find input sensitivities for execution in a complete system test lab
  - Generation of random or statistically meaningful inputs (combinatorial testing)
  - Generation of object-oriented, data-driven, or structured analysis
  - MBT inputs to test coverage

- Allow generation and assessment of IoT outputs

  - IoT qualification or acceptance testing based on functionality
  - Design oracle for verification
  - System oracle for validation

Automate, Automate, Automate

IoT test plans should include automation when and where it is beneficial. Automation has the following benefits and must be implemented carefully to ensure success:

> ➤ Make better use of resources (time and money)
- Avoid human mistakes in traditional testing over millions of tests
- Support agile testing in the whole lifecycle from coding to system validation
- Run more tests with a fast turnaround of new tests
- Achieve faster regression testing
- Drive testing with models, which are then fed into automation tools
- Use keywords with simulation and emulation in the areas of hardware, software, and systems

Simulation Test Tools Needed to Support IoT

When a tester says "test tools," everyone typically thinks of an automated test execution tool. And while this may be part of the story, when I say the word "tool," I mean anything that helps me do better testing. Tools can be software, such as capture-playback tools, but a tool can also be as simple as a checklist that supports manual testing. Tools, model-based testing, automation, environments, and data analytics are covered throughout the book and specifically in Part 4.

I recommend white box and black box testing, including analysis tools, such as static code analysis. These levels and approaches allow testing, verification, and validation throughout the lifecycle. Also, these combinations are complementary, increasing the likelihood that errors will be found. Finally, many IoT embedded projects may benefit from the use of models and mathematical analysis, which, in my experience, more progressive organizations will use. Additional complementary ideas are reflected in test plans, IV&V [2], test automation, and AI.

Evolve into Automation Tooling for IoT Success

As software systems grow in complexity and size, test execution automation will become expected and, sometimes, the only viable option. This situation will be seen in IoT and large-scale, complex IoT systems of systems.

IoT testers may start with manual testing, but doing manual testing for every phase may not be a good survival or cost-effective strategy. For project test planning, testers need to be thinking about situations where the manual approach is not the best option. Too often, many testers default to manual testing because it is what they know. Still, given that there are some situations where automated test execution must be included in planning, testers need *not* be trapped in that manual thought pattern.

In my book *Software Test Attacks to Break Mobile and Embedded Devices* [3], I detailed the need for test automation in the test environment. To date, the key rationale and points for this have not changed. However, even with automation, teams still need thinking and skilled testers. The ability to recognize the need for automation is a skill worth building and one of the factors that will make automation successful.

In IoT projects, there are several items critical to success. First, the test team needs skilled programmers and testers who understand the keywords and the system under test. The project needs dedicated people in these areas. Next, in test project planning, management must allocate a proper schedule to develop the test system and allow the training of people to use the system. These efforts require resources that must be included in project-level planning. During test automation, projects that do not allocate a proper schedule and appropriate resources are more likely to fail. To assure these allocations, all levels of project management need to support the test execution automation efforts. Finally, the IoT hardware and software efforts need to evolve and integrate with the test development.

Test teams considering automation should take the risk of actually trying it. My personal experience with automation and tooling is that out of every ten attempts, two or three failed, four or five were just okay, and the final two or three were great successes. I highlighted and built on the successes, improved the "just okay," and abandoned the failures. Remember the old saying, "Nothing ventured. Nothing gained."

There are many lessons to learn in the test automation of IoT systems. The first and most important is that automation is only one part of the story for IoT testing since you still need thinking testers who can do rapid manual testing and use tools when needed. *Automation does not always save money but can help to yield IoT products that delight customers.* Many managers believe they can get rid of test staff and just let the computers do the work, thinking it saves budget. However, computers – even those running AI – need skilled, thinking, creative humans to drive the testing of hardware, software, and the overall system.

My favorite source of lessons learned on automated testing comes from Dorothy Graham [4]. Dot's book addressed

➢ Tool management.
• Agile tooling.
• Database automation (essential for IoT data storage and analysis at the edge and cloud).
• Cloud tooling.
• The test tool creator becomes more automated.
• Framework automation.
• Automation in a complex system (like IoT).
• Device simulation and emulation (very applicable to IoT).
• Automated reliability testing with hardware interfaces.
• Model-based testing (MBT – very applicable to IoT test automation).
• Regression test automation.
• Case studies in a variety of environments which in many cases relate to IoT.

Continue your reading on automated testing there.

Evolve IoT Testing with AI Tools

AI is addressed in several parts of this book (Chapters 13, 20, and 22). AI implies the use of tooling, and, as much as I could, I have avoided specific tool or company recommendations, opting instead for general tool categories. However, in this case, I would like to refer to [5]. I know the company and have played with their tools. I recommend a search in general for AI test tools since there are many. As of the writing of this book, the test area of AI is growing and changing at blinding speed. Given the challenges covered in this book, I believe AI will be a significant player in many aspects of IoT. Readers should follow developments on and with AI and do their homework in this area.

Summary

In recent years, support by vendors for embedded, mobile, and IoT testing has been increasing. Classic test execution automation will support IoT issues, such as testing device configurations and capture/playback. Teams working on IoT are advised to conduct tool trade studies and searches to find the best tools for their project's context.

Using tools and automation does *not* mean that all testing should be automated. The progressive groups mix automated and manual tests, mainly guided exploratory testing, with tours and attack patterns. This idea follows the concept of having complementary approaches and activities to guide the testing.

In IoT systems, testers must include the hardware as well as the software. This chapter provided a very short overview of SIL tool configurations and recognized that there are many options for specific vendor tools. The range of tool customization is extensive and based on each IoT project. Categories of tools were identified with pointers to other references on test automation since this is an entire book topic in itself. The next chapter addresses environments to support independent testing and IV&V for large IoT systems.

References

1. "IEEE/ISO/IEC 29119, ISO/IEC/IEEE International Standard – Software and systems engineering – Software testing" – Parts 1 to 5 series, particularly Part 5 for keyword testing
2. "Software Verification and Validation: An Overview" by D. Wallace and R. Fujii. IEEE Software, May 1989
3. *Software Test Attacks to Break Mobile and Embedded Devices* by Jon D. Hagar, CRC Press, 2013
4. *Experiences of Test Automation: Case Studies of Software Test Automation* by Dorothy Graham and Mark Fewster, Addison-Wesley, 2012
5. https://test.ai/ – accessed winter 2022

Chapter 21
Environments for Independent Testing and IV&V on Large IoT Systems

The previous two chapters addressed the test SIL, environments, and tools. Large IoT systems may be composed of multiple IoT devices, edges, layers, and central processing. I have mentioned the issue of ownership of such large IoT systems. There may be many vendors of IoT devices, different stakeholder organizations looking to be using the IoT system, and no clear-cut "owner" of testing the whole system. In this chapter, I offer advice on using test independence and IV&V to support such large IoT systems. This chapter will interest procuring stakeholders, as they recognize the need for test ownership of the system or system of systems.

This chapter focuses on test independence in detail and references closely to the IEEE 1012 V&V standard [1]. IV&V can be applied to any software system with the different factors of Part 1 of this book, if a stakeholder or customer is willing to pay for the IV&V. The best factors to consider are the ones given in IEEE 1012. In the case of large and complex IoT systems and systems of systems, it may be that the various vendors and even a prime contractor for the whole are *not* fully prepared to do all of the testing and evaluation. When tens or even hundreds of IoT elements have been tested, the ownership of assessing the whole needs to be addressed, but the procuring stakeholder organization often cannot do the verification and validation (V&V) and testing. Whether private industry or government, such a case calls for independent testing or independent V&V (IV&V).

Getting the Most Out of Independent Testing and IV&V

This chapter provides an example of IV&V history and why for many larger IoT systems IV&V may be important to implement. Independence in testing and IV&V is not without some controversy. *NASA studies concluded that its effectiveness is low* [2] *when poorly practiced and funded.* However, IEEE 1012 and my experience conclude that V&V/IV&V/testing can be successful if software testing is funded and conducted from a user's and/or system's perspective to assess functions, performance, usability, quality, and reliability. This system view of the software test process considers success information and classes of errors that can be practically detected in software. To accomplish this, software evaluation must be done at a V&V level by engineers skilled in more than just the software disciplines.

IV&V should be focused on

- Negative testing, in which finding errors is the goal
- Positive quality(ies) assessment, particularly critical items including functionality, performance, and security
- Looking at the whole of software, hardware, system, and operations

© Jon Duncan Hagar 2022
J. D. Hagar, *IoT System Testing*, https://doi.org/10.1007/978-1-4842-8276-2_21

IV&V should cover these topics by having all levels of testing (unit to entire system-software qualification), multiple tools (simulations, instrumentation, oracles, etc.), and highly knowledgeable engineering people. IV&V/test should be viewed as serving in place of the customer performing detailed acceptance testing, as defined in ISO 9001 [3].

I have found that the key to obtaining reliability in the IV&V method is using domain experts to define, conduct, and analyze the testing products. *When doing IV&V*, I have found that experts from a variety of non-software engineering domains devise tests and identify faults during analysis that exceed what is found during routine testing by the development staff. Further, such an independent team avoids the development blind spot or prejudice about the overall nature of the system and software. These blind spots often afflict engineers after constant contact and review of material related to the software. The fresh view of an independent, multidisciplined test team avoids this biased situation. Finally, this multidisciplined team voids what I believe is the mistake of having the test team be composed of pure software engineers, testers, or programmers. Many software engineers have limited knowledge about the entire IoT problem being solved. Their skills lie in the use of computer systems and software implementation issues. All too often, they only know the requirements as a "passed down" artifact, coming from a system analysis domain expert. Unfortunately, this myopic view can incorporate abstraction errors into the software and/or introduce deductive errors during the test.

V&V ON A SMART IOT VEHICLE CONTROL

A vehicle control software was under IV&V. This software had the following characteristics: real time; vehicle control; minimal human intervention possible; and numerically intensive calculations of movement, dynamics, vehicle body characteristics, and moving targets. Programs were small, under 100,000 source lines of code, but critical to the control and success of the overall system. The IV&V staff had engineers in electronics, system controls, guidance and navigation, systems engineering, math, physics, hardware, software engineering, and management areas. Additionally, the IV&V project had a mix of experienced or senior engineers since the new staff sometimes brought new viewpoints.

The concentration of these experts is on creating test cases and analyses designed to provide information to stakeholders and detect faults. And, by successful completion of all identified V&V, testers demonstrated that the software program accomplished its mission. This real-world, system-based testing allowed evaluation of the software above what the development team did. Failures in software at this level received the most visibility and publicity for the users. [4] argues that the goal in V&V is beyond simple correctness but includes the detection of errors. This team and story proved this point and practiced testing consistent with it.

Verification

Verification concentrates on the detection of programming and abstraction errors. Programming faults have two subclasses: computational logic errors and data errors. The team practiced white box or structural verification at a very low level of the computer, including assembly language and binary files. Automated tools and computer probes allowed monitoring of individual instructions, logic/data flow, configuration management, modeling, and comparing results to predetermined and computed success criteria. Verification is conducted primarily by software engineers or computer scientists with some aid from the other members of the team. The software system under test is subject to the deductive class of errors. The transformation of concept-to-requirements, requirements-to-design, design-to-code, and hardware-to-software interface all can experience deductive class errors that can result in failures.

Validation

Validation is conducted at several levels of black box or functional testing. Validation is conducted at the design, specification, and system concept levels. Validation is where the largest concentration of human resources, time, and testing is expended. Engineers from each of the domains define analysis and testing appropriate for the lifecycle level of the software.

SUCCESSFUL VALIDATION TEAM

In this case, a team used extensive computer simulation models to serve as oracles. Each simulation or model is specifically designed to concentrate on one error class (deductive or abstraction) at high levels of the system. Multiple runs on the hardware/software and tools were used with comparisons and cross-checks between results. These cross-checks serve as good oracles and constitute a multiple version (see the section "N-Version Testing Supporting Independence and IV&V") based on testing methodology. Additionally, since the expert engineers had internal knowledge of what the software should be doing from a design or concept standpoint, they could detect faults associated with missed or poorly abstracted requirements. This was accomplished by detailed inspection and review of analysis results.

The project lasted for years. Over 4000 issues were found and corrected. An independent analysis of this IV&V project found eight issues that saved stakeholders of this system money – at least two of these issues would have resulted in a total system loss.

N-Version Testing Supporting Independence and IV&V

The IV&V approach for large IoT systems can be compared to a multiple version (or N-version) based testing methodology. This approach is because the team had simulations and partial implementations of the system generated separately from that of the software development team and used in V&V.

Figure 21-1 depicts the different levels commonly found on an IV&V project. The team has the hardware test environment (shown at the bottom of the figure). But, as shown in the top left of the figure, the team has software programs based on requirements, including several different levels of requirements documentation. The IV&V team creates design-based simulation tools that replicate a function or groups of software functions. These form the equivalent of an N-version programming system. These tools are compared at different points, as shown in Figure 21-1. One project example had a minimum N=3 and a maximum of N=8 during the test phases. In N-version testing, the testers compare and check points of "N" different programs or levels.

N-version programming has been criticized for not significantly aiding reliability [5]. Brilliant's, Knight's, and Leveson's experiments indicate that different teams make the same mistakes in "independent" software versions. The underlying assumption of N-version is that independently coded programs fail independently, and Leveson's experiment proved this to be incorrect in the general case. This has been and continues to be a concern for IV&V teams. However, my use of the N-version program-based testing had successfully detected errors. I follow [6, 7] guidance, where the judge of the value of tools and IV&V is their ability to detect errors and provide helpful IoT product information such as qualities.

Problems were found by the N-version testing approach in all phases of the software's maturity. The approach found the following specific kinds of errors:

- Compiler-generated logic problems
- Differences between requirement specification and code equations

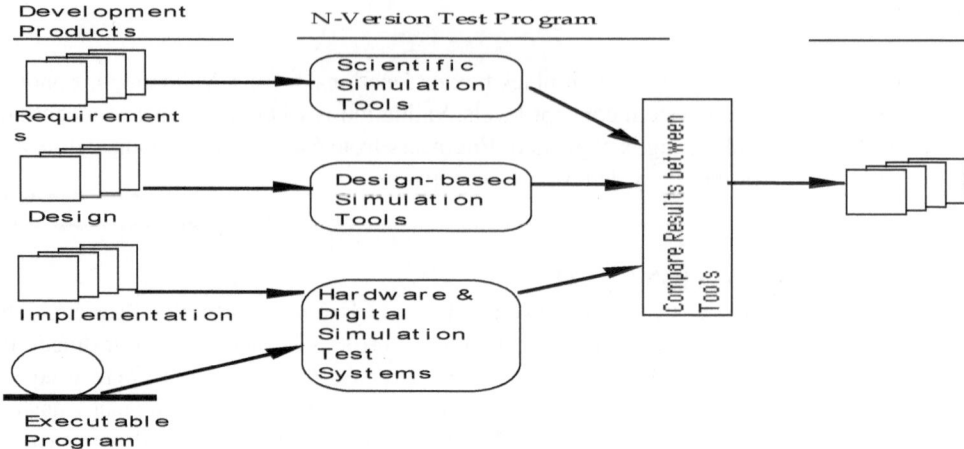

Figure 21-1. IV&V tool levels

- Differences in logic between requirements and control logic
- Differences in the numeric representation of floating-point values
- Differences in requirements-to-design deduction
- Problems between development products and test products
- Incorrect data load and initialization values
- Differences in tools between development and test groups

Lessons Learned in Testing, in Independence, and IV&V

After years of IV&V, the following represents my main lessons learned from various IV&V programs. Several of these bullets I see as mandatory on any large, complex, highly critical IoT development project. Also, this could be considered a mandatory review checklist for every project:

- The IV&V team should be involved in the software development cycle at the earliest possible time. For established programs, the developer should provide extensive technical information and documentation of the system design to the test team.
- The IV&V scope must be established and agreed to by all involved parties. The management scope would include a statement of work (SOW), reporting structure, interface requirements for standard deliveries and support, and final deliverable software procedures.
- It is best and most independent if a separate budget, controlled by the IV&V contractor, is established to negotiate changes in scope and deliverable products. (Beware of scope creep, which can kill a budget and schedule.)
- External to the software development group, a customer should be established for the IV&V effort. For the IV&V effort to influence the software design, findings and recommendations should be reported to an interested third party with organizational responsibility for the software development group.
- A documented data exchange list should be created between the development (developer organization) and the IV&V organization, as well as a format to control the interchange of documentation and products.
- Establish combined schedules for all tasks and maintain active reviews with notification to impacted activities of any schedule impacts. This is a management function.

- Accomplish active planning for all tasks, including review of schedules, metrics (trending and projection), and deliveries. This is reported to management.
- Establish a forum for exchanging technical, cost, schedule, and other information regularly. Weekly seems to work best.
- There should be an independent development of a requirements traceability matrix shared between development and IV&V teams to ensure all requirements are testable, all changes are tracked, and any requirement closures. CM and SCM can help here with their tools or other project tools.
- The IV&V team should implement complete configuration management (CM and SCM) of all delivered products. Hence, everyone knows what they are testing, and any "unknown" or suspicious software or hardware is *not allowed* (this could be a security issue).
- Increasingly with the aging of a project, a primary focus of testing should be to validate the interfaces and integration between software and hardware elements. Compatibility problems are hard to find and often remain even after extensive testing and IV&V. *Good interface definitions help, but systems of any complexity hide interface problems.*
- *There is a hidden cost associated with using reuse and heritage software.* Typically, the requirements documentation and the interfaces can have problems that complicate their use.
- The cost and time to develop the test software, process, and tools are often underestimated, resulting in the failure of these efforts. Also, part of the cost that is often overlooked is the actual training of people to do the work.
- *Any requirement that only states what to do or is unclear and incomplete presents problems.* For critical or safety-related code, no lines of code exist for which there is no corresponding requirement definition information or objective evaluation criteria (design, story, or use case). The software requirement must be stated to develop tests and test requirements so that a test or IV&V engineer can infer how the code is supposed to function.
- If present, it is vital that the software quality assurance group (SQA) be staffed with experts in software development engineering, since they need to make decisions about the validity of tests, tools, processes, and IV&V efforts. For large, complex, or highly critical IoT projects, SQA should be mandatory.
- Ensure the CM/SCM personnel establish a common problem reporting (issue) and tracking database with proper use and access procedures for all project users and possibly stakeholders (context dependent). CM/SCM records from this database will prove invaluable to project management. Allow CM/SCM to hold change board meetings as regularly as needed to allow smooth daily operations of development and IV&V and to ensure changes to any configuration are adequately evaluated before implementation, as well as tested or verified/validated before closing out a problem/issue in the tracking database.

The preceding lessons have limitations in an entire lifecycle, so added testing levels and actions should be considered, including

- Security holes from historic code may be missed, so specialized security attacks are conducted.
- Performance testing of the full system is complex, so added testing with long waits (seconds vs. milli- or microseconds), loads, slow network, dropped connections, etc., is done.
- Over time, the viability and completeness of recorded data meant these areas were assessed during operations.
- Usability of the system with the new connection(s) and expanded use requires assessment in operations.
- Coupling impacts from the new logic, new hardware, and the system requires assessment in operations.
- Monitoring off-the-shelf software issues is an ongoing development and operations task.

Table 21-1. Standard Sampling of IV&V Tools

Activity	Tool	Function	Benefit
Verification	Commercial	Coverage	Measurement of test
Scientific validation	Engine utility program	Control system simulation	Assessment of data values from testing
Validation	Hardware test environment	Execution of software system	Realistic assessment of software

Tooling Example for Testing Large IoT Systems and IV&V

Many independent/IV&V testing concepts originated in IEEE 1012–based programs where high reliability is needed. Table 21-1 has some examples of IV&V tool configurations that have worked for teams.

In verification, the team tests to demonstrate compliance of the code-to-design, design-to-requirements, and even an executable configuration to its source files. Testers treat the design product as "truth" and test it to be correctly transformed into the next level. Validation, on the other hand, tests that the requirements, design, or code does what "works" for a stakeholder. This is a much harder question and requires the human expert to quantify "works." For example, in validation, testers could look to see if an IoT vehicle control system has sufficient authority for road conditions, given vehicle and environment characteristics (snow, rain, traffic, etc.).

The IV&V team uses extensive computer simulations to analyze requirements in scientific validation which then serve as oracles for the actual "black box" system test results. Each simulation or model should be specifically designed to concentrate on one error class (deductive or abstraction) or system level. For example, the simulations are higher-order, nonreal-time models of the software or aspects of the system. The team's validation efforts start at what would be considered hardware-software integration testing. At this level, the team's simulations are design-based tools, and they simulate aspects of the system but lack some functionality of the total system. These tools allow the assessment of software for these particular aspects individually.

Simulations can be used in a holistic system fashion or on an individual lower-level functional basis. A simulation may model the entire system, while another simulation may model the specifics of how a control function of the engine is required to work. This approach allows system evaluation from a microscopic level to a macroscopic level. Identical startup condition tests on the actual hardware/software can be compared to these modeling tools to cross-check the results. The IV&V team uses these scientific validation tools as oracles for the test environment as well as "stand-alone" analysis functional aids.

Finally, in the other major validation aspect of this approach, the team develops a comprehensive test environment of hardware, software, and the system. This IV&V environment is crucial, and every attempt to replicate some or all of the system should be made. Such an IV&V environment can be costly to create, as the cost figures are directly dependent on the complexity and size of the system. However, this is the best way to test the software in a realistic environment. Some test facilities I have seen included ground operation systems, ground cabling, the real world, and actual vehicle configurations. However, sometimes there can be aspects of the critical systems that cannot be wholly duplicated in a test environment and thus must be simulated. For example, in some testing one might need thousands of IoT devices, hundreds of users, interface systems, many edges and fog configurations, and public clouds. These would be best done in simulation.

When to Consider IoT Test Independence and IV&V Environments

IV&V is not suitable for every IoT project. Comprehensive, fully independent IV&V is justified when the risk of software system failure is high. A risk can be from cost and quality factors, including security, functionality, safety, and dependability. IV&V must also be considered from the viewpoint of the customer and management. If customers and management are willing to do active engineering with software, where other people can provide "outside eyes of independence," then less rigorous IV&V may be acceptable. IV&V certainly can find errors, but if the cost of the errors being found does not offset the cost of the IV&V, then benefits can be questioned. IV&V can be considered in IoT, IIoT, or high-risk software programs. IEEE 1012 offers more details on integrity levels for IV&V.

The use of independence partially mitigates risk in testing that the development project cannot fully address. The consuming stakeholder must pay for the added testing/V&V. These are often government or large corporate entities where IoT failure is not an option. Over pure full IV&V, there are other ways to get independence and certification of IoT systems, including

1. Crowd testing where the IoT system is made available to "for hire" freelance testers and test teams paid via a bounty system (e.g., Utest [8])
2. Internal IV&V teams running inside of the corporate structure
3. Freely independent alpha and beta testers [9] where people test for "fun" or bounties

There are varying cost issues and ways to pay for independent testing. But remember, there is no single correct answer.

Models, chaos engineering, and model-based testing are often crucial parts of the testing with IV&V. Independent labs may take these test approaches [6]. These organizations may go with certification (cert.) labs, UL qualified parts, security labs, test companies, or IT test facilities. There may even be questions about how such architectural environments are funded and staffed.

Summary

This chapter addressed the considerations and environments that support independent testing and IV&V. Such types of testing may become necessary for large systems made up of many IoT devices and support systems where the "owner" of the testing may be the procuring group. This group may need to contract with outside or third-party testing or IV&V groups since the procuring group lacks the knowledge to do testing. In such IV&V environments and general testing, data analytics in IoT will become an advantage that should be considered, as defined in the next chapter.

References

1. "IEEE Standard for System, Software, and Hardware Verification and Validation," in *IEEE Std 1012-2016 (Revision of IEEE Std 1012-2012/Incorporates IEEE Std 1012-2016/Cor1-2017)*, vol., no., pp. 1–260, 29 Sept. 2017, DOI: 10.1109/IEEESTD.2017.8055462
2. F. McGarry and G. Page, "Performance Evaluation of an Independent Software Verification and Integration Process," Tech. Report SEL 81-110, NASA Goddard Space Flight Center, Greenbelt, MD., Sept. 1982
3. ISO 9001:2015 Quality Management Systems – Requirements
4. W. Howden, "Program Testing versus Proofs of Correctness," Journal of Software Testing Verification and Reliability, vol. 1, issue no. 1, 1991
5. S. S. Brilliant, J. C. Knight, and N. G. Leveson, "Analysis of faults in an N-version software experiment," in *IEEE Transactions on Software Engineering*, vol. 16, no. 2, pp. 238–247, Feb. 1990, DOI: 10.1109/32.44387

6. "How to Build a 20-Year Successful Independent Verification and Validation (IV&V) Program for the Next Millennium," by Jon Hagar and Lisa Boden, Quality Week Conference, 1999
7. D. Wallace and R. Fujii, "Software Verification and Validation: An Overview," IEEE Software, May 1989
8. Utest example – www.utest.com/ – accessed winter 2022
9. www.geeksforgeeks.org/difference-between-alpha-and-beta-testing/ – accessed spring 2022

Chapter 22
Self-Organizing Data Analytics (SODA): IoT Data Analytics, AI, and Statistics

After IV&V, testing, test environment, and STA concerns of the previous chapters, a final development and test support environment will be the tools and analysis used to understand the IoT data in support of engineering and particularly of testing (Part 1). In this chapter, I address the idea of self-organizing data analytics (SODA) [1]. Many testers and engineers have limited data analytics skills or experience, yet SODA can offer a positive path for them. This chapter outlines an environment to address this limitation but leads to better testing and better products and sales.

There is nothing magical about the concept of SODA for test data analysis. I choose it because it represents an activity I have done for many years. I suspect that many other testers and engineers will be mentally applying it, though maybe not consciously. Some testers are not aware they are learning testing patterns. SODA test environments relate to the pattern classification identified in other sections. I formally grow my test pattern activities into this section's SODA model.

SODA Model Examined

This section defines the process flow and details of inputs and outputs of SODA. Testers have moved into the age of IT, big data, clouds, and rapid project development changes, including testing. I present the idea of SODA, starting first with crawling before walking into automation and finally running with it using AI. Indeed, not every person and most testers will not be AI or data analytic experts with full SODA. However, I believe many advancing testers will be able to do the basics of crawling with SODA. Testers need to *use data* to run and improve our jobs. This situation is true for sales, management, development, and testers.

The idea of SODA is to have a system of self-organizing data analytics (hence the name) that can be set up by a tester and then used across a project. Some testers can set up SODA by themselves. You may need to hire an expert in statistics, AI, or analytics to help set up SODA on big projects. I recommend self-organization because I find the task of analyzing and projecting data has elements of repetition. This work takes the time that is best saved for automation. As a tester or manager, what I was interested in was not the raw data, which in modern systems can be overwhelming, but what I wanted *from the data*: the trends, patterns, graphics, and maybe "the answer." So, when I was doing data analysis and analytics, I moved into automation that would parse and do much of the work for me. This work freed me actually to do the analytic thinking. Now modern AI systems are becoming available that may be able to do more and more analytics for me.

© Jon Duncan Hagar 2022
J. D. Hagar, *IoT System Testing*, https://doi.org/10.1007/978-1-4842-8276-2_22

This section examines how to apply SODA. I use examples of inputs into SODA that might interest testers and illustrate how to get started. Also, I consider examples of the outputs that testers might find helpful. This section is not a complete cookbook that can be followed without thinking and doing some work. I realize many people just want "the answer," but I find that answers usually must be given within a project context, and I find almost an infinite number of contexts in the real world. I do try to give examples and sample use cases for SODA. Finally, I fully expect the ideas on SODA to change over time. The forces driving SODA's change include AI advancements, human user understanding of the concepts, and the tools to do analytics. Anyone trying to use the SODA concept successfully will have different inputs, outputs, and processes than my examples. This situation happens with most software projects as tools and programming styles change. Go with the flow of change.

All levels of IoT projects shown in Figure 22-1 should consider SODA.

BUG HIDES IN DATA ANALYTICS

A test team had Terabytes of data over hundreds of tests. A bug was hiding in three test data results. However, the data trend line in the three tests was in the wrong direction yet never tripped the failure threshold because the test did not run long enough, since most tests were short-duration runs. The system's first use almost did "run" long enough, and a fatal error almost happened as users watched the realtime data. However, with three seconds remaining at the system shutdown, a complete system failure did not happen by the luck of time. However, Ops engineers could see the failure was almost about to happen. They were not happy. Management wanted to know why the test team had never seen the failure in the data. The error was buried in a Terabyte of data, and the test team failed to use automated organized analysis processing. Management demanded better analytics to recheck all of the existing Terabytes of data. It was a long few months while the test team put in better data analytics and called it self-organizing data analytics (SODA).

Defining SODA

I have done development, QA, testing, and management for many decades. I learned to be successful, and the one toolset I needed was data analysis and analytics. I used these tools to communicate with my team and management.

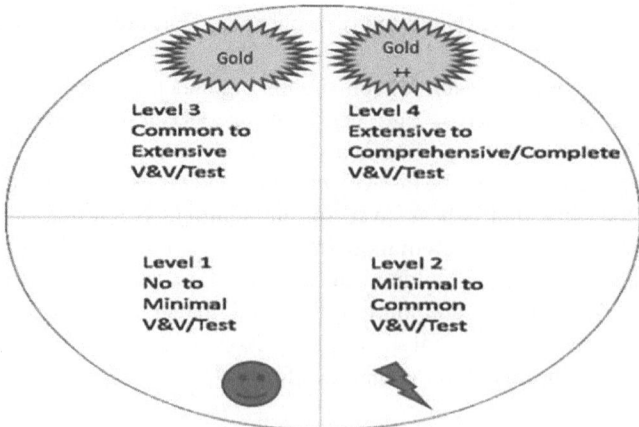

Figure 22-1. IoT project levels

In the SODA model world, I define analysis as the activity of using math, graphics, simulations, statistics, and AI to collect and process raw data into a form that I, as a human, can better understand. From Wikipedia [4], testers have "Data analysis" as a process of inspecting, cleansing, transforming, and modeling *data* to discover useful information, informing conclusions, and supporting decision-making. All of the above are varieties of "data analysis."

Consider a data system with ten billion individual points. No human can keep that many points in their head at once. However, if I plot those billions of points into a graphic, I can see historical trends. Further, I can feed the ten billion points into other math, statistical tools, and AI, which may yield even more understanding.

My definition of analytics is a little different. From Wikipedia again, testers have "Analytics" as the discovery, interpretation, and communication of meaningful patterns in *data*. Digital *analytics* is a set of business and technical activities that *define*, create, collect, verify, or transform digital *data* into reporting, research, etc. Here, with analytics, I use some of the same tools as in analysis, but I "look forward" and look for things such as patterns. I am looking for things like "when is the next bug likely to happen" or "what types of bugs should I be testing for and where should I be looking?" *There is no guarantee that future bugs will happen. There is no guarantee that future bugs won't happen either.* Still, in the cases where the future risk of software failure is sufficiently high, it may justify more work in testing using information from analytics.

As I said, the name SODA defines creating an idea or concept for data analytics rather than a specific tool system. Testers should feel free to make up their own name and change the activities, inputs, and outputs I define here. My experience is that analytics and analysis of data are how management communicates. Testers are often asked to justify why they are needed, and the costs, and indeed they are asked to report on the schedule of where they are and when they might be done. As a test manager, every management team I have worked with expected me to provide "real" status data. Further, SODA leads to better testing and better products, both of which management and stakeholders end up liking.

SODA Implementation Options Using Stats, Taxonomies, and AI

While many people may want to jump to AI and advanced analytics, my experience is that these concepts take knowledge, skill, and experience to implement successfully. So, in this section, I present an implementation going from a simple crawling stage to the advanced running stage with AI.

To start with analysis and analytics, I began using spreadsheets, mind maps, and simple graphics (see examples in Figure 22-2). Most engineers and testers have experience with these tools, but I am always surprised by how few testers use these simple tools. Instead, they seem to want progressive ideals like AI without building the skills of "crawling" with the data. They then do not understand why the advanced topics "fail" for them. We humans always forget that there are no panaceas or even easy answers, regardless of the field testers are working in; *everything is complex*. The three items in Figure 22-2 present simple analysis information: a pie chart, a mind map, and a taxonomy spreadsheet. The details and names are not important; what is important is the analysis concept name.

The first concept is the pie chart in drawing A. Here, one can see the "larger" piece of the pie is where I might want to focus my analysis attention. For example, the orange problem area in requirement stories could have been written better to support testers. In drawing B, the mind map (Part 3, Chapter 10) contains analysis data points and associations linked to the green circle. This gives me places to focus on new test cases. And finally, I used the spreadsheet in drawing C to create a failure taxonomy analysis, which yielded the kinds of data for the information on the sidebar.

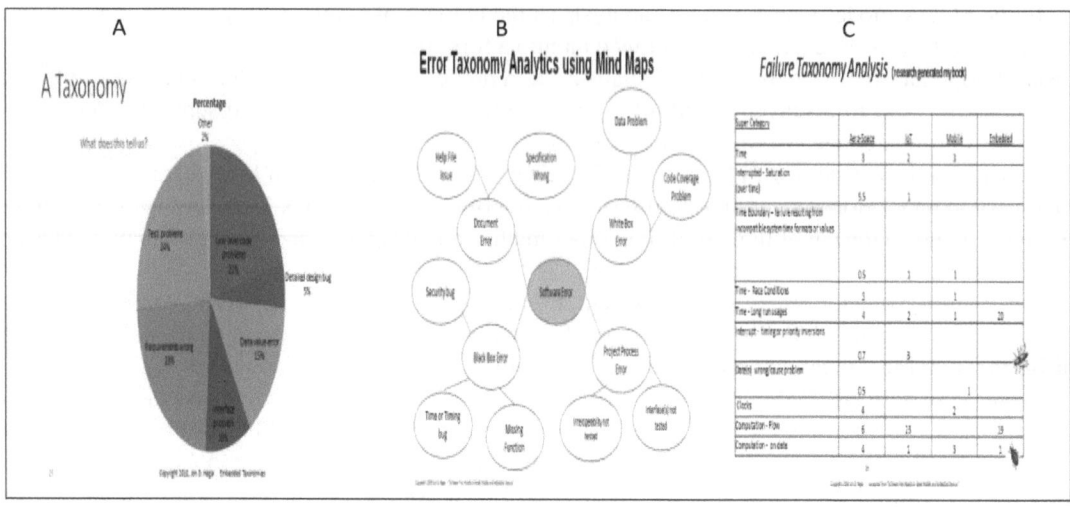

Figure 22-2. Simple analysis examples (A, B, C)

Note: On these SODA charts, the exact data values and information are unimportant for our discussion here.

Reference: Jon Hagar

REAL ERROR TAXONOMY DATA POINTS
To create parts B and C in Figure 22-2, I had to become a data geek based on the following:

- IV&V has filed over 1600 error reports on one project. This number represented about 60 percent of all errors found by the project. I used this to produce taxonomies and analyses manually.
- 30 percent of all software problems were found by IV&V on tested products.
- 20 percent of IV&V defined problems resulted in code or requirement changes and included errors that could have resulted in degraded mission performance.
- One error would have lost the *full system* with a cost of multimillion dollars.
- Over 80 percent of problems found by IV&V were in the documentation. While not impacting code, finding these errors reduced maintenance costs and helped in project process improvements related to concepts, such as the Capability Maturity Model Integration (CMMI) and ISO audits.
- The IV&V logic errors found included software configuration management problems, code standards violations, incorrect data parameters, logic faults, design faults, and incorrect requirements that impacted the design and code. These were equally distributed over the remaining 20 percent of error reports.
- Over 75 percent of the total errors were found during initial development. However, after years of operations and maintenance on one program, 30 error reports were filed during a significant system upgrade to create errors even in Ops.

These simple "tools" were my starting point for understanding my test data analytically. The SODA started with such data points and allowed me to improve my test planning.

Next, in becoming a test data analytics person, I learned more analysis techniques by building data analysis skills and knowledge from "walking leading into running." In walking SODA, I used tools and concepts such as error taxonomies, more advanced models with prediction, complicated math and statistics, or complete simulations, which could start helping me to look forward with patterns and do analytics and future prediction SODA, as seen in Figure 22-3.

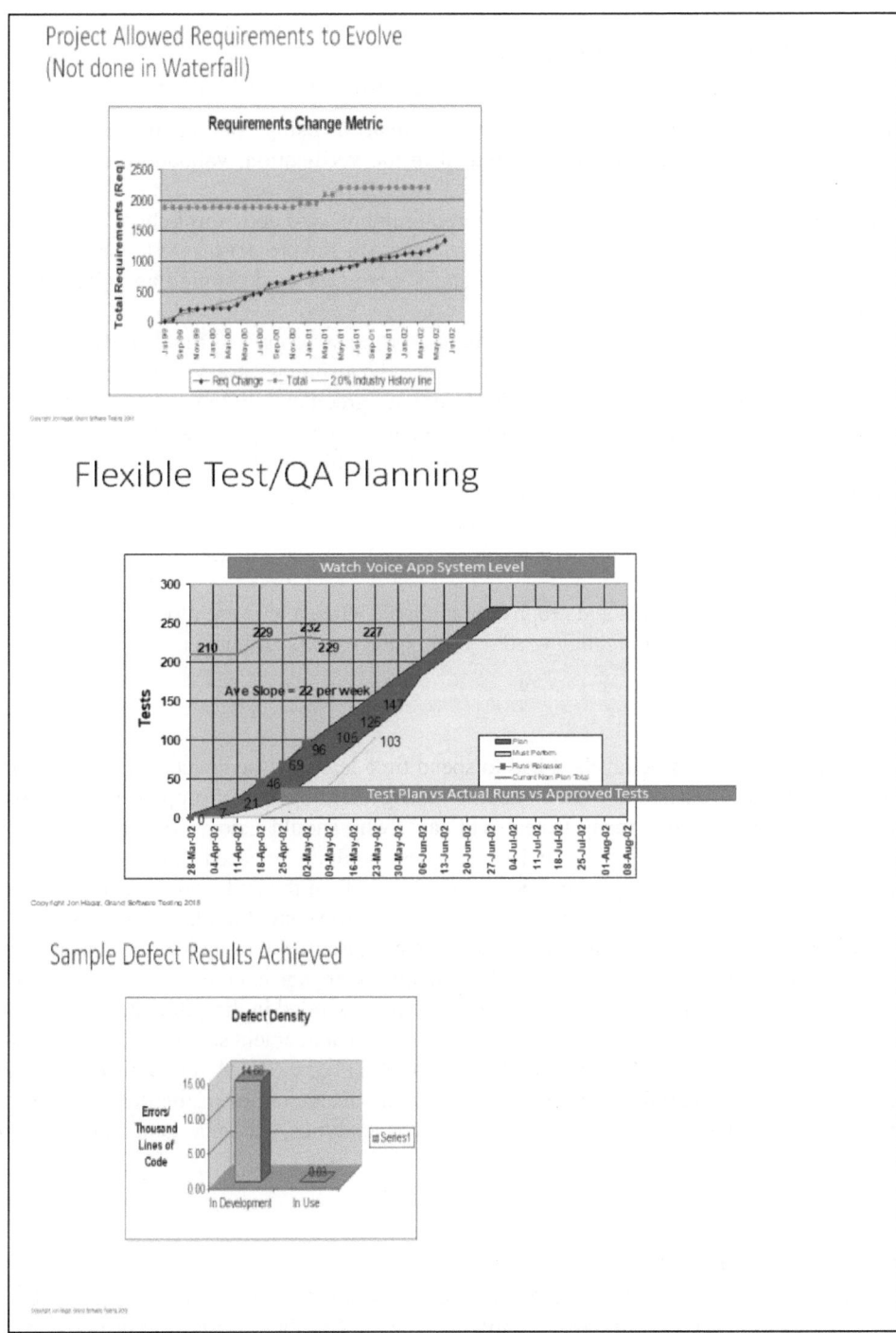

Figure 22-3. Simple analytics graphics A, B, and C

Note: On these SODA charts, the exact data values and information are unimportant for our discussion here.

Reference: Jon Hagar

USING THE EXAMPLE DATA CHARTS

On a startup development project, another data set yielded 5000 issue reports:

- Problems were found by inspection, testing, and analysis using both manual and automated approaches. Problems are equally distributed in the verification, validation, and scientific validation areas.
- The problems found resulted in changes to requirements and code throughout the lifecycle. In the old way of thinking, changing requirements over the project lifecycle were deemed a significant risk, but on this project, the changes were managed through an agile approach (Figure 22-3A).
- The agile test team used monthly planning and progress reports to report testing status to management (Figure 22-3B). Management wanted the testing progress (considered red lines) and predictions of when testing would be complete (green and yellow schedule areas). The colors represent changes to the respective management data, in this example, schedules.
- During the lifecycle, the high number of issues and requirement changes worried the upper management organization. Still, the issues found in the field (Figure 22-3C) were minor (right side of the graph) compared to the number on the left side of the graph. Hence, the team concluded they had found issues before the product was released into the field, which was the desired result.
- Today, many of these reports and graphs are available when teams use modern CM/SCM and management tools, but testers may need to ask for the reports and understand how to use the data provided.

To implement SODA at these levels, I had to spend time learning the concepts myself, or in some cases, I hired "experts" to assist me. When I had no management or project support, I worked on these concepts on my own time. Yes, many testers will say, "why should I do work for free?" From my perspective, it helped me be successful and get the best products released. Success was noticed by management, and I advanced in my career. So, I did get "paid" in the end, and the company stayed in business. This approach of extra work and learning is not for everyone. I did have some projects where management supported my analysis and analytics, often because they had seen my teams be successful before and understood what I was doing. Here, I would hire or engage experts in an area where I lacked skill. For example, testers built complex simulation models that aided in the automated realtime testing of embedded flight systems. Again, more success encouraged management support of my SODA efforts.

Finally, after winning a management marathon by running basic SODA on my project, I upgraded by engaging with AI and neural networks (NN). Again, the exact values of the items in Figure 22-4 are important, so I will not share much of the specific data analytics, except as general concepts found in Figure 22-4A and B.

RELIABILITY AND AI/NN

Figure 22-4 uses an error estimation tool that takes the actual error arrival rate and estimates the remaining errors to be found. In this example, Figure 22-4A, the red line is a project, and the blue line is what was achieved. I used this data in drawing to convince project management that the product was ready to ship. In Figure 22-4B, I programmed a neural network to identify trends in software errors. I used actual data, which projected that the better the comments programmers wrote, this resulted in much better code. While interested in getting programmers to do better work, I could also identify areas to focus test improvements.

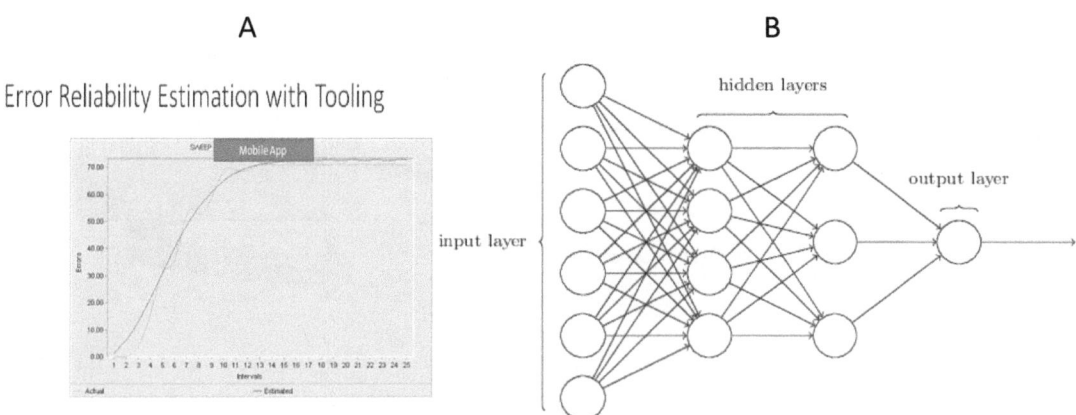

Figure 22-4. Reliability growth modeling estimation (A) and AI/NN SODA matrix (B)

Reference: Jon Hagar

Everyone wants to get to an end game of AI and NN, but my experience is that it is easier to make progress over the continuum of "crawling to running" than running first. I can tell you that the initial efforts are what I would term "promising," and I fully expect other substantial efforts in AI/NN to aid testers greatly in the future. Some projects and companies will, in many cases, reach AI/NN analytics. However, these are advanced topics and will take time and exceptional skills to work on projects. I look forward to seeing something like IBM's Watson for software testers. Some projects may already exist that I have not seen published work on yet. I would expect such an AI "tester" system to be of great interest in the marketplace. Further, I have heard of a few testing tools using the AI buzz phrase, but again I have not seen many AI and SODA results from projects that I can genuinely call a success. However, these concepts are still relatively new.

AI/NN systems are expensive and still evolving in software test spaces. The "crawling and walking" stages have worked for me for decades. If things get complicated, statistics, math, and simulation models are the best support tools. However, these tools will require people with knowledge and skill in those areas. I trained in math in college, but I contracted with an expert for accurate statistical analysis.

The use of graphics to display and organize complex data is the most significant hint for success in this area. People, particularly managers, are visually oriented. For such users, information needs to be easily digested at a glance when displayed in graphics rather than reviewing long paragraphs of text and numbers. It may take time to explain the specific graphic you create, but once others understand the layout of information, it is my experience that they will want to see that kind of graphic repeatedly. Several of the preceding charts were produced for management frequently and represented both analysis and analytics. They were provided throughout the test project.

The newer CM/SCM tools may offer a graphical analysis of the raw data extracted from the tool, as does Microsoft Excel spreadsheets. In the best tools, data can be exported from the CM/SCM tool to Microsoft Excel where a graphical picture in various forms can be saved or copied to other graphical tools such as Microsoft PowerPoint. This is handy to share across enterprise tools in various forms.

Finally, in SODA, testers should remember KISS (keep it simple, sir). Just as I cannot process ten billion data points in my head, I can get lost in hundreds of analyses and analytic charts or tool outputs. I once heard a phrase I like (not sure whom to give credit to): "a fool with a tool is still a fool."

AI/NN and testing analytics (I call it SODA) will be the future for many testers, and if you are an adventuresome person, maybe now is the time to build your skills and tools in these areas. I predict a big AI/NN analytics business for testers and software engineers/developers. Start with "crawling" and practice toward (running) the complete AI.

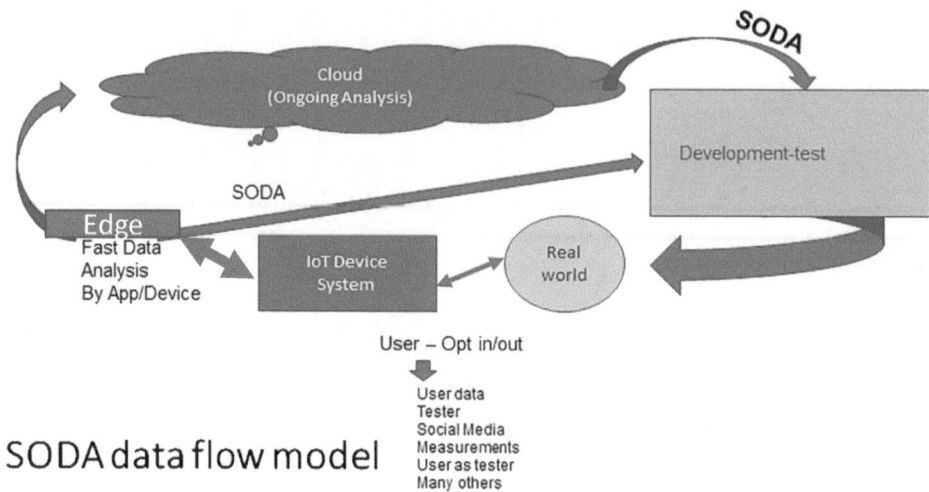

Figure 22-5. SODA with edge system data flow

Reference: Jon Hagar

Use Case Example for SODA and Edge Data Flow

The data flow model of Figure 22-5 starts in the development test phase. There will be ongoing data analysis from a test perspective within the Dev-test gray box during whatever lifecycle is being followed. Many testers will be familiar with aspects of SODA during the development test efforts. However, testers need to consider SODA after development in the IoT world. In the data flow view of the SODA model, there can be at least three time points (development test, user analysis at the edge, and cloud ongoing analysis) where self-organization of the data will be helpful. Once the product is out in the "real world," it *will* generate data. The predictions are for massive data streams. The data available on the edge systems and cloud will be massive. Some of the analysis will likely be done on the edge device itself. This is the "fast data analysis by app/device." Information can flow in two directions. One is into the cloud, where it can be used after SODA. More interesting for some test teams where they may be following the development-operations lifecycle, for example, is realtime SODA. Here, the development and test teams will use SODA to solve the problem in real time. For example, imagine seeing a software bug happening in real time using SODA and then determining a fix that is pushed to the device before things "get worse." There are both risks and benefits with such realtime edge SODA updates, but it is worth considering for some IoT systems.

The other SODA path is to use information in the cloud in a nonreal-time set of analytics. Cloud mining will be where many teams look for project process improvement ideas, but given the data sizes in the cloud, some SODA model concepts will be needed. Examples include classic error taxonomies and pattern identification to improve future projects.

Figure 22-6 considers a sample Unified Modeling Language (UML) use case diagram for SODA. The tester is associated with test input data in the analytic system. The running analytic systems use the data. The "system under test (SUT)," which is hardware-software, both provides and can use the data from the analytic system. For example, the SODA output might be new test cases for the SUT (right side of Figure 22-6) to run on the output side. The engineering teams (right side of Figure 22-6) include testers, teammates, and other teams. These "users" will cyclically consume the analytic system outputs over iterations during the lifecycle.

After some time and iterations, analytic test predictions will come from the test team in this use case. Examples of these test predictions are test progress, test completion estimates, and fitness of the product

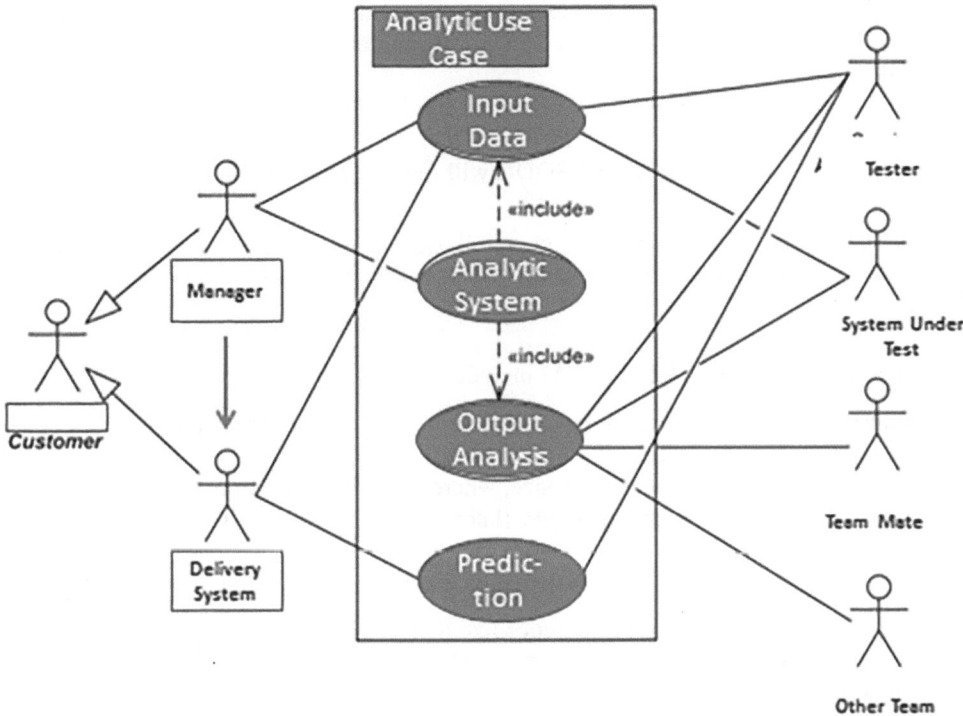

Figure 22-6. SODA use case sample

Reference: Jon Hagar

for delivery. The manager will use the information directly from the analytic system to control the project. As the product matures, the product's fitness, provided by tests and other analytic information, will be used by the manager to direct delivery of the system or to release the product to a customer. Notice that the tester is not the "gatekeeper" of product release in this use case. Still, they provide analytic system information and prediction that management uses to make the "ready for delivery" call. Finally, it is expected that management should be the direct line of communication to customers – not the engineering/test staff. Having a single point of contact (POC) with any customer always works best, which include a POC with knowledge of management an critical stakeholders. Only in a compliance scenario might the technical staff be interfacing directly with a customer on the release of any product.

Leveraging SODA for IoT Testing

I have used these analytics and analysis examples to do many things, including

- Selling testing to stakeholders (increasing testing efforts and justification)
- Improving my test team and efforts
- Identifying problems in test and development processes and products
- Improving the project products and processes
- Supporting risk and opportunity analysis
- Knowing when it is time to deliver the software (or not)
- Providing status and information to management of testing efforts and predictions of completion
- Measuring qualities of the product(s) under test
- Conducting future project/product planning and cost/schedule estimations

For most test teams, it is helpful to understand the past with analysis and prediction for the future with analytics. There are many other uses for SODA information. The examples cited are items I cared about or needed to provide.

There is a concern for SODA when it is being used in IoT testing. *The concern is bias in the data and analysis during SODA, even with AI. Bias in SODA and AI can cause algorithm and data risks and errors* [2]. Addressing bias in AI and SODA will require addressing human, systemic, and collected data biases.

> *If we develop trustworthy AI systems (including SODA), we need to consider all the factors that can chip away at the public's trust in AI. Many of these factors go beyond the technology itself to the impacts of the technology.*
>
> —Reva Schwartz, principal investigator for AI bias

Bias factors that can impact SODA and AI include

- Algorithm issues where the logic used within the tools, AI, or software is faulty.
- Wrong tools or methods are used.
- Data sampling and bias mistakes include size, where the data comes from, aging of data, etc.
- Prejudice is when the engineers misjudge (knowingly) the source(s) of the data used in the analysis.
- Measurement and analytics where the wrong data are unknowingly used or misapplied.

There are many cases of bias in data analytics like SODA and AI. Teams using these must be cognizant of the bias factors, including testing the SODA and AI information for bias risks. The bias consideration leads testers to the AI SODA future. Readers moving to SODA should consider good references on statistics [3]. Statistical tools and math are not for everyone, but they are very powerful tools to IoT teams.

AI SODA – A Near Future for IoT/Edge/Cloud Data Analytics

The next extension of SODA will include AI in the test lab with data coming back from the field. The AI SODA will extend into the edge and fog. The AI SODA will have advantages such as dealing with massive amounts of data, realtime processing, and better user feedback.

I have used some AI data analytics in a neural network (NN) to analyze test and code complexity metrics. The data was from an actual project, and the results were unexpected and exciting.

SODA GOES AI

I trained the NN using thousands of error reports from a project. The report was cross-correlated to the level of testing, technique, and code areas. There were hundreds of code modules. The project also had code complexity measurements from a tool, including lines of code (LOC), cyclomatic complexity, Halstead complexity, numbers of comments in a module, and a calling tree. I trained the NN with 1000 reports and complexity measures. I believed that errors would correlate to higher code complexity in history. I used the remaining reports to check the prediction of where modules would have errors. There was no good prediction. There was no correlation between the error rates and module complexity. There was an unexpected analytic observation from the NN, which was that models that had more and better comments had fewer errors. This finding was unexpected, but some research in the literature shows similar observations by other researchers. Moral of the story: AI SODA may learn an unexpected thing, but that does not mean it is right or wrong. So, testers need to apply solid critical test thinking to the story the AI and the data tell you.

Artificial intelligence (AI) is used for good during testing in the following ways:

- AI can analyze data patterns and identify anomalies and threats.
- AI can help testers hack or attack a system during security testing.
- AI can be used to protect test privacy during security attacks.

As the previous story indicates, AI data analytics may be good at supporting concepts in testing, including that of SODA and product understanding. AI can find patterns in data and test information that may escape most testers. However, I am concerned that AI will scare some testers away or cause them to reject AI and the unexpected things it may learn or have to teach them. Testers should approach SODA and AI with open minds but with a reasonable degree of healthy skepticism.

Much more to this story will unfold in the coming years. Organizations such as IEEE and ISO are currently producing AI standards (in work pointer references IEEE Working Group on IEEE p7003 Algorithmic Bias Considerations and ISO/IEC JTC 1/SC 42/JWG 2). Note: As of 2022, these are active working groups studying AI and bias issues that the author directly supports. Dates for publishing have not been determined.

Summary

SODA offers a final development and test support environment, which, if used, will be the tool to support testing analysis and understanding of the IoT data. SODA and data can lead to better testing. Many testers and engineers have limited data analytics skills and experience. This chapter outlined an environment to address their limitation while leading to better testing, better products, and sales.

The Appendixes of this IoT testing book are annexes on supporting interface standards, careers, and "fast start" checklists.

References

1. How Analytics Can Drive Software Test Architectures and Advanced Support Environments, Jon Hagar, Softec Asia Conference, 2018
2. www.nist.gov/news-events/news/2022/03/theres-more-ai-bias-biased-data-nist-report-highlights
3. *Statistics for Dummies*, by Deborah J. Rumsey, Wiley Publishing, 2nd Edition, 2011
4. https://en.wikipedia.org/wiki/Data_analysis – accessed spring 2022

IoT Supporting Interface, Hardware, Platform, and Protocol Standards

There are various support products and process standards that may be of interest to testers. These are just for reference as many of them may change or evolve rapidly. Example product standards are currently presented in the following lists as a starting point for readers.

Interface examples:

- Wi-Fi (802.11)
- 802.15.4: Zigbee, 6LoWPAN (RFC 6282)
- Bluetooth Smart (formerly low energy)
- Bluetooth
- Ethernet and PoE

Hardware standard examples:

- Raspberry Pi and Beaglebone
- Atheros AR9331: Arduino Yun, WeIO, Black Swift, Onion
- TI CC3200 and CC3100
- ESP8266 (see Hackaday, Arduino IDE port, Nodemcu)
- Electric Imp
- Spark: Core, Photon, Electron
- Intel Edison

Platform standard examples:

- Electric Imp (Imp001, Imp002, Imp003)
- Spark (Core and Photon)
- Thingsee
- TinkerForge
- SmartThings
- WICED (Broadcom)
- Cosino
- littleBits

© Jon Duncan Hagar 2022
J. D. Hagar, *IoT System Testing*, https://doi.org/10.1007/978-1-4842-8276-2

Protocol examples:

- MQTT
- ZeroMQ
- Thread (6LoWPAN on 802.15.4)
- Protocol Buffers
- HTTP and WebSockets, often with JSON
- CoAP (RFC 7252)

Careers in IoT Testing

Testers tend to focus on building skills, learning, providing information, and getting the job done on time. They want careers, reasonable compensation, and the ability to find work as needed. Of course, these are important considerations, but they miss an area that surveys only occasionally touch on, which is job satisfaction. I will not give you long lists of tester skills and things to learn in this appendix. I only ask that you think about your career and job satisfaction.

I extend the satisfaction theme further. I believe that testers should have fun, find their jobs challenging, and look forward to an exciting career. I say this because when I talk about my life's path in testing, people always ask how I did it.

I believe that the main point in life is to have fun, love, and learn. This path does not mean every workplace task must make you happy; there will be stress and maybe even some pain, but the big picture of how you work should result in happiness, support your learning, and lead to love of at least your career. If it does not, consider changing something.

How do you change your future test path career?

The high-tech world of computers and software means those of us working in it must *always be learning, building skills through practice, and looking to the future.* The tools, ideas, and systems of 40 years ago (when I started) were vastly different from today, and tomorrow's systems will be even more different. Are you at risk of being caught in a skills-practice gap that will lead to unhappiness?

To keep my career and skills on track while having fun, I have always had a career plan covering what I want to do this year and what I'd like to be doing in the next few years. I learned and built skills by going to conferences, reading/finding the appropriate sources which would further my skills and knowledge, and taking risks to practice new ideas. These challenges made me happy, so I had fun and learned to love my job (while keeping close to my family).

Specifically, I learned languages designed to support testing. Next, I learned artificial intelligence (AI), where I used neural networks (NN) to do data analytics on error reports. This led me to an interest in taxonomies and understanding patterns to improve software testing. In turn, I wrote a conference award-winning paper on this subject, which led me to publish books on testing embedded, mobile, and IoT systems. This was my plan and path to practice and happiness. Yours will be different, but I have always found fun challenges in software and testing. You, too, can do this.

Here are a few things you can try or work on to improve your IoT testing:

- Become more artistic (tech is not everything and you may find some surprises in the arts).
- Take a risk at learning new things.
- Learn more engineering concepts in differing areas (e.g., civil engineering, medical (all branches and subfields), electronics, mechanical, psychology, biology, etc.).

– Establish an extensive personal reference library of books and materials (I have over 500 books and reference sources).

Note
Do not automatically trust everything found on the Internet or any reference. Find the credible and trustworthy sources and decide how to save that pertinent information to have as a ready reference, that is, links fail.

– Work on building skills in new areas that you are not familiar with (more risks will open your eyes).
– Learn about IoT as it moves into the metaverse.
– Work product improvement and safety engineering by learning standards such as

• ISO/IEC/IEEE 12207 Systems and software engineering – Software lifecycle processes
• ISO/IEC/IEEE 15288 Systems and software engineering – System lifecycle processes
• ISO/IEC/IEEE 32430 Software engineering – Software nonfunctional sizing measurements
• IEEE 982.1 Standard for Measures of the Software Aspects of Dependability
• ISO/IEC/IEEE 29119 Software Testing – All parts

Note
Standards are only one way to improve your testing, and they may not be for every IoT team, but these are standards that I have supported and learned from as well as other materials referenced in this book.

– Become an entrepreneur.
– Work at being more of a risk and challenge taker, within reason.
– Have a career plan, near term and lifelong.
– Work on tech skills, always (there are always more tools and more technology to learn).
– Work on physical skills (I ski, bike, sail, and do yoga).

Have some fun, or change your job, career, or path in life. Use your power of choice wisely.

IoT Testing Startup Checklist

I am a big fan of checklists, even if you have to modify them, which you should do for your local team and device. Checklists help remind testers of things to do, check, and test. This is a starting point, but you can tailor it for your context and project.

Getting Started with IoT Testing

1. Start with paper prototype testing – Even before implementation or design, draw out the user interface (UI) and the architecture on a piece of paper and analyze if the concept actually works. Listen to the tester. A tester's point of view may vary quite a bit from a developer's point of view.
2. Test competitor's app – Do this to understand what they have gotten right and how your IoT could be better, different or unique. This groundwork is essential to build a comprehensive device covering all essential functionalities.
3. Set up a testing schedule – After the IoT devices enter the development phase, it's better to start testing at least once a week to check all risky functionalities and discover how everything works together. Then, ask yourself how different your original understanding was from the actual functionality. Were there some surprises? Did you get "backdoored?"
4. Do test planning per Part 2 – Plan, strategize, and create an STA to understand boundaries, risks, and tailoring.
5. Have you planned the right level of security, attack, tour, quality, technique, and framework testing? Do you have stakeholder consensus?
6. Have you applied the OWASP tables of Part 3, Chapter 14?

Usability Testing

This area focuses on the user's mobile app usage. If users find it challenging to perform basic functions on your app, they will uninstall it. These usability tips aim to craft the best user experience possible for mobile app users.

Test the basic wireframes of the device – It's better to initially test the UI of the application and how the navigation works. What can the user do perhaps that they shouldn't be able to do given the current design? Testing the UI and its navigation helps elucidate if any efficient design structures might help reduce the number of taps and swipes, making navigation simple and desirable, as well as preventing users from doing something they shouldn't (security risks).

© Jon Duncan Hagar 2022
J. D. Hagar, *IoT System Testing*, https://doi.org/10.1007/978-1-4842-8276-2

Compatibility and Integration Testing

Many smart IoT devices are available in various sizes, platforms, interfaces, communication lines, etc. Have testers identified IoT devices that may need integration testing, and if so

1. Create a checklist or matrix for all hardware, software, communication, and user interfaces.
2. Assess critical checklist items.
3. Test integrations with combinatorial testing.

Touch-screen, input, and button functionality – An aspect of the IoT features makes the functionality easy and extremely user-friendly (or not):

1. Test each button and function to specification.
2. Stress each button and input (hold down to make a sticky input test).
3. Test combinations of inputs at the same time.
4. Are button sizes large enough and meet industry and legal requirements (ADA)?

App Localization Issue

1. A lot of IoT devices are used in multilingual settings. Ensure there are no encoding issues, display truncation issues, or any UI changes due to variations in character length and language.
2. Social app integration

   a) Today, social app integration features are a must-have option. Many devices will allow users to log in using their social media account logins (such as Facebook or Google); this feature is essential. If this test fails, it is critical for many IoT devices. However, in the embedded world, this may be a poor requirement choice (due to security risks).
   b) In the case of phone banking, should there be login or interface capability to apps such as Facebook or Google?

External Connectivity Issues

1. Test to check the connectivity on all of the following:

   • Bluetooth
   • Wi-Fi
   • USB
   • NFC
   • 2/3/4/5/6G networks
   • GPS functionality, etc.

2. Check IoT switching ability between different connections.
3. Test with dropout and dip-outs (intermittent network changes).
4. Test with required hardware and software.
5. Check for bad connectivity messages on failure.

Sensor/actuator checks (note: may need support from systems or hardware staff)

1. Test that sensor readings recorded are accurate. This may mean specialized STA/environment requirements to stimulate and calibrate input.
2. Test that actuator outputs are accurate. This may mean specialized STA/environment requirements to sample and record outputs for analysis.
3. If there are multiple sensors and actuators, they must be tested in stand-alone and combinations.

Interrupt Testing

1. Test for IoT device interrupts from communication, battery/power status, memory issues, or other interrupt notifications that should occur without disturbing the IoT performance.
2. Identify interrupt levels and actions so that each one is tested.

Operational Testing

1. Test operational recovery events in case the IoT device/system crashes while users are on it. The system should recover or degrade gracefully (this should be defined in the specification or in use case stories).
2. Test if the IoT retrieves backed up data after recovery or degrades (does it recognize bad input data).
3. Test all user stories or use cases in positive and negative (failure) modes.
4. Run tests with low memory, low or degraded power, slow networks, or unstable sensors/actuators.

Installation Test

1. Check if the support apps get installed or upgraded entirely and within a reasonable time.
2. Check if IoT software version updates are added without hesitation and assess prompt to users for necessary actions.
3. Check if the installation follows user guides – step by step.
4. Test or check for uninstall – successful, no junk files with data, and user confirmation.
5. Check that needed historic data and information (e.g., settings) are not overwritten improperly.

Miscellaneous

1. Test user guides/documentation for accuracy.
2. Ensure regression is planned and designed; revisit as needed with updates/upgrades.
3. Check the compatibility of the IoT edge-cloud-server-client relationship to performance.

Example of an IoT System of Systems

Appendix D gives a larger version of a figure used in several chapters of the book. The figure shows a series of IoT systems that form a system of systems (of everything). It is an example to show how large a set of IoT devices and networks could become. Test teams may be needed at the device, system and system of systems levels to assure quality and success to stakeholders.

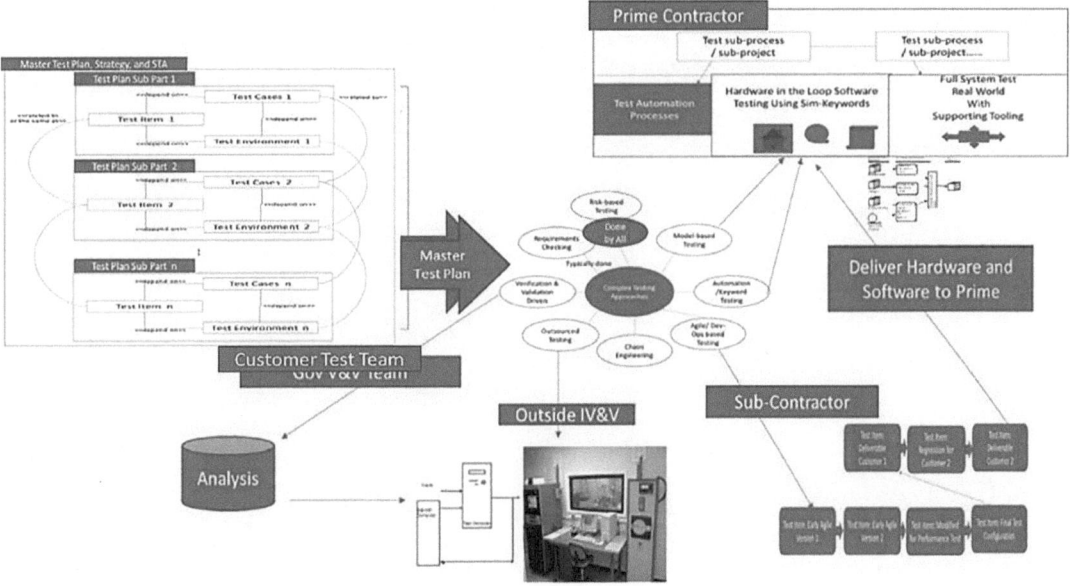

J. D. Hagar, *IoT System Testing*, https://doi.org/10.1007/978-1-4842-8276-2

Index